Functional Cement-Based Composites for Civil Engineering (Volume II)

Functional Cement-Based Composites for Civil Engineering (Volume II)

Guest Editor

Jonathan Oti

MDPI

Basel • Beijing • Wuhan • Barcelona • Belgrade • Novi Sad • Cluj • Manchester

Guest Editor
Jonathan Oti
Faculty of Computing,
Engineering and Science
University of South Wales
Pontypridd
UK

Editorial Office
MDPI AG
Grosspeteranlage 5
4052 Basel, Switzerland

This is a reprint of the Special Issue, published open access by the journal *Materials* (ISSN 1996-1944), freely accessible at: https://www.mdpi.com/journal/materials/special_issues/LJZW47H9NT.

For citation purposes, cite each article independently as indicated on the article page online and as indicated below:

Lastname, A.A.; Lastname, B.B. Article Title. *Journal Name* **Year**, *Volume Number*, Page Range.

ISBN 978-3-7258-6436-2 (Hbk)
ISBN 978-3-7258-6437-9 (PDF)
https://doi.org/10.3390/books978-3-7258-6437-9

Contents

About the Editor

Jonathan Oti

I am an academic and researcher in civil engineering with extensive expertise in cement science, concrete technology, and soil stabilization. I have published over 80 peer-reviewed journals and conference papers and authored technical research reports for government and industry stakeholders.

My research focuses on sustainable construction materials, including the development of waste-derived and geopolymer binders, as well as the mitigation of sulphate-induced swelling in stabilized soils. I have supervised doctoral and postdoctoral researchers, served as an examiner for UK and international research degrees, and contributed actively to teaching, academic governance, and national and international technical committees.

materials

MDPI

Editorial

Functional Cement-Based Composites Materials for Future Applications

Jonathan Oti

School of Engineering, Faculty of Computing, Engineering and Science, University of South Wales, Pontypridd CF37 1DL, UK; jonathan.oti@southwales.ac.uk

1. Introduction

The global demand for cement-based materials continues to rise, particularly in emerging economies where infrastructure and housing development are accelerating. Cement remains one of the most widely used materials in the built environment, and its production is expected to grow despite increasing environmental regulations aimed at reducing carbon emissions [1–5]. Cement production alone contributes approximately 8% of global CO_2 emissions, primarily due to the calcination of limestone and the energy-intensive nature of the process [6,7].

In response to these challenges, the construction industry is undergoing a transformation toward sustainability. This Special Issue brings together innovative and economically viable materials derived from alternative cement replacement strategies. It highlights interdisciplinary approaches to developing functional cement-based composites with reduced environmental impact. These include the use of waste plastic as coarse aggregate in geopolymer concrete, granite powder as partial cement replacement, and ground glass powder as a supplementary cementitious material.

Recent advancements have demonstrated the potential of geopolymer concrete (GPC) as a sustainable alternative to traditional Portland cement [8,9]. GPC offers superior mechanical properties, high thermal resistance, and up to 80% reduction in CO_2 emissions [10]. Innovations such as nanomaterial integration, hybrid binder systems, and 3D printing are expanding its applications and performance [11]. Moreover, recycled cement technologies have shown promise in reducing emissions by up to around 61% while maintaining strength comparable to conventional cement [12].

These developments underscore the importance of continued research and collaboration across academia, industry, and policy. The papers in this Special Issue provide valuable insights for civil engineers, material scientists, and sustainability practitioners seeking advanced techniques and alternative approaches to low-carbon construction. By integrating performance-based assessments, life cycle analysis, and digital optimization tools, the industry can move toward net-zero emissions while maintaining the durability and versatility of cement-based materials [13]

2. An Overview of Published Articles

Huang et al. [14] presents an experimental study on dovetail profiled steel concrete composite shear walls (DPSCWs) connected using self-tapping screws, aiming to improve seismic resilience and construction efficiency. Traditional welded connections in thin-walled steel structures often suffer from deformation and are unsuitable for prefabrication. By introducing self-tapping screws, the study fills a gap in functional cement-based composites by offering a reliable, flexible, and damage-resistant connection method. The research evaluates hysteretic behavior, failure modes, ductility, and energy dissipation, and proposes a design method based on effective strip theory. The findings demonstrate that screw-connected DPSCWs meet seismic design standards, enhance deformation capacity, and simplify construction, contributing to the development of intelligent, resilient structural systems in civil engineering. The study concludes that self-tapping screw-connected DPSCWs show reliable seismic performance, meeting code requirements. Increasing screw quantity improves ductility and energy dissipation, while higher axial compression boosts stiffness but reduces deformability.

The article by Rogojsz and Rudnicki [15] explores the impact of various mineral additives, such as white and compacted microsilica, limestone flour, glass flour, basalt dust, and glass granulate, on the alkaline reactivity of aggregates in cement-based composites. The study addresses an important gap in civil engineering: the limited usability of reactive aggregates in concrete due to alkali–silica reactions (ASR), which compromise durability. By replacing 10% and 20% of cement with these additives, the authors demonstrate that certain materials, especially white microsilica, can significantly reduce ASR, enabling the use of lower-grade aggregates in infrastructure applications. This contributes to sustainable construction by expanding aggregate sources and reducing cement usage, aligning with environmental goals. The findings offer practical solutions for enhancing the performance and longevity of functional cement-based composites in road and bridge engineering. The work concludes that white microsilica significantly reduces aggregate reactivity, enabling use of reactive materials in road concrete. Other additives like glass flour and limestone also help. Compacted microsilica, however, increases reactivity and is unsuitable.

The third article is from Rudnicki [16], the study introduces a novel method for designing self-compacting concrete (SCC) with reduced cement content by optimizing aggregate packing. It addresses important gaps in functional cement-based composites for civil engineering by proposing a simplified, performance-driven design approach that minimizes voids and enhances durability. The study systematically evaluates 36 mix designs using the blocking method to achieve optimal aggregate proportions, reducing cement from 500 to 350 kg/m^3 without compromising strength or freeze–thaw resistance. This contributes to sustainable construction by improving material efficiency, reducing environmental impact, and expanding SCC applicability in complex structural designs. The study concludes that optimizing aggregate packing in SCC design enables reduced cement use while maintaining strength and durability. The blocking method proves effective, offering sustainable, high-performance concrete with minimal voids and enhanced flow properties.

The fourth text published in this Special Issue is Zhang et al. [17], the study presents a novel approach to enhancing vegetation concrete, a cement-based composite used in ecological slope protection, by incorporating plant protein foaming agents. Traditional vegetation concrete suffers from consolidation, limiting root growth and nutrient uptake. The study fills this gap by introducing foam to improve porosity and reduce weight, while maintaining sufficient mechanical strength. Through physical, mechanical, and biological tests, the authors demonstrate that a foam volume of 20–30% optimizes plant growth and structural integrity. This innovation advances functional cement composites by balancing ecological performance with engineering requirements, offering a sustainable, lightweight, and porous solution for slope stabilization. It contributes to civil engineering by integrating environmental functionality into structural materials, addressing both ecological restoration and material efficiency. The study concludes that foamed vegetation concrete with 20–30% plant protein foam improves porosity, reduces weight, and supports plant growth while maintaining strength for slope protection. It offers a sustainable, efficient solution for ecological restoration.

Costafreda et al. [18] study, investigates a newly discovered natural zeolite-rich tuff in the Los Frailes Caldera, focusing on its mineralogical, chemical, and pozzolanic properties. Through advanced characterization techniques (XRD, SEM, XRF, TGA), the authors demonstrate that the material—rich in mordenite—exhibits high reactivity and mechanical strength when blended with Portland cement. The research addresses an important gap in sustainable civil engineering by offering a locally sourced, high-performance supplementary cementitious material (SCM) that enhances durability and reduces CO_2 emissions. Unlike conventional SCMs, this zeolite-rich tuff shows superior purity and pozzolanicity, contributing to long-term strength gains in mortars and concretes. The findings support its use in functional cement-based composites, promoting eco-efficient construction and extending the life of existing mineral reserves. The main conclusions were that newly discovered zeolite-rich tuff shows excellent pozzolanic properties, high silica content, and strong mechanical performance, making it a viable supplementary cementitious material and promising for expanding Spain's industrial reserves.

The work by Rudnicki and Stałowski [19] investigates on the performance of cement concrete pavements using low-emission cements (CEM II, III, V) to address environmental and durability challenges in civil engineering. It fills a key gap in functional cement-based composites by demonstrating that multi-component cements not only reduce carbon footprint—up to 39%—but also meet or exceed mechanical and durability standards required for road infrastructure. Through comprehensive testing (strength, frost resistance, water penetration, and air void analysis), the study confirms that these sustainable mixes are viable alternatives to traditional CEM I cement, offering enhanced functionality for modern, eco-conscious civil engineering applications. The study concludes that low-clinker cements (CEM II, III, V) significantly reduce concrete's carbon footprint—up to 39%—while meeting strength and durability standards. These eco-friendly mixes are viable for sustainable road infrastructure and heavy traffic applications.

Bouchikhi et al. [20] explores the use of ground glass powder (GP) as a supplementary cementitious material to enhance the functional performance of cement-based composites, specifically for zinc (Zn) stabilization. It addresses an important gap in civil engineering materials research: the limited understanding of how SCMs like GP contribute to the immobilization of heavy metals in cement matrices. By comparing a 30% GP–70% Portland cement blend with a conventional 100% cement binder, the study demonstrates that GP significantly improves Zn immobilization over time, especially after 90 days of curing. Through microstructural, mechanical, and chemical analyses, the study reveals new fixation mechanisms involving GP, such as Zn–Si and Zn–C–S–H interactions, offering insights into designing more durable, sustainable, and pollution-resistant cement composites for infrastructure applications. The work concludes that glass powder enhances zinc stabilization in cement binders after 90 days, forming stable Zn–Ca phases. Though Zn delays hydration, GP promotes long-term immobilization, making it a viable solution for heavy metal remediation in concrete.

Rafieizonooz et al. [21] focuses on the integration of Carbon Nanotubes (CNTs) into Ultra-High-Performance Concrete (UHPC) to enhance its mechanical properties, chloride resistance, and service life. It addresses important gaps in functional cement-based composites by exploring how nanoscale reinforcement via CNTs can improve durability and structural performance, especially under aggressive environmental conditions. The study fills a research void by evaluating CNT compatibility with other advanced materials like artificial lightweight aggregates and micro hollow spheres, and by analyzing microstructure through SEM, XRD, and chloride diffusion tests. It also incorporates life service prediction modeling, offering practical insights for long-term infrastructure resilience in civil engineering. The study concludes that CNTs at 0.025–0.05% improved UHPC's compressive strength, chloride resistance, and service life. CNT3 showed reduced performance due to poor dispersion. CNT1 and CNT2 mixes offered optimal durability and structural benefits.

Sanjuán et al. [22] investigates the carbonation resistance of ternary Portland cements incorporating silica fume and limestone, addressing key gaps in sustainable cement-based composites for civil engineering. Traditional Portland cement contributes significantly to CO_2 emissions, and while blended cements offer environmental benefits, their carbonation resistance has been a concern. This study fills that gap by evaluating the long-term performance of ternary mixes under natural carbonation conditions. It demonstrates that with proper curing, these cements meet durability standards and enhance CO_2 uptake, offering a viable, low-carbon alternative for structural applications. The findings support performance-based design approaches for future sustainable infrastructure. The work concludes that ternary cements with silica fume and limestone show better-than-expected carbonation resistance. With proper curing, they meet durability standards and enhance CO_2 uptake, supporting their role in sustainable, low-carbon concrete solutions.

Suarez-Navarro et al. [23] study evaluates the radioactive content in ashes from recycled fuels used in cement production, ensuring safety in co-processing practices. It addresses an important gap in civil engineering by linking chemical composition to radiological behavior, enabling safer integration of recycled materials into functional cement-based composites. By using gamma spectrometry and principal component analysis, the study identifies correlations between radionuclide concentrations and metal oxides like Fe_2O_3, supporting predictive assessments. The findings confirm low radiation exposure for workers and promote sustainable, circular economy practices in cement manufacturing, advancing the development of safe, eco-friendly construction materials. The work concludes that co-processing recycled waste in cement production poses no radiological risk. Sewage sludge showed the highest natural radionuclide levels, but worker exposure remained below safety limits. Future studies should explore radionuclide leaching.

Huang et al. [24] study explores the incorporation of granite powder (GP), a byproduct of stone processing, into mass-manufactured sand concrete as a partial cement replacement. Replacing 5–10% of cement with GP enhanced 28-day compressive (up to +17.6%) and flexural strength (+20.9%), reduced autogenous shrinkage (−19.7%), and lowered hydration heat (−7.2%), mitigating thermal cracking risks in mass concrete. Microstructural analyses (SEM, XRD, MIP) revealed that GP improved pore structure and acted as nucleation sites without altering hydration products. The study addresses important gaps in functional cement-based composites by promoting sustainable reuse of industrial waste, optimizing GP dosage for performance, and providing mechanistic insights into GP's role in enhancing durability and mechanical properties—critical for large-scale civil engineering applications. The work concludes that granite powder enhances mass-manufactured sand concrete by improving strength, reducing shrinkage and hydration heat, and refining microstructure. Optimal benefits occur at 5–10% replacement, beyond which performance may decline.

Adeleke et al. [25] study addresses an important gap in the development of functional cement-based composites for civil engineering by exploring the under-researched potential of using polylactic acid (PLA) waste plastic as a coarse aggregate in geopolymer concrete. While previous studies have focused on plastic as fine aggregates or fibers in ordinary Portland cement (OPC) concrete, this research uniquely investigates the performance of geopolymer concrete incorporating coarse recycled plastic aggregates. It evaluates both mechanical and microstructural properties, demonstrating that geopolymer concrete with up to 70% plastic replacement maintains superior strength compared to OPC mixes. The study also highlights improved interfacial bonding and reduced porosity in geopolymer matrices, offering a sustainable, high-performance alternative to traditional concrete. The work concludes that using waste plastic as coarse aggregate in geopolymer concrete, will potentially achieve notable strength and sustainability benefits. It supports greener alternatives in construction.

Abbey et al. [26] study investigates the durability and microstructural behavior of expansive subgrade soils treated with sustainable cementitious waste materials—specifically ground granulated blast furnace slag (GGBS) and brick dust waste (BDW). Through wetting–drying cycle tests, California Bearing Ratio (CBR) evaluations, and microstructural analyses (SEM and EDX), the study demonstrates that combining GGBS and BDW enhances the strength, durability, and resilience of treated subgrades under cyclic environmental stress. The formation of calcium silicate hydrate (CSH) gels and increased alumina/silica content from BDW contribute to improved interparticle bonding and reduced mass loss. The study addresses an important gap in civil engineering: the limited durability of cement-based composites under fluctuating moisture conditions. Traditional binders like cement and lime, while effective in strength enhancement, often degrade under cyclic wetting and drying. By introducing BDW—a highly pozzolanic and sustainable material—the study offers a viable alternative that not only improves mechanical performance but also aligns with environmental sustainability goals. The work concludes that Expansive subgrades treated with GGBS and BDW demonstrated enhanced durability and strength under wet–dry cycles. The pozzolanic synergy between these materials reduced mass loss and improved microstructural integrity, making them sustainable alternatives for resilient road construction in fluctuating environmental conditions.

The fourteenth text published in this Special Issue is a review by Li and Cao [27] on the recent developments on the effects of micro- and nano-limestone on the hydration process, products, and kinetics of cement. The study presents a comprehensive review of the role of limestone-based materials in cement hydration. The authors focus on three distinct forms of calcium carbonate—limestone powder (LP), calcium carbonate whiskers (CWs), and nano-calcium carbonate (NC)—and examine their effects on hydration mechanisms, product formation, and kinetics. This work is particularly relevant to the development of functional cement-based composites in civil engineering, where performance, sustainability, and durability are critical. The study begins by acknowledging the environmental challenges associated with cement production, notably its high carbon footprint and energy consumption. In response, researchers have explored supplementary cementitious materials (SCMs) to partially replace cement in concrete. Among these, limestone has emerged as a promising candidate due to its chemical compatibility, abundance, and cost-effectiveness. However, the mechanisms by which limestone influences hydration—especially at micro and nano scales—have not been fully understood or quantified. This study addresses that gap by systematically analyzing the physical and chemical effects of limestone on cement hydration. The work concludes that micro- and nano-limestone improve cement hydration via nucleation and chemical effects, while larger particles mainly dilute. Their use enhances strength, reduces CO_2 emissions, and supports sustainable concrete development.

3. Conclusions

This Special Issue showcases a diverse and forward-thinking collection of research that advances the development of functional cement-based composites for civil engineering. Each contribution addresses critical gaps in sustainability, durability, and performance, offering innovative solutions that align with global efforts to reduce carbon emissions and promote eco-efficient construction. From the integration of recycled materials and nanotechnology to the optimization of mix designs and the use of alternative binders, the studies collectively highlight the transformative potential of interdisciplinary approaches in cement science [28,29].

The findings underscore the importance of leveraging locally available resources, enhancing material efficiency, and adopting performance-based design strategies. Whether through the use of geopolymer concrete, low-emission cements, or supplementary cementitious materials like glass powder and zeolite-rich tuff, the research demonstrates that sustainable construction is not only feasible but also capable of meeting stringent engineering requirements [10,29,30].

As the civil engineering sector continues to evolve in response to environmental and societal demands, the innovations presented here provide a roadmap for future research, policy development, and industrial application. By embracing these advancements, stakeholders can contribute to a resilient, low-carbon built environment that supports both infrastructure growth and ecological stewardship [31].

Future research should compare self-tapping screw and welded connections in DP-SCWs under seismic conditions, optimize screw layouts for ductility, and refine design via parametric FE studies. Studies on cement composites with mineral additives like white microsilica should assess long-term durability, shrinkage, and mechanical performance. SCC design using crushed aggregates (basalt, granite) and fillers (slag, limestone flour) needs expanded testing and mix model refinement. Foamed vegetation concrete should be tested in varied climates and soils to validate field performance. Zeolite–cement ratios beyond 25% and scalability for industrial use warrant exploration. Low-emission concrete research should focus on field performance, recycled aggregates, and rapid construction mix optimization. GP–cement composites need durability studies, pollutant-specific GP content optimization, and multi-metal stabilization. CNT dispersion, durability beyond 90 days, smart sensing, and economic feasibility require attention. Carbonation models for ternary cements need refinement under diverse conditions. Radionuclide leaching from Fe_2O_3-rich ashes and predictive modeling should be studied. Granite powder concrete's durability, structural applications, and synergy with other SCMs should be explored. Geopolymer concrete with waste plastic needs admixture optimization and durability studies. BDW–GGBS blends require field validation, mix optimization, life cycle analysis, and modeling. Micro/nano-limestone research should quantify nucleation area, optimize dispersion, and study hydration kinetics.

Conflicts of Interest: The author declares no conflicts of interest.

References

1. Belaïd, F. How does concrete and cement industry transformation contribute to mitigating climate change challenges. *Resour. Conserv. Recycl. Adv.* **2022**, *15*, 200084. [CrossRef]
2. Nilimaa, J. Smart materials and technologies for sustainable concrete construction. *Dev. Built Environ.* **2023**, *15*, 100177. [CrossRef]
3. Barbhuiya, S.; Kanavaris, F.; Das, B.B.; Idrees, M. Decarbonising cement and concrete production: Strategies, challenges and pathways for sustainable development. *J. Build. Eng.* **2024**, *86*, 108861. [CrossRef]
4. Scrivener, K.L.; John, V.M.; Gartner, E.M. Eco-efficient cements: Potential economically viable solutions for a low-CO_2 cement-based materials industry. *Cem. Concr. Res.* **2018**, *114*, 2–26. [CrossRef]
5. Imbabi, M.S.; Carrigan, C.; McKenna, S. Trends and developments in green cement and concrete technology. *Int. J. Sustain. Built Environ.* **2012**, *1*, 194–216. [CrossRef]
6. Andrew, R.M. Global CO_2 emissions from cement production. *Earth Syst. Sci. Data* **2018**, *10*, 195–217. [CrossRef]
7. Aday, A.; Richardson, K.; France, D.; Olavarria, J.; Schaidle, J.; Cozart, T. *State of Innovation 2024: Paving the Way for Low-Carbon Cement and Concrete*; NREL: Golden, Colorado, 2024.
8. Oti, J.; Adeleke, B.O.; Mudiyanselage, P.R.; Kinuthia, J. A Comprehensive Performance Evaluation of GGBS-Based Geopolymer Concrete Activated by a Rice Husk Ash-Synthesised Sodium Silicate Solution and Sodium Hydroxide. *Recycling* **2024**, *9*, 23. [CrossRef]
9. Ndigui, B.; Oti, J.; Kinuthia, J. Using silica fume based activator in sustainable geopolymer binder for building application. *Constr. Build. Mater.* **2021**, *275*, 122177. [CrossRef]
10. Khasawneh, M.A. Geopolymer concrete in construction projects: A review. *Discov. Civ. Eng.* **2025**, *2*, 124. [CrossRef]

11. Yang, M.; Chen, L.; Lai, J.; Osman, A.I.; Farghali, M.; Rooney, D.W.; Yap, P.-S. Advancing environmental sustainability in construction through innovative low-carbon, high-performance cement-based composites: A review. *Mater. Today Sustain.* **2024**, *26*, 100712. [CrossRef]
12. Shivaprasad, K.N.; Yang, H.; Singh, J.K. Path to carbon neutrality in construction: An overview of recent progress in recycled cement usage. *J. CO_2 Util.* **2024**, *83*, 102816. [CrossRef]
13. Marandi, N.; Shirzad, S. Sustainable cement and concrete technologies: A review of materials and processes for carbon reduction. *Innov. Infrastruct. Solut.* **2025**, *10*, 408. [CrossRef]
14. Huang, Z.; Tan, Y.; Zhong, Z.; Zhang, S.; Guo, L.; Wang, Y. Experimental Study on Seismic Performance of Dovetail Profiled Steel Concrete Composite Shear Walls with Self-Tapping Screw Connections. *Materials* **2025**, *18*, 49. [CrossRef]
15. Rogojsz, G.; Rudnicki, T. The Influence of Mineral Additives on Aggregate Reactivity. *Materials* **2025**, *18*, 7. [CrossRef]
16. Rudnicki, T. Design of Self-Compacting Concrete with Reduced Cement Content by Aggregate Packing Method. *Materials* **2025**, *18*, 4. [CrossRef] [PubMed]
17. Zhang, T.; Li, T.; Xu, H.; Wang, M.; Lu, L. Preparation and Performance Study of Novel Foam Vegetation Concrete. *Materials* **2024**, *17*, 6295. [CrossRef] [PubMed]
18. Costafreda, J.L.; Martín, D.A.; Sanjuán, M.A.; Costafreda-Velázquez, J.L. New Discovery of Natural Zeolite-Rich Tuff on the Northern Margin of the Los Frailes Caldera: A Study to Determine Its Performance as a Supplementary Cementitious Material. *Materials* **2024**, *17*, 4430. [CrossRef] [PubMed]
19. Rudnicki, T.; Stałowski, P. Performance Research of Cement Concrete Pavements with a Lower Carbon Footprint. *Materials* **2024**, *17*, 3162. [CrossRef]
20. Bouchikhi, A.; Safhi, A.e.M.; Maherzi, W.; Mamindy-Pajany, Y.; Kunther, W.; Benzerzour, M.; Abriak, N.-E. Advancements in Heavy Metal Stabilization: A Comparative Study on Zinc Immobilization in Glass-Portland Cement Binders. *Materials* **2024**, *17*, 2867. [CrossRef]
21. Rafieizonooz, M.; Kim, J.-H.J.; Kim, J.-S.; Jo, J.-B. Effect of Carbon Nanotubes on Chloride Diffusion, Strength, and Microstructure of Ultra-High Performance Concrete. *Materials* **2024**, *17*, 2851. [CrossRef]
22. Sanjuán, M.Á.; Menéndez, E.; Recino, H. Carbonation Resistance of Ternary Portland Cements Made with Silica Fume and Limestone. *Materials* **2024**, *17*, 2705. [CrossRef]
23. Suarez-Navarro, J.A.; Sanjuán, M.Á.; Mora, P.; Alonso, M.d.M. An Evaluation of the Radioactive Content of Ashes Obtained from the Use of Fuels from Recycled Materials by Co-Processing in the Cement Industry. *Materials* **2024**, *17*, 2287. [CrossRef]
24. Huang, J.; Xu, G.; Chen, S.; Yu, D.; Fu, T.; Feng, C.; Wang, Y. Enhancing Mechanical Properties and Microstructures of Mass-Manufactured Sand Concrete by Incorporating Granite Powder. *Materials* **2024**, *17*, 2234. [CrossRef]
25. Adeleke, B.O.; Kinuthia, J.M.; Oti, J.; Pirrie, D.; Power, M. Mechanical and Microstructural Investigation of Geopolymer Concrete Incorporating Recycled Waste Plastic Aggregate. *Materials* **2024**, *17*, 1340. [CrossRef]
26. Abbey, S.J.; Amakye, S.Y.O.; Eyo, E.U.; Booth, C.A.; Jeremiah, J.J. Wet–Dry Cycles and Microstructural Characteristics of Expansive Subgrade Treated with Sustainable Cementitious Waste Materials. *Materials* **2023**, *16*, 3124. [CrossRef] [PubMed]
27. Li, X.; Cao, M. Recent Developments on the Effects of Micro- and Nano-Limestone on the Hydration Process, Products, and Kinetics of Cement. *Materials* **2024**, *17*, 2133. [CrossRef] [PubMed]
28. Vignesh, J.; Ramesh, B.; Xavier, J.R. Recent advances in nano-engineered cement composites for sustainable and smart infrastructure. *Nanotechnol. Environ. Eng.* **2025**, *10*, 81. [CrossRef]
29. Gatti, G.; Singh, R.K.; Kumar, V.P. Performance study of cement concrete containing natural zeolite powder and glass powder as partial replacement of cement. *Int. J. Emerg. Res. Eng. Sci. Manag.* **2024**, *3*, 48–56. [CrossRef]
30. Virtanen, A.; Makinen, E. Low-carbon materials in construction: Opportunities and challenges for civil engineers. *Int. J. Hydropower Civ. Eng.* **2024**, *5*, 34–39. [CrossRef]
31. Barbhuiya, S.; Das, B.B.; Adak, D.; Kapoor, K.; Tabish, M. Low carbon concrete: Advancements, challenges and future directions in sustainable construction. *Discov. Concr. Cem.* **2025**, *1*, 3. [CrossRef]

materials

MDPI

Article

Experimental Study on Seismic Performance of Dovetail Profiled Steel Concrete Composite Shear Walls with Self-Tapping Screw Connections

Zhenfeng Huang [1,2], Youwen Tan [1,2], Zheng Zhong [3], Sumei Zhang [1,2,*], Lanhui Guo [4] and Yunhe Wang [5]

[1] School of Civil and Environmental Engineering, Harbin Institute of Technology, Shenzhen 518055, China; huangzhenfeng@stu.hit.edu.cn (Z.H.); tanyouwen@stu.hit.edu.cn (Y.T.)

[2] Guangdong Provincial Key Laboratory of Intelligent and Resilient Structures for Civil Engineering, Harbin Institute of Technology, Shenzhen 518055, China

[3] School of Science, Harbin Institute of Technology, Shenzhen 518055, China; zhongzheng@hit.edu.cn

[4] School of Civil Engineering, Harbin Institute of Technology, Harbin 150090, China; guolanhui@hit.edu.cn

[5] School of Civil Engineering, Taiyuan University of Technology, Taiyuan 030024, China; wangyunhe@tyut.edu.cn

* Correspondence: smzhang@hit.edu.cn

Abstract: To achieve the assembled connection between dovetail profiled steel sheets and the boundary members in dovetail profiled steel concrete composite shear walls (DPSCWs), self-tapping screws were employed. Three DPSCW specimens connected with self-tapping screws were tested under combined axial and cyclic lateral loads to evaluate their hysteretic response, focusing on the influence of the number of self-tapping screws and the axial compression ratio. The self-tapping screw-connected DPSCWs exhibited a mixed failure mode, characterized by shear failure of the profiled steel sheets and compression-bending failure of multiple wall limbs divided by ribs on the web concrete. Except for slight deformation at the screw holes located on the profiled sheets at the corners of the wall, the connections exhibited minimal visible damage. The yield drift ratio of the DPSCW specimens in the test ranged from 1/286 to 1/225, and the ultimate drift ratio ranged from 1/63 to 1/94, both meeting the relevant deformation standards specified in the "Code for Seismic Design of Buildings. Increasing the number of self-tapping screws enhanced the development of local tensile fields on the profiled steel sheets, thereby improving the wall's load-carrying, deformation, and energy dissipation capacities. However, increasing the axial compression ratio improved the initial stiffness of DPSCWs but reduced their load bearing and deformation capacity. Moreover, a design method for the self-tapping screw connections in DPSCWs was proposed.

Keywords: composite shear wall; dovetail profiled steel sheet; seismic performance; self-tapping screw connection; design method

1. Introduction

Dovetail profiled steel concrete composite shear walls (DPSCWs) are an innovative form of composite shear wall, comprising double layers of dovetail profiled steel sheets filled with concrete (Figure 1). The unique form of the ribs of the profiled steel sheets enables the sheets to work together with concrete without additional shear connectors, avoiding the welding or hole-opening associated with the installation of additional shear connectors in traditional double plate concrete shear walls. Meanwhile, the vertical concealed slits formed by the dovetailed ribs in the web concrete can prevent the penetration of diagonal

 https://doi.org/10.3390/ma18010049

cracks in the web concrete, thus avoiding the occurrence of brittle shear damage in the web concrete and improving the deformation capacity of the shear wall. Additionally, the dovetail profiled steel sheet has greater out-of-plane stiffness compared with the flat steel plate, which can improve the out-of-plane stability of the dovetail profiled steel sheet under the action of vertical force and reciprocating horizontal load, significantly improve the flexural bearing capacity of the dovetail profiled steel sheet, but also ensures that, during the construction of the shear wall, only a small amount of support can be set up to meet the requirements, which reduces the difficulty of construction. Additionally, profiled steel sheets exhibit significantly higher out-of-plane stiffness than flat steel sheets. This increased stiffness improves their stable load-bearing capacity under vertical and cyclic lateral forces. Furthermore, it enables shear wall construction to achieve out-of-plane stability with minimal support, simplifying the construction process. Therefore, DPSCWs with superior mechanical properties, simple structure, and efficient construction have good application prospects as a lateral force-resisting member in low- and mid-rise buildings.

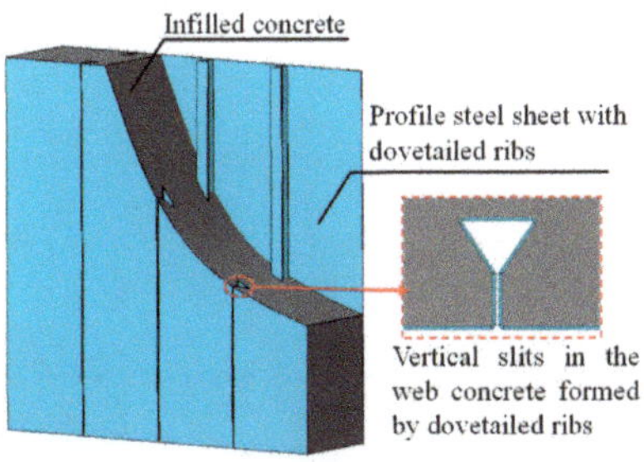

Figure 1. Dovetailed profiled steel concrete composite wall.

Uy and Wright et al. [1,2] conducted early experimental studies on the axial compression and compression-bending performance of DPSCWs and performed a theoretical analysis of the compressive buckling behavior of the profiled steel sheets. Based on this, a compression bending capacity calculation formula was proposed considering the post-buckling strength of profiled steel sheet. Tong et al. [3,4] further investigated the overall axial compressive stability, out-of-plane bending, and in-plane eccentric compression performance of DPSCWs using the finite element method based on the experimental data in the above studies, which supplemented and improved the corresponding design methods for closed dovetail profiled steel sheets concrete combined shear walls. Accordingly, the corresponding design method of DPSCWs was supplemented and optimized. Xu et al. [5] conducted experimental research on the seismic performance of DPSCWs with dovetailed ribs arranged in a staggered pattern. The effects of parameters such as axial compression ratio, concrete strength, steel sheet thickness and reinforced connection structure on the seismic performance of DPSCWs were studied. Consequently, the shear bearing capacity calculation formula for the shear wall was developed. Huang and Zhang et al. [6–9] conducted experimental research on the axial compression and hysteretic performance of DPSCWs. The results showed that the composite action between profiled steel sheets and web concrete can be achieved under the anchorage action provided by the dovetailed rib embedded in the concrete. Combined with finite element analysis, the axial compression and seismic failure mechanism of DPSCWs were revealed, along with the effects of parameters such as thickness ratio, shear span ratio, and axial compression ratio on lateral performance. A simplified calculation method and design recommendations for the axial compression and lateral bearing capacity of DPSCWs were provided.

In the above-mentioned study, the dovetail profiled steel sheets in DPSCWs were welded to the boundary members, and the results confirmed that this welding method provided a reliable connection. However, because the dovetail profiled steel sheets are thin-walled steel sheets with a thickness range of 0.7–2.0 mm, welding operations often cause obvious residual welding deformation near the welded area. Additionally, on-site welding is unsuitable for the requirements of prefabricated construction. Self-tapping screws, known for their high stiffness, bearing capacity, and deformation capability, offer advantages such as flexible installation, ease of operation, and low precision requirements, making them a preferred connection method for thin-walled steel plates [10–21]. This paper proposes using self-tapping screws to connect dovetail profiled steel sheets to the boundary members (Figure 2), providing an assembled and effective connection approach for DPSCW. This method addresses the residual deformation issues caused by welding in thin-walled structures and enhances construction flexibility.

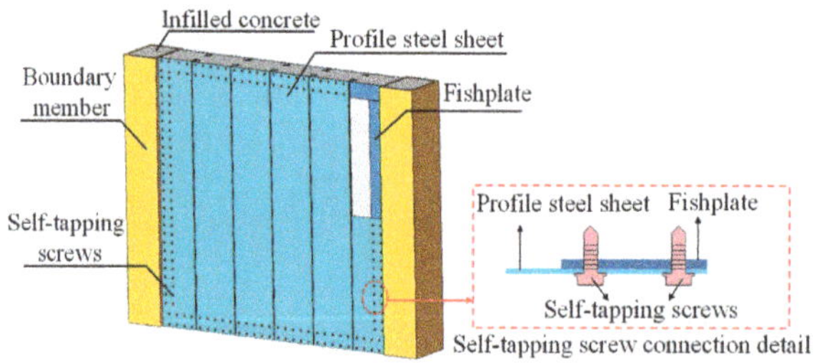

Figure 2. Self-tapping screw-connected DPSCWs.

Currently, many scholars have conducted studies on the mechanical properties of self-tapping screw connections [10–16] and thin-walled steel components connected by self-tapping screws [17–21], analyzing the force mechanisms involved and proposing design methods for these connections. However, no research has been conducted on the mechanical performance of DPSCWs connected by self-tapping screws, and their failure mechanism has not yet been clarified, especially the failure characteristics of self-tapping screw connections between thin sheets and thick substrates. Consequently, three DPSCW specimens connected by self-tapping screws were designed and fabricated for quasi-static tests. The influence of the number of self-tapping screws and the axial compression ratio on the failure mode, bearing capacity, strength and stiffness degradation, energy consumption, and deformation capacity of DPSCWs connected by self-tapping screws was analyzed to provide a reference for the seismic design of self-tapping screw-connected DPSCWs.

2. Test Program

2.1. Specimen Design

To ensure that the profiled steel sheets of the shear wall experienced shear failure, which would be the most unfavorable condition for testing the reliability of the self-tapping screw connections, three low shear-span ratio DPSCW with self-tapping screw-connected specimens, labelled DPSCW-S-1~DPSCW-S-3, were designed and tested to examine the cyclic behavior of DPSCW under combined axial and cyclic lateral loads. The parameters studied were the number of self-tapping screws and the axial compression ratio of the wall (Table 1). Figure 3 illustrates the specimen geometry and sectional details. All specimens had the same overall geometry, and each specimen had a barbell-shaped cross-section with an overall width of 1350 mm, which included a 1050 mm wide wall panel composed of double DPSs and concrete infill, as well as two 150 mm × 200 mm rectangular boundary concrete-filled steel tubes (CFSTs). It should be noted that the specimen dimensions were determined with consideration of the capacity of the loading equipment. The wall thickness was set at 160 mm, with 8 mm diameter U-shaped bars welded to the boundary steel tubes to facilitate shear transfer along the interface between the boundary CFSTs and the web concrete, ensuring coordinated deformation and preventing premature failure of the connections. In addition, three HRB400 longitudinal rebars (28 mm in diameter) were placed on each side near the outer web of the boundary CFSTs to enhance the bending capacity of the wall and to ensure that the boundary members did not fail before the wall web.

Table 1. Details of specimens.

Specimen	Number of Self-Tapping Screws	Design Axial Compression Ratio	Applied Axial Force N (kN)
DPSCW-S-1	56 (2 rows)	0.20	2000
DPSCW-S-1	78 (3 rows)	0.20	2000
DPSCW-S-3	78 (3 rows)	0.50	5000

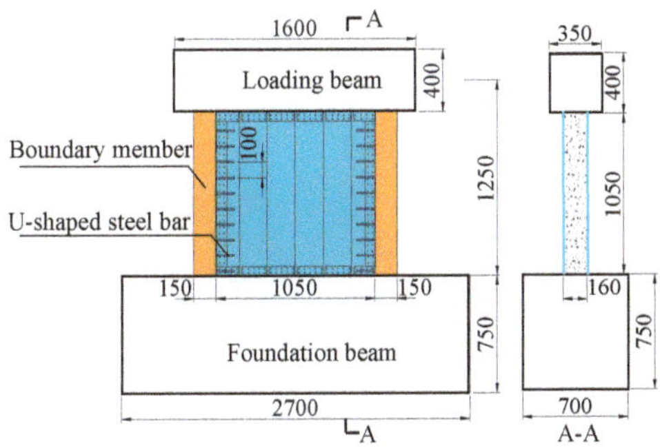

(a)

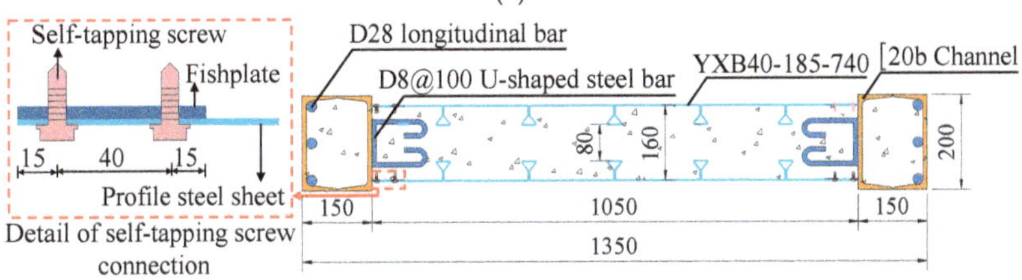

(b)

Figure 3. Configurations of specimen. (**a**) Overall Dimensions Schematic. (**b**) Cross-section. (unit: mm).

In the specimens, commercially available YXB40-185-740 (Xinshijie Coloured Steel Products Co., Wuxi, China) profiled steel sheets with a plate thickness of 1.2 mm (Figure 4) were selected and connected to the 3-mm-thick fishtail plates welded to the boundary members using ST4.8 self-drilling self-tapping screws (Figure 5). It is worth noting that these screws were carefully selected through precise design to ensure they would not fail before the connected steel plates. The technical specifications provided by the manufacturer for the ST4.8 screws are as follows: a diameter of 4.8 mm, a tensile strength of ≥550 MPa, a yield strength of ≥400 MPa, and a hardness range of HRC 20 to 28. The ST4.8 screws are suitable for connecting steel plates with a thickness of up to 3 mm and are available in various lengths to meet specific application requirements. The number of self-tapping screws (56 on one side) for specimen DPSCW-S-1 was determined based on the shear capacity of the profiled steel sheet according to in-plane shear yielding, as shown in Equation (1). These screws were arranged in two rows in accordance with the relevant placement requirements (Figure 5a). Considering the self-tapping screws are also subjected to a certain amount of tensile force after the profiled steel sheets buckling, the number of self-tapping screws of specimen DPSCW-S-1 is increased by approximately 1.4 times and are arranged in three rows according to the plum-shaped arrangement, thus obtaining the arrangement scheme (78 on one side) of specimens DPSCW-S-2 and DPSCW-S-3 (Figure 5b).

$$n = t_{\mathrm{sp}} l_{\mathrm{d}} f_{\mathrm{vp}} / N_{\mathrm{v}}^{\mathrm{f}} \tag{1}$$

where, t_{sp} denotes the thickness of the profiled steel sheet; f_{vp} denotes the shear strength of the profiled steel sheet; l_{d} indicates the width of wall web; 0.8 is the reduction factor applied to self-tapping screws when connecting the ends of the profiled steel sheets to supporting members; and $N_{\mathrm{v}}^{\mathrm{f}}$ presents the shear strength for individual self-tapping screws, which can be calculated according to Equation (2) with reference to the "Technical Specification for Cold-Formed Thin-Walled Steel Structures" [22].

$$N_{\mathrm{v}}^{\mathrm{f}} = 0.8 \times 2.4 t_{\mathrm{sp}} d f_{\mathrm{up}} \tag{2}$$

where d denotes the self-tapping screw diameter and f_{up} denotes the tensile strength of the profiled steel sheet.

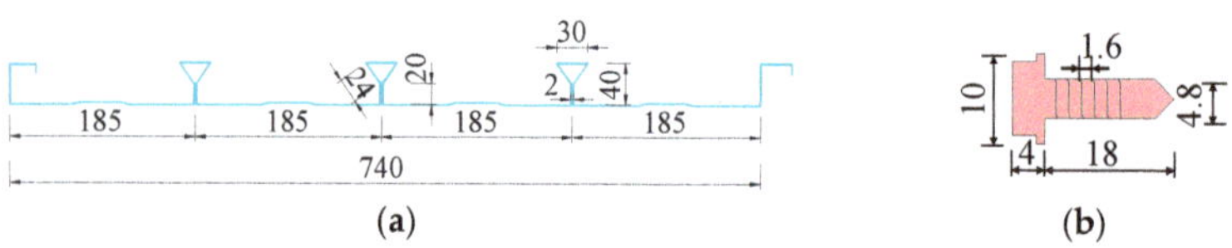

Figure 4. Details of YXB40-185-740 profiled steel sheet and self-tapping screw. (**a**) YXB40-185-740 profiled steel sheet. (**b**) Self-tapping screw. (unit: mm).

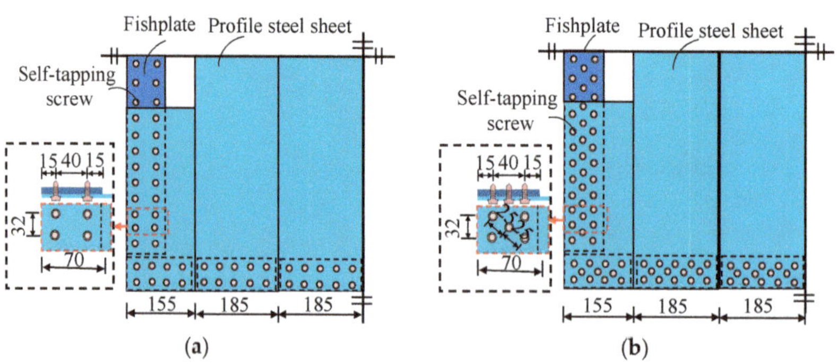

Figure 5. Layout of self-tapping screws. (**a**) Two rows of self-tapping screws. (**b**) Three rows of self-tapping screws. (unit: mm).

The designed axial compression ratio of 0.2 and 0.5 was adopted in this study. The relationship between the applied compression load in test and designed axial compression ratio can be seen in Equation (3):

$$n_{\mathrm{d}} = \frac{0.8 \times 2.4 t_{\mathrm{sp}} d f_{\mathrm{up}}}{f_{\mathrm{c}} A_{\mathrm{c}} / 1.4 + (f_{\mathrm{yp}} A_{\mathrm{p,eff}} + f_{\mathrm{yt}} A_{\mathrm{st}} + f_{\mathrm{yr}} A_{\mathrm{sr}}) / 1.11} \tag{3}$$

where n_d represents the design axial load ratio, f_c denotes the compressive strength of infilled concrete, and f_{yp}, f_{yt}, and f_{yr} indicate the yield strengths of the profiled steel sheet, boundary steel tube, and longitudinal rebars placed in the boundary members, respectively. A_c, A_{st}, and A_{sr} correspond to the cross-sectional areas of the concrete, boundary steel tubes, and longitudinal rebars, respectively, and $A_{sp,eff}$ refers to the effective cross-sectional area of the profiled steel sheets after considering the post-buckling strength, which can be determined using the formula provided in Ref. [7]. Finally, 1.4 and 1.11 are the partial safety factors for concrete and steel materials, respectively.

A reinforced concrete foundation beam with a cross-section of 750 mm × 600 mm was cast integrally with the wall, securing the specimen to the reaction floor. A reinforced concrete loading beam, measuring 400 mm × 350 mm, was also cast to apply vertical and lateral loads to the wall. The effective height (H) of the lateral force application point from the top of the foundation beam (Figure 3a) was 1250 mm. Consequently, the shear-span ratio (λ) of these specimens was 0.93, defined as the ratio of the specimen's effective height to the section width [23,24].

2.2. Specimen Fabrication

The fabrication steps for the specimens are as follows: (1) Weld U-shaped steel bars and fishtail plates onto the side of the boundary steel tube that connects to the profiled steel sheets. Position and assemble welded U-bars and fishtail plate boundary steel tubes with profiled steel plates. Then, secure the profiled steel sheets to the fishtail plates welded to the boundary steel tubes using self-tapping screws, thereby completing the steel component of the wall. (Figure 6a). (2) Tie the reinforcement cages for the loading and foundation beams, and weld sufficient studs onto the sections of the steel wall components embedded within the loading and foundation beams to prevent pull-out. Subsequently, position the steel components within the reinforcement cages and secure them in place (Figure 6b). (3) Pour commercial concrete in the wall after formwork support of the load beams and foundation beams (Figure 6c). The concrete for all specimens was poured in layers during a single session, with a vibrating rod used during the process to ensure proper compaction and density. After pouring, the specimens were cured outdoors under natural conditions. The completed specimens are shown in Figure 6d.

2.3. Material Properties

The specimens were made of commercial concrete with a strength grade of C30. During concrete casting, three cubic specimens (150 mm side length) and three prismatic specimens (150 mm × 150 mm × 300 mm) were reserved for each specimen. In accordance with the Chinese standard GB/T 50081-2019 [25], these concrete specimens were tested simultaneously with the main specimen using a 200-ton press. The concrete cubic strength (f_{cu}), prism strength (f_{cp}), concrete modulus of elasticity (E_c), and Poisson's ratio (ν_c) of each specimen were measured and are listed in Table 2. It should be noted that the measured strength of concrete is lower than the nominal strength due to factors such as low winter temperatures during concrete curing.

Table 2. Concrete properties.

Specimen	f_{cu} (MPa)	f_{cp} (MPa)	E_c (GPa)	υ_c
DSCW-S-1	28.25	26.33	27.9	0.19
DSCW-S-2	30.07	28.87	29.8	0.20
DSCW-S-3	27.11	22.63	26.9	0.20

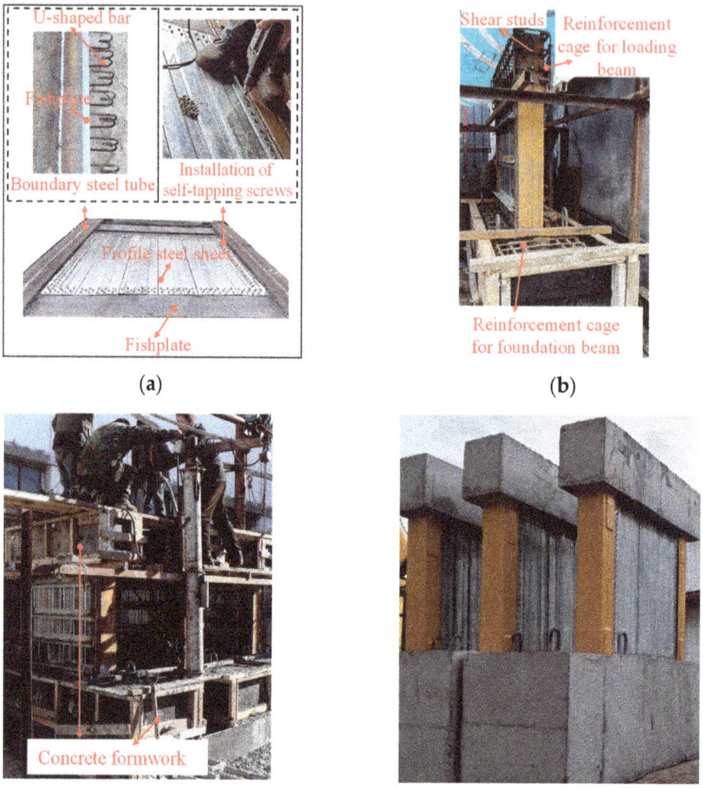

(a) (b) (c) (d)

Figure 6. Layout of self-tapping screws. (**a**) Fabrication of steel components. (**b**) Specimen assembly. (**c**) Formwork support and concrete pouring. (**d**) Completed specimens.

Three standard tensile coupons were prepared for the profiled steel sheet, fishplate, channel steel, and rebar used in the DPSCW specimens, and the main mechanical property indexes of steel were measured in accordance with the provisions of "Tensile Test of Metallic Materials Part 1: Room Temperature Test Methods" (GB/T228.1-2010) [26]. The average measured results of thickness or diameter, yield strength (f_y), yield strain (ε_y), ultimate strength (f_u), Young's modulus (E_s), and Poisson's ratio (ν_s) of these coupons are presented in Table 3.

Table 3. Mechanical properties of steel elements.

Components	Measured Thickness or Diameter (mm)	f_y (MPa)	ε_y	f_u (MPa)	E_s (GPa)	ν_s
YXB40-185-740	1.17	331	0.0018	382	200	0.28
Fishplate	2.96	325	0.0017	372	195	0.29
Channel steel	8.95	341	0.0018	490	199	0.26
Longitudinal rebar	27.40	467	0.0023	648	182	-

2.4. Test Set-Up and Instrumentation

The test loading was performed on a multifunctional loading device at the Structural Engineering Laboratory of Harbin Institute of Technology (Figure 7). The device had a vertical loading capacity of 30,000 kN and a lateral loading capacity of 3000 kN. The specimen's foundation beam was anchored to the test platform with ground anchor bolts, and two hydraulic jacks were placed at both ends to prevent sliding. The specimen loading beam was connected to the 300 t horizontal actuator via end plates and lead screws. To ensure uniform distribution of the vertical load across the wall section, a rigid steel beam was positioned between the specimen's loading beam and the vertical actuators. The vertical actuator, connected to the reaction frame with polytetrafluoroethylene (PTFE) slide plates, was free to move horizontally to accommodate the specimen's lateral displacement.

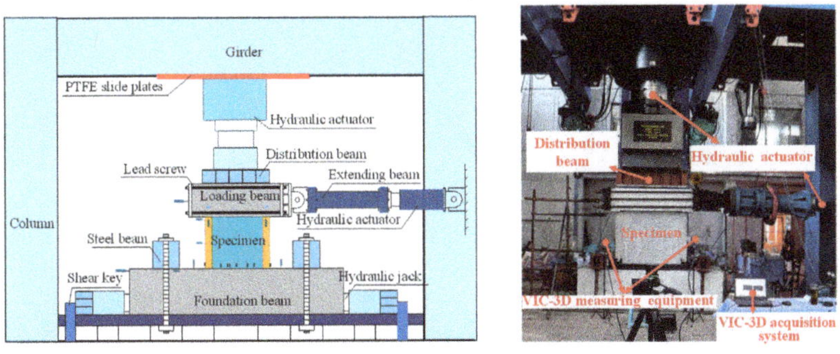

Figure 7. Test setup. (**a**) Schematic of the loading device. (**b**) Photo of loading device.

The measurement system was designed to capture the structural responses of the specimens during the whole loading process (Figure 8). The VIC-3D 8 non-contact full-field strain measurement system, produced by Correlated Solutions, Inc. (CSI), Columbia, SC, USA, was employed for full-field observations of strains and displacements, allowing for the analysis of buckling development in the profiled steel plate. As shown in Figure 8a, the entire front steel surface of the specimens was painted white to enhance the contrast, with random black dots speckled on the steel surface. Linear variable differential transformers (LVDTs) were used to measure displacements at key locations. As displayed in Figure 8b, four LVDTs labeled as H1~H4 were positioned at different heights of the specimens to measure their lateral displacements. Additionally, four LVDTs, labeled as V1~V4, were placed on the loading beam and foundation beam to detect the rotation of the specimens. Strain gauges were installed at important positions to measure the strain development of the steel plate.

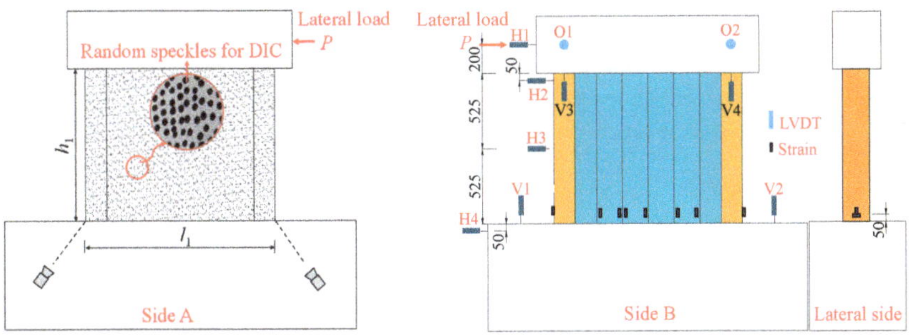

Figure 8. Measurement scheme. (**a**) VIC-3D measurement system. (**b**) Displacement and strain gauges arrangement. (unit: mm).

2.5. Loading Scheme

The loading process followed the Chinese specification JGJ/T 101-2015 [27]. Preloading was first carried out to verify the proper functioning of the loading systems. Compression loads were then applied, beginning at 50% of the expected forces and gradually increasing to 100% in accordance with the axial compression ratio. The compression load was maintained at a constant level during the subsequent lateral cyclic loading phase. The initial loading cycle was force-controlled, with a lateral load of 200 kN. Once the drift ratio reached approximately 1/400, the control mode switched to displacement control. The yield deformation of the specimen was taken as the step length, and each subsequent step consisted of three cycles until the actuator force decreased to 85% of the peak load.

3. Test Results and Discussion

3.1. Experimental Phenomenon

The specimens exhibited similar failure characteristics under both push and pull loading directions. For clarity, the failure phenomena described here are based on experimental observations during positive loading, when the actuator was pushed to the left (Figure 7).

For the specimen DPSCW-S-1 with two rows of self-tapping screws, no significant changes were observed during the axial compression phase or the initial stage of lateral loading (lateral drift ratio θ less than 1/500). When the lateral load P reached approximately 1000 kN (0.58 P_{m}, where P_{m} is the peak lateral load), the VIC-3D system captured slight shear buckling on the strip of the steel sheets (Figure 9a), and the sound of bond-slip between the concrete and profiled steel sheets was heard. At this point, the angle between the buckling wave and the loading direction currently is about 45°~50°; after unloading, the buckling wave deformation mostly recovered. As the load increased, new buckling waves continuously formed, the original buckling wavelength extended, and the height of the waves grew. At a lateral load of 1474 kN (0.86 P_{m}, θ = 1/254), 2–3 buckling waves appeared on most strips, except for the third strip from the left. Additionally, slight warping of the steel plate was observed between the two self-tapping screws at the upper right corner of the wall. When the load was reduced to zero, visible residual deformations were observed on the plate strips (Figure 9b), indicating that the specimen had entered a distinct elastic-plastic stage. At 1709 kN (θ = 1/170), the specimen reached the peak lateral load, and the previously formed buckling wave extended into a tensile band, with the angle between the tensile band and the loading direction increasing to 60–65°. Meanwhile, slight warping was observed on the profiled steel sheet between the two self-tapping screws in the upper left corner (Figure 9c). After the peak load, the cyclic lateral load caused further damage and plastic deformation, leading to a gradual reduction in the specimen's bearing capacity. The tension fields extended and the gaps between the dovetailed ribs widened as the drift ratio increased, but the inclination angle remained stable. Additionally, warping of the profiled steel sheets between the screws at the lower right corner was noted (Figure 9d). During the whole test, the specimen experienced a progression of failure from local shear buckling of the profiled steel sheet strips, followed by the development of tension fields bounded by the ribs embedded in the concrete, no disengagement or shearing of the self-tapping screws was observed, and no loosening was detected when the screws were twisted by hand.

The failure processes of specimen DPSCW-S-2 and specimen DPSCW-S-3 were similar to that of specimen DPSCW-S-1 and will not be repeated here. Notably, the shear buckling of the profiled steel sheets for specimen DPSCW-S-3 was captured by VIC-3D at the load level of 800 kN, indicating that higher axial compression ratio causes the profiled steel sheets of DPSCW to experience shear buckling earlier. In addition, tearing was observed at the turning point of the local tensile field in specimen DPSCW-S-2 and DPSCW-S-3 during the load decrease period.

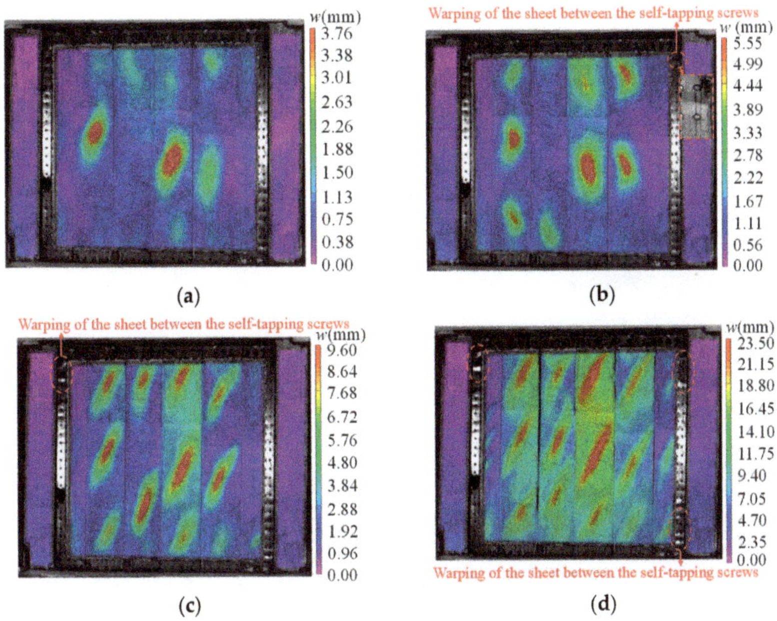

Figure 9. Experimental observations of specimen DPSCW-S-1. (**a**) Initial buckling of DPS. (**b**) Obvious residual deformation. (**c**) Peak load. (**d**) Ultimate displacement.

3.2. Failure Mode

The overall failure modes of the three specimens are shown in Figure 10. Obvious residual localized tension fields deformation was observed on profiled steel sheets for each specimen, exhibiting significant shear damage. Warping of the steel sheet between the two self-tapping screws was observed around the whole steel sheet, especially at the corners. The warping of the specimen DPSCW-S-1 with two rows of self-tapping screws was more serious than that of the specimens DPSCW-S-2 and DPSCW-S-3 with three rows of self-tapping screws. Additionally, tearing occurred at the turning points of the local tension fields in specimen DPSCW-S-2 and DPSCW-S-3. Apart from the localized warping of the profiled steel sheets between the self-tapping screws, the failure modes of the self-tapping screw-connected profiled steel plates were generally consistent with the failure modes of welded profiled steel plates reported in the existing literature [9]. After removing the steel sheets, there were small, inclined cracks in the concrete column between the ribs, but these inclined cracks were not continuous and did not form a diagonal crack, and concrete crushing was observed at the upper and lower ends of each concrete column between the two ribs. Upon removing the self-tapping screws, it was discovered that the nail holes in the profiled steel near the areas of warping were slightly enlarged, while the holes in the fishtail plate showed no significant deformation. The removed self-tapping screws remained intact.

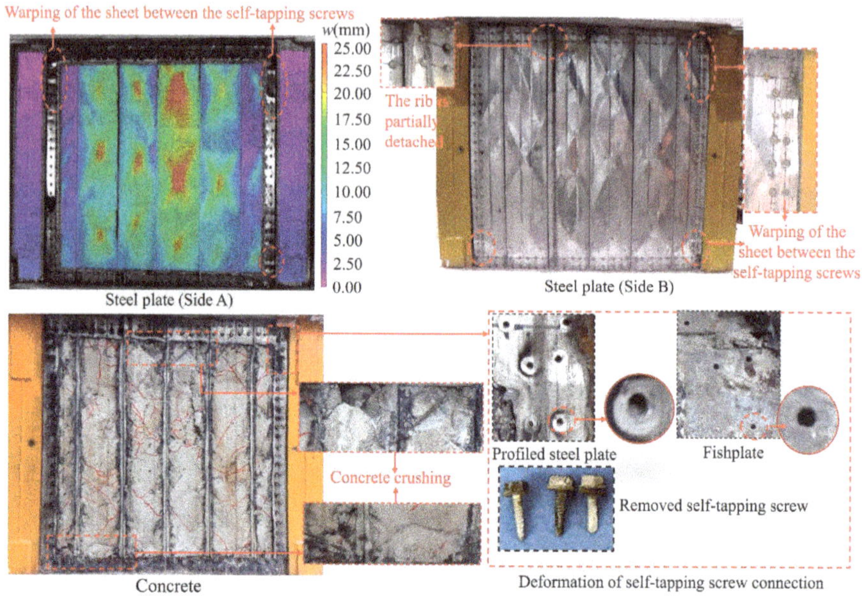

(a)

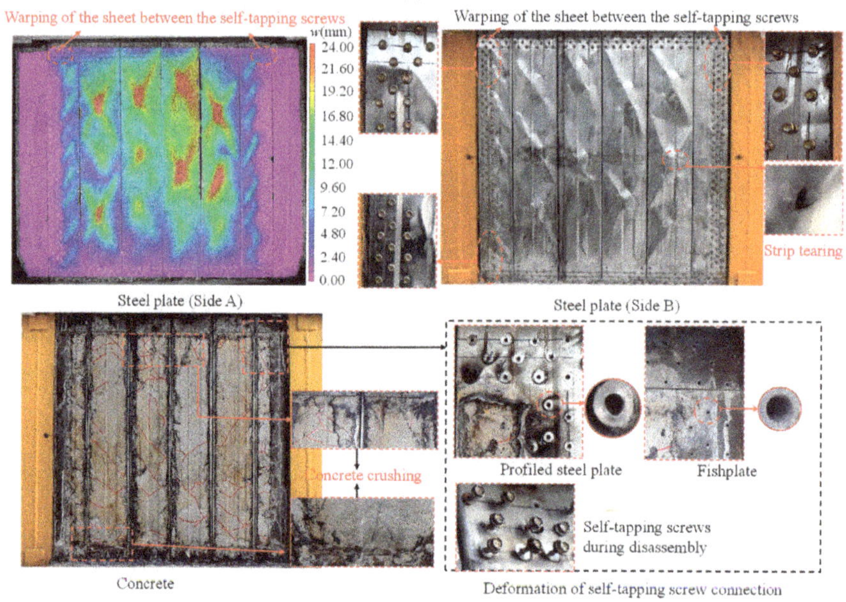

(b)

Figure 10. *Cont.*

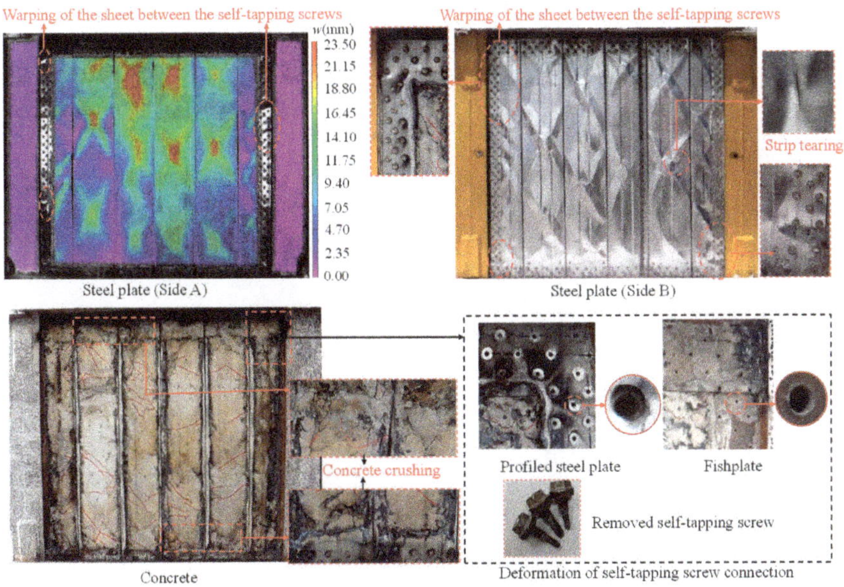

(**c**)

Figure 10. Specimen failure mode. (**a**) Specimen DPSCW-S-1. (**b**) Specimen DPSCW-S-2. (**c**) Specimen DPSCW-S-3.

3.3. *Lateral Load-Displacement Hysteretic Response*

The lateral load-displacement hysteresis curves of the specimens generally exhibit a reverse "S" shape (Figure 11), with some pinching observed, and the positive and negative loading curves were basically symmetrical. The characteristic points of the first buckling of the DPS, the overall yield of the specimen, the peak load of the specimen, and the yield of the boundary CFSTs for each specimen were pointed at their skeleton curves, where the definition of specimen yielding point is determined from equivalent elastoplastic energy method [28,29]. Before the shear buckling of profiled steel sheets were generated and observed, the specimens showed the linear deformation characteristics, for which the lateral load-displacement curves of loading and unloading were coincident basically with few energy dissipations. With the generation and development of the shear buckling of profiled steel sheets, accompanied by the increases of lateral load, the stiffness of the three specimens decreased obviously, the residual deformation increases, and the area of the hysteresis loop gradually increases. As the horizontal load continued to increase, numerous shear buckling waves were generated and developed on the strips, indicating that the profiled steel sheets exhibit the profound shear deformation under combined axial compression and lateral cyclic loads. Besides, due to the development and closure of concrete cracks under lateral reciprocating loads and the flattening of the buckling wave during unloading and reverse reloading, the hysteresis curve gradually exhibits pinching. After reaching the peak load points, all specimens exhibited a gradual decline in lateral bearing capacity as lateral displacement increased, indicating good deformation capacity.

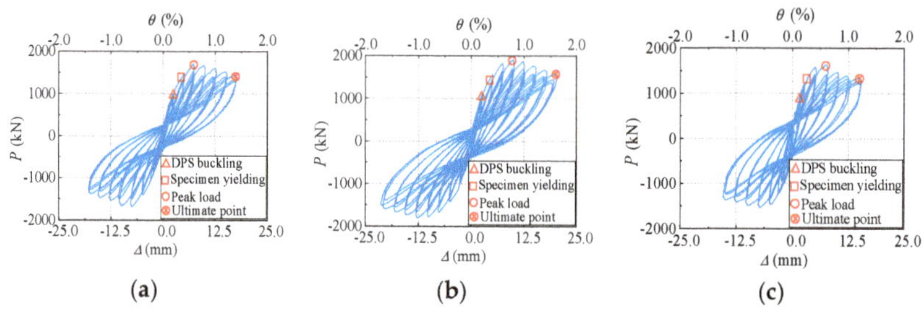

Figure 11. Hysteresis curves. (**a**) Specimen DPSCW-S-1. (**b**) Specimen DPSCW-S-2. (**c**) Specimen DPSCW-S-3.

By connecting the peak points of the first cycle at each increment, the skeleton curves for the hysteresis curves were established (Figure 12). The skeleton curves of each specimen are basically symmetric under different loading directions, with the load differences under push and pull loading being at the same displacement level within 5%. Comparative analysis of the envelope curves of specimens DPSCW-S-1 and DPSCW-S-2 showed that the lateral load capacity of DPSCW-S-2 with more self-tapping screws increased from 1695 kN to 1843 kN, which is an increase of 9%, compared to DPSCW-S-1 with fewer screws. However, their initial stiffnesses were similar at 1044 kN/mm and 1066 kN/mm, respectively. For specimens DPSCW-2 and DPSCW-3, which had the same number of self-tapping screws, the axial compression ratios were 0.2 and 0.5, respectively. Increasing the axial compression ratio from 0.2 to 0.5 raised the initial stiffness of the shear wall from 1066 kN/mm to 1289 kN/mm, a 21% increase. However, the load capacity decreased from 1843 kN to 1581 kN, and the load capacity decreased more quickly after the peak load.

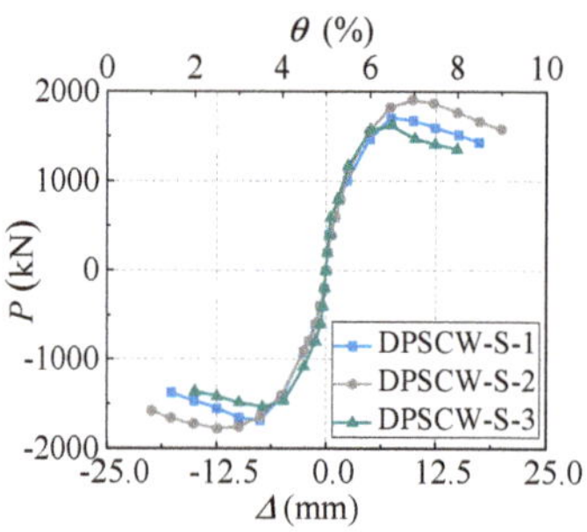

Figure 12. Skeleton curves.

3.4. Ductility

The ductility of specimens can be evaluated by the displacement ductility ratio (μ), which can be calculated through $\mu = \Delta_u/\Delta_y$; Δ_y represents the lateral displacement of yielding and Δ_u represents the lateral displacement at which the lateral resistance of specimens decreases by 85% of the peak load. Table 4 lists the horizontal load, drift ratios, and displacement ductility coefficients corresponding to the three characteristic points of yield point, peak point, and limit point of each specimen. The deformation capacity of the self-tapping screw-connected specimen is slightly worse than that of the welded specimen. The average yield drift ratios for specimens DPSCW-S-1, DPSCW-S-2, and DPSCW-S-3 are 1/257, 1/225, and 1/286, respectively, and the average ultimate drift ratios are 1/78, 1/63, and 1/94, respectively. These values exceed the requirements specified in the "Code for Seismic Design of Buildings" (GB 50010-2010) [30], which states that the elastic inter-story drift angle for shear walls should not exceed 1/1000, and the inelastic inter-story drift angle should not exceed 1/120. The ductility coefficients for specimens DPSCW-S-1, DPSCW-S-2, and DPSCW-S-3 are 3.31, 3.51, and 3.09, respectively, all greater than 2, meeting the criteria for moderate ductility failure as defined in the literature [31].

Table 4. Comparison of experimental results.

Specimen	Load Direction	Initial Stiffness	Yield of DPSCW			Peak Lateral Load			Ultimate Displacement			Ductility Coefficient
		K_0/(kN/mm)	P_y/kN	Δ_y/mm	θ_y	P_m/kN	Δ_m/mm	θ_m	P_u/kN	Δ_u/mm	θ_u	$\mu = \Delta_u/\Delta_y$
DPSCW-S-1	Puch	1063	1454	4.80	1/263	1709	7.30	1/172	1453	16.18	1/78	3.38
	Pull	1025	1438	4.94	1/250	1680	7.52	1/167	1428	15.97	1/78	3.23
	Average	1044	1446	4.87	1/257	1695	7.41	1/170	1441	16.07	1/78	3.31
DPSCW-S-2	Puch	1072	1617	5.48	1/227	1876	9.94	1/125	1628	19.52	1/64	3.56
	Pull	1061	1507	5.89	1/213	1810	12.62	1/99	1587	20.35	1/61	3.45
	Average	1066	1562	5.69	1/225	1843	11.28	1/111	1608	19.94	1/63	3.51
DPSCW-S-3	Puch	1340	1385	4.14	1/294	1627	7.5	1/167	1391	13.05	1/94	3.15
	Pull	1238	1355	4.49	1/278	1535	7.09	1/175	1348	13.58	1/93	3.02
	Average	1289	1370	4.32	1/286	1581	7.30	1/167	1369	13.32	1/94	3.09

Increasing the number of self-tapping screws by 40% led to a 14% increase in elastic drift ratio and a 23% increase in inelastic drift ratio, with the ductility coefficient rising from 3.31 to 3.51, suggesting improved deformation capacity and ductility with more screws. However, increasing the axial compression ratio from 0.2 to 0.5 decreased the elastic drift ratio by 22%, the inelastic drift ratio by 33%, and the ductility coefficient from 3.51 to 3.09, reflecting diminished deformation capacity. Although the load-bearing capacity of the unconnected specimens did not significantly decrease, its deformation capacity failed to meet the relevant code requirements.

3.5. Stiffness and Strength Degradation

The development of cracks as well as the accumulation of inelastic deformations due to the development of local tensile fields in the profiled steel sheet under cyclic loading can result in the gradual deterioration of stiffness and strength, which can assess the damage degree of the DPSCWs connected by self-tapping screws. Referring to Chinese code JGJ/T 101-2015 [27], the secant stiffness K_j, which is defined in Equation (4), is used to depict stiffness degradation (Figure 13). At the initial loading stage, all specimens experienced significant stiffness reduction, which followed an exponential distribution as lateral displacement increased. After yielding, the stiffness degradation of each specimen gradually slowed and stabilized. Under the same axial compression ratio, the stiffness degradation curves of specimen DPSCW-S-2, with more self-tapping screws, nearly coincided with those of specimen DPSCW-S-1, which had fewer screws, up to the peak load. Beyond the peak load, DPSCW-S-2 exhibited a slower rate of stiffness degradation compared to DPSCW-S-1. With a higher axial compression ratio, the initial stiffness of specimen DPSCW-S-3 increased from 1066 kN/mm to 1289 kN/mm, a 21% improvement. However, the cyclic stiffness degradation rate was faster and more pronounced in specimens with a higher axial compression ratio.

$$K_j = \frac{|+P_j| + |-P_j|}{|+\Delta_j| + |-\Delta_j|} \quad (4)$$

where $+P_j$ and $-P_j$ represent the peak lateral loads at j_{th} loading level under push and pull loading, respectively; $+\Delta_j$ and $-\Delta_j$ represent the corresponding lateral displacement associated with $+P_j$ and $-P_j$, respectively.

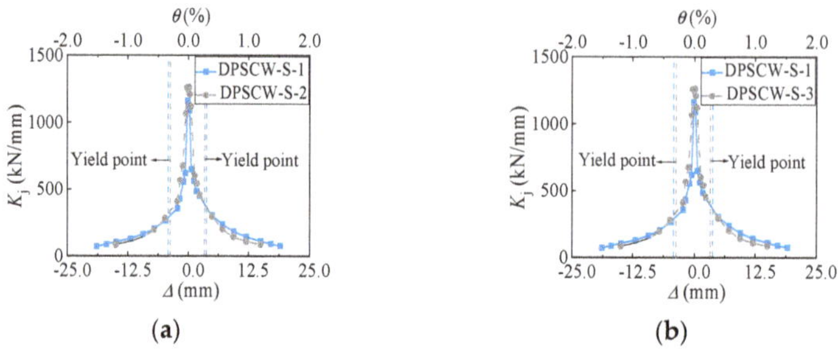

Figure 13. Stiffness degradation curves. (**a**) Specimen DPSCW-S-1. (**b**) Specimen DPSCW-S-2.

Referring to Chinese code JGJ/T 101-2015 [27], determination of the strength degradation ratio λ_j can be calculated using Equation (5), where a lower coefficient value indicates greater strength loss. As shown in Figure 14, before peak load, the strength degradation index of all specimens gradually decreased from 0.97–1.0 to approximately 0.84–0.88. After the peak load, the strength degradation index exhibits fluctuations but all larger than 0.9. This indicates that the load-bearing capacity of self-tapping screw-connected DPSCWs is generally stable, and that the self-tapping screw connections can ensure cooperative performance between the profiled steel sheets and boundary members. Additionally, it was observed that the strength degradation index during the third cycle of repeated loading was higher than that during the second cycle, indicating that the damage caused by the third loading cycle was less severe than that of the second at the same drift angle.

$$\lambda_j = \frac{P_j^{i+1}}{P_j^{i}} \tag{5}$$

where P_j^{i} represents the peak lateral load of i_{th} load cycle under the j_{th} load level and P_j^{i+1} is the peak lateral load of $(i + 1)_{th}$ load cycle under the j_{th} load level.

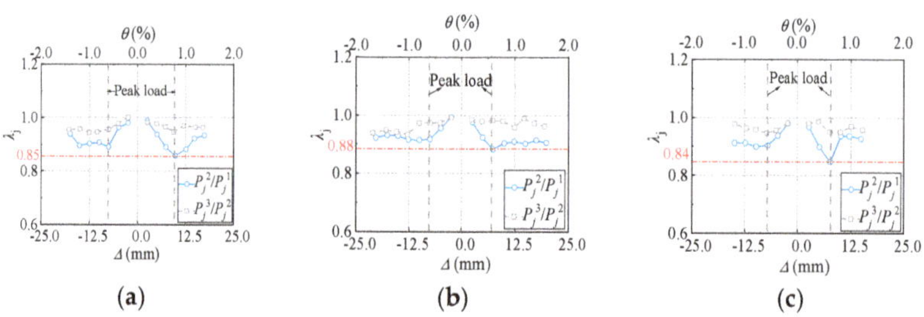

Figure 14. Strength degradation curve. (**a**) Specimen DPSCW-S-1. (**b**) Specimen DPSCW-S-2. (**c**) Specimen DPSCW-S-3.

Overall, the DPSCW specimens connected by self-tapping screws show relatively stable strength and stiffness degradation behavior, indicating that the self-tapping screw connection ensures effective cooperation between the profiled steel plate and the boundary members under seismic loads. The reliability of the self-tapping screw connection has been validated.

3.6. Energy Dissipation

The enclosed area (E_j^{i}) of hysteresis curves can be used to evaluate the energy dissipation of DPSCWs under cyclic loadings. Specifically, at the same deformation level, the larger the area enclosed by the average single-cycle hysteresis curve ($E_j^{\,i} = \sum_{i=1}^{n} E_j^{\,i}/n$), the greater the energy dissipated per cycle, indicating higher energy dissipation efficiency. The greater the total area of all hysteresis loops ($\sum E_j = \sum_{j=1}^{m}\sum_{i=1}^{n} E_j^{\,i}$) of the specimen before failure, the larger the total energy dissipated, reflecting the specimen's stronger energy dissipation capacity. Higher energy dissipation efficiency and capacity are beneficial for improving the seismic performance of self-tapping screw-connected DPSCWs. Figure 15 illustrates the average energy dissipation per cycle (E_j) and the cumulative energy dissipation($\sum E_j$) at each displacement level of the test specimens. Initially, both the average energy dissipation per cycle and cumulative energy dissipation of the specimen are small. As the local buckling of the profiled steel plate occurred, the specimens entered the elastoplastic stage, significantly increasing energy dissipation. At the same displacement level, DPSCW-S-2, with three rows of self-tapping screws, exhibited 55% higher cumulative energy dissipation (2486 kN·mm) compared to DPSCW-S-1, which had two rows (1605 kN·mm). This indicates that increasing the number of self-tapping screws around the profiled steel sheet enhances the restraining effect of the screws on the sheet, leading to more fully developed localized tension filed along the sheet strips, and the deformation of the specimen and the ability to dissipate energy were improved. Specimen DPSCW-S-3, which had a higher axial compression ratio, exhibited greater average and cumulative energy dissipation than specimen DPSCW-S-2, which had a lower axial compression ratio, at the same displacement level. However, due to the poorer deformability of specimen DPSCW-S-3 with a larger axial compression ratio than that of specimen DPSCW-S-2, the final cumulative total energy dissipation of DPSCW-S-3 (1399 kN·mm) is 44% less than that of specimen DPSCW-S-2 (2486 kN·mm). This suggests that although increasing the axial compression ratio enhances per-cycle energy dissipation, it diminishes overall deformability, resulting in lower total cumulative energy dissipation. Therefore, the axial force applied to concrete composite shear walls with self-tapping screw-connected dovetail profiled steel sheets should be carefully managed.

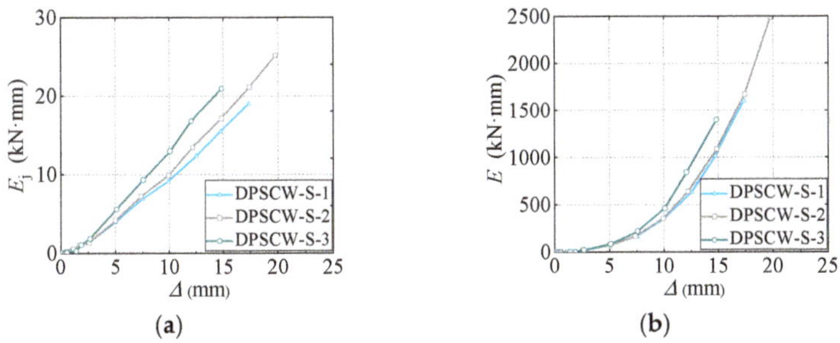

Figure 15. Energy dissipation curve. (**a**) Dissipated energy per load cycle. (**b**) Cumulative dissipated energy.

4. Design Method for the Self-Tapping Screw Connections in DPSCWs

Based on the test analysis results of DPSCWs connected using self-tapping screws, the failure mode of specimen DPSCW-S-1, designed according to the shear capacity of the profiled steel sheet, was similar to that of specimen DPSCW-S-2, in which the number of self-tapping screws was increased by approximately 1.4 times compared to DPSCW-S-1 and arranged in a plum blossom pattern across three rows. However, the deformation of the self-tapping screw holes on the profiled steel sheet of the specimen DPSCW-S-1 is more obvious, and the ribs of the sheet are partially detached. Furthermore, the load-bearing capacity and deformation ability of specimen DPSCW-S-1 were both inferior to those of specimen DPSCW-S-2. This indicates that the number of self-tapping screws calculated using the design method adopted during the specimen design should be insufficient. Therefore, the design method for the lateral resistance and connections of self-tapping screw-connected profiled steel plates in DPSCWs should be developed based on their failure mechanisms.

Experimental studies on DPSCWs with self-tapping screw connections showed that failure of the dovetail profiled steel sheets is characterized by local buckling and the development of extended tension fields on the strips. Additionally, within the studied range of axial compression ratios, calculations based on the von Mises yield criterion show that the influence of axial stress on the shear capacity of the profiled steel sheets is less than 5.2%. Therefore, the effect of axial stress on the profiled steel sheets is negligible, and the overall shear capacity of the profiled steel sheets can be approximated as the shear response of multiple independent strips. Due to the periodic occurrence of buckling waves along the length of the profiled steel strips, the shear behavior of each strip can further be approximated as the shear response of a single tension band segment (Figure 16).

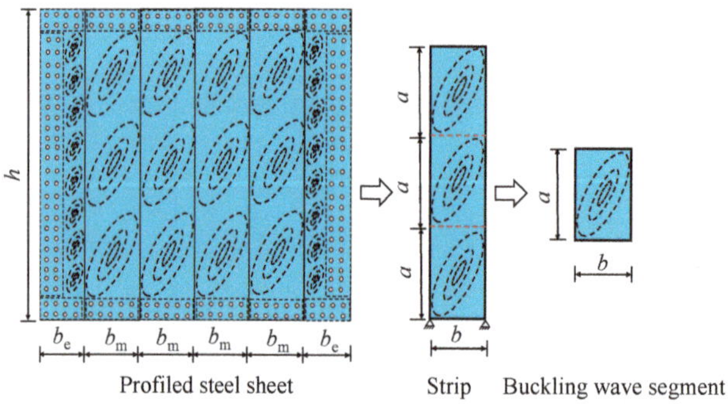

Figure 16. Simplified analytical model for profile steel sheet.

Referring to the North American specification AISI S400-15 [32,33], the effective strip method was employed to calculate the shear capacity of the tension band segment. Figure 17a illustrates the effective strip method model for a single tension band segment. In Figure 17a, a represents the height of a single tension band segment on the profiled steel sheet's strip and b denotes the width of the tension band segment; the length a is twice the width b, as determined from the local tension bands observed in the test. The angle α, defined as the angle between the lateral and diagonal directions of the tension band segment, is given by $\alpha = \arctan(a/b)$. T_u represents the load carrying capacity of the effective strips. W_e denotes the width of the effective strips, which can be calculated according to the following formula [34,35]:

$$W_e = \begin{cases} W_{max}, & \lambda \leq 0.0819 \\ \rho W_{max} & \lambda > 0.0819 \end{cases} \tag{6}$$

$$\rho = \frac{1 - 0.55(\lambda - 0.08)^{0.12}}{\lambda^{0.12}} \tag{7}$$

$$\lambda = \frac{1.736\alpha_1\alpha_2}{\beta_1\beta_2{\beta_3}^2\psi} \tag{8}$$

where W_{max} represents the maximum effective strip width, calculated as $b/\sin\alpha$. ρ is the discount factor of effective strip width. λ is the correction factor that accounts for component thickness, material properties, self-tapping screw spacing, aspect ratio, and other factors of the profiled steel sheet. α_1, α_2 are the correction factors considering the difference of materials, $\alpha_1 = f_{up}/310.3$, $\alpha_2 = f_{uf}/310.3$, with f_{up} and f_{uf} being the ultimate tensile strength of the material of the profiled steel sheet and the fishtail plate, respectively. β_1, β_2 are the correction coefficients considering the difference in the thickness of the members, defined as $\beta_1 = t_{sp}/0.457$, $\beta_2 = t_{sf}/0.457$, with t_{sf} being the thicknesses of the fishtail plate; β_3 is the correction coefficient considering the spacing and rows of the tapping screws, given by $\beta_3 = (2R \times s)/152.4$, where R and s are the rows and self-tapping screw spacing of the self-tapping screws, respectively; ψ represents the height-to-width ratio of individual tension strip segments.

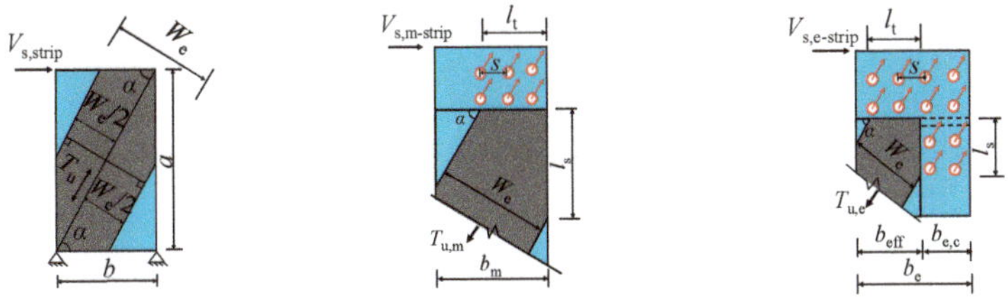

Figure 17. Simplified analytical model for self-tapping screw connections of profile steel sheets. (**a**) Effective strip method model. (**b**) Self-tapping screw force analysis in middle strip. (**c**) Self-tapping screw force analysis in edge strip.

Based on the effective strip model for the single tension band segment, the analytical model of the self-tapping screw force within the effective strip width on the middle strips and edge strips was given in Figure 17b,c, so that the shear bearing capacity of the intermediate and edge strips can be obtained as:

$$V_{s,m-strip} = T_{u,m} \times \cos\alpha \tag{9}$$

$$V_{s,e-strip} = T_{u,e} \times \cos\alpha \tag{10}$$

where $T_{u,m}$ and $T_{u,e}$ are the bearing capacity of the effective strip for the middle and edge strips, respectively, as calculated in Equations (11) and (12):

$$T_{u,m} = \min(P_{n,m}, \frac{W_e}{2} t_{sp} f_{yp}) + \frac{W_e}{2} t_{sp} f_{yp} \tag{11}$$

$$T_{u,e} = \min(P_{n,e}, \frac{W_e}{2} t_{sp} f_{yp}) \tag{12}$$

where $P_{n,m}$ and $P_{n,e}$ represent the shear bearing capacities of the self-tapping screws within the effective strip width of the middle and edge strips, respectively.

Based on the arrangement and stress analysis of self-tapping screws within the effective strip width presented in Figure 17, the formulas for calculating the shear bearing capacities $P_{n,m}$ and $P_{n,e}$ of these self-tapping screws in the tension band sections on the intermediate and edge strips are as follows:

$$P_{n,m} = (\frac{W_e}{2s\sin\alpha} R - \xi) N_{v,b}^f \tag{13}$$

$$P_{n,e} = (\frac{W_e}{2s\sin\alpha} R - \xi) N_{v,b}^f + (\frac{W_e}{2s\cos\alpha} R - \xi) N_{v,b}^f + n_c N_{v,b}^f \tag{14}$$

where $N_{v,b}^{f}$ is the shear bearing capacity of a single self-tapping screw, which can be calculated using Equation (2); ξ is equal to 0 when the self-tapping screws are arranged side by side, and 1 when arranged in plum blossom shape; n_c represents the number of self-tapping screws at the corners of the edge strips with the effective strips.

Adopting Equations (9)–(14), the formula for calculating the shear capacity of self-tapping screw-connected profiled steel sheets in DPSCWs can be obtained:

$$V_s = 2V_{s,e-strip} + mV_{s,m-strip} \tag{15}$$

where m is the number of middle strips.

The shear capacity of the profiled steel sheets for the test specimen can be determined by subtracting the loads carried by the web concrete and boundary members from the total wall capacity obtained from the test. The lateral resistance of the web concrete (V_{cn}) is calculated using the simplified model shown in Figure 18, as described in reference [8], and is given by Equations (16)–(18).

$$V_c = 2\frac{V_{c,r}b_e}{h} + m\frac{V_{c,r}b_m}{h} \tag{16}$$

$$V_{r,c} = 0.15f_c t_{rib} h \tag{17}$$

$$V_{cn} = 0.15f_c t_{rib} l_d + 0.13N \tag{18}$$

where $V_{r,c}$ is the shear strength of the connecting concrete; h is the height of the wall; b_e and b_m are the widths of the middle and edge concrete wall limbs divided by dovetailed ribs, respectively; and t_{rib} is the thickness of the rib-area-concrete. N is the compressive axial load applied to the wall, which should be less than $0.2f_c A_w$, and A_w is the cross-sectional area of the wall.

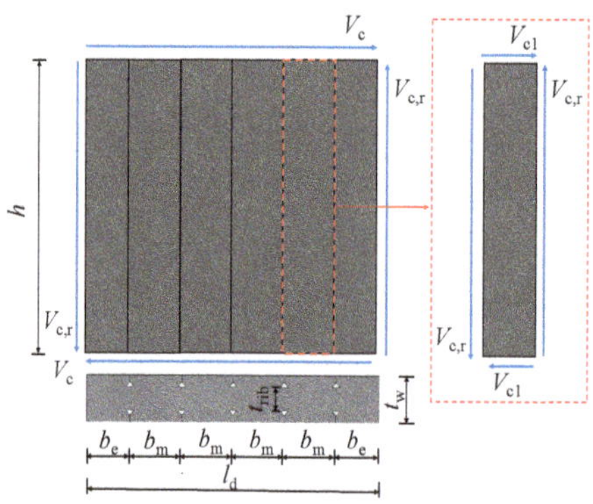

Figure 18. Analytical model for web concrete.

For the boundary members in shear walls, which primarily resist the overturning moment of the wall [8], their lateral resistance is mainly provided by the boundary member on the compressive side, as specified in Code for Design of Composite Structures (JGJ 138-2016) [36]:

$$V_b = 0.5f_t l_b t_{bw} + 0.4f_{yst} A_{st} \tag{19}$$

where f_t and f_{yst} is the tensile strength of the concrete and the yield strength of the boundary tube, respectively, l_b and t_{bw} are the width and thickness of the boundary member, respectively, and A_{st} signifies the cross-sectional area of a single boundary steel tube.

In summary, the shear capacity of the profiled steel sheets for the test specimen can be calculated using the following formula:

$$V_{s,exp} = P - 0.15 f_c t_{rib} l_d - 0.13N - 4 f_{v,b} l_b t_{sb} \tag{20}$$

The shear capacity $V_{s,cal}$ of the profiled steel sheets was calculated for each specimen using Equation (19) and is presented in Table 5, along with the test shear capacity values $V_{s,exp}$ obtained from Equation (16). The error between the calculated and test results is within 12%, demonstrating that the proposed calculation method effectively predicts the shear capacity of profiled steel sheets in the DPSCWs connected by self-tapping screws.

Table 5. Comparison between the calculated and tested shear capacity of profiled steel sheets.

Specimen	Tested Lateral Load P_m (kN)	Tested Shear Capacity of Profiled Steel Sheets $V_{s,exp}$ (kN)	Calculated Shear Capacity of Profiled Steel Sheets $V_{s,cal}$ (kN)	$V_{s,cal}/V_{s,exp}$
DPSCW-S-1	1695	359	331	0.92
DPSCW-S-2	1843	437	389	0.89
DPSCW-S-3	1581	346	389	1.12
			Average	0.98

Based on the above analysis, to ensure a reliable connection of the profiled steel sheets, the number of self-tapping screws within the effective strip width of the middle and edge strips must satisfy the following requirements:

$$n_{m-strip} \geq \frac{W_{e,max} t_{sp} f_{yp}}{2 N_{v,b}^{f}} \tag{21}$$

$$n_{e-strip} \geq \frac{W_{e,max} t_{sp} f_{yp}}{N_{v,b}^{f}} \tag{22}$$

Based on the calculated number of self-tapping screws needed within the effective strip width, the screws should be arranged within the effective width range (as shown in the l_t and l_s ranges in Figure 16) according to the relevant requirements. For design purposes, the arrangement of self-tapping screws outside the effective width range should be consistent with the arrangement within the effective width range.

5. Conclusions

This study experimentally investigated the seismic behavior of self-tapping screw-connected DPSCWs, focusing on the effects of varying the number of self-tapping screws and axial compression ratios. The following conclusions were drawn:

(1) The self-tapping screw-connected DPSCW specimens exhibited a failure progression starting with local shear buckling of the strips, followed by inclined tension fields anchored by the concrete-embedded ribs. The wall concrete experienced compression-bending failure in several concrete columns, while the self-tapping screw connections remained intact with no significant damage.

(2) The specimens achieved yield drift ratios of 1/286 to 1/225, ultimate drift ratios of 1/63 to 1/94, and ductility coefficients of 3.09 to 3.51, all meeting the deformation capacity requirements in the "Code for Seismic Design of Buildings". This confirms the effectiveness of self-tapping screw connections in ensuring reliable force transfer between the profiled steel plates and boundary members.

(3) Increasing the number of self-tapping screws by 40% slightly increased peak load capacity by 9.5%, ultimate drift ratio by 23%, and ductility factor by 6%, with minimal impact on initial stiffness.
(4) Increasing the axial compression ratio from 0.2 to 0.5 resulted in a 13% increase in initial stiffness. However, the peak load capacity, ultimate drift ratio, and ductility coefficient decreased by 14%, 33%, and 12%, respectively.
(5) An analytical model for the shear capacity of profiled steel sheets connected by self-tapping screws in DPSCWs was developed based on the effective strip method, with the error between the calculated and experimental results within 12%.

In future studies, a comparative analysis will be conducted on the seismic performance of self-tapping screw-connected DPSCWs and welded-connected DPSCWs to evaluate the adaptability and applicability of different connection methods. Additionally, finite element analyses will be performed on the mechanical behavior of self-tapping screw-connected DPSCWs, accompanied by systematic parametric studies. Based on the results of the finite element analyses, the proposed design methodology will be refined to enhance its accuracy and reliability.

Author Contributions: Conceptualization, Z.H., S.Z. and Y.T.; methodology, Z.H., S.Z., Y.T. and L.G.; software, Z.H. and Y.W.; validation, Z.H., Z.Z. and S.Z.; formal analysis, Z.H., S.Z., Y.T. and L.G.; investigation, Z.H., Y.W. and Y.T.; resources, Z.H.; data curation, Z.H. and Y.W.; writing—original draft preparation, Z.H.; writing—review and editing, Z.H., Z.Z. and S.Z.; visualization, Z.H.; supervision, Z.Z. and S.Z.; project administration, S.Z.; funding acquisition, S.Z. All authors have read and agreed to the published version of the manuscript.

Funding: This study is funded by China National Natural Science Foundation (51678193).

Institutional Review Board Statement: Ethical approval is not required.

Informed Consent Statement: Not applicable.

Data Availability Statement: The original contributions presented in this study are included in the article. Further inquiries can be directed to the corresponding author.

Conflicts of Interest: The authors declare no conflicts of interest.

References

1. Uy, B.; Wright, H.; Bradford, M. Combined axial and flexural strength of profiled composite walls. *P. I. Civ. Eng-Str. B.* **2001**, *146*, 129–139. [CrossRef]
2. Wright, H. Local stability of filled and encased steel sections. *J. Struct. Eng.* **1995**, *121*, 1382–1388. [CrossRef]
3. Tong, J.; Pan, W.; Shen, M. Performance of double-skin composite walls with re-entrant profiled faceplates under eccentric compression. *J. Build. Eng.* **2020**, *28*, 101010. [CrossRef]
4. Yu, C.; Tong, J. Compressive behavior of slender profiled double-skin composite walls. *J. Constr. Steel Res.* **2021**, *182*, 106657. [CrossRef]
5. Xu, L.; Guo, Z.; Guo, Y. Experimental study on seismic behavior of low-aspect-ratio concrete filled funnel-shaped profiled steel sheet composite shear walls. *J. Build. Struc.* **2023**, *44*, 119–128. (In Chinese)
6. Huang, Z.; Sun, Z.; Zhang, S. Mechanical behavior of dove-tail-shaped profiled steel concrete composite shear walls under axial compression. *J. Build. Struc.* **2019**, *40*, 84–90. (In Chinese)
7. Zhang, S.; Huang, Z.; Guo, L. Performance of dovetail profiled steel concrete composite sandwich walls under axial compression. *Constr. Build. Mater.* **2021**, *309*, 125090. [CrossRef]
8. Zhang, S.; Huang, Z.; Guo, L. Hysteretic performance of dovetail profiled steel-concrete composite sandwich shear walls. *Eng. Struct.* **2022**, *256*, 113994. [CrossRef]
9. Huang, Z.; Zhang, S.; Guo, L. Cyclic behavior of low shear-span ratio dovetailed profiled steel-concrete composite shear walls. *J. Build. Eng.* **2022**, *59*, 105094. [CrossRef]
10. Li, Y.; Shuai, Y.; Shen, Z. Experimental study on tension behavior of self-drilling screw connections for cold-formed thin-walled steel structures. *J. Build. Struc.* **2015**, *36*, 143–152. (In Chinese)

11. Roy, K.; Lau, H.; Tina, C. Experiments and finite element modelling of screw pattern of self-drilling screw connections for high strength cold-formed steel. *Thin Wall Struct.* **2019**, *145*, 106393. [CrossRef]
12. Huynh, M.; Cao, H.; Hancock, G. Design of screwed connections in cold-formed steels in shear. *Thin Wall Struct.* **2020**, *154*, 106817. [CrossRef]
13. Lu, L.; Wang, D.; Wang, W. Shear bearing capacity of Self-drilling screw group connections of CFS sheets. *Structures* **2022**, *35*, 160–171. [CrossRef]
14. Zhao, J.; Wang, Z.; Qian, F. Finite element analysis of the shear capacity of stainless-steel screw connections. *Structures* **2022**, *41*, 957–968. [CrossRef]
15. Liu, W.; Deng, L.; Zhong, W. Parametric study on the pull-out performance of screw connections in cold-formed thin walled steel structures. *Eng. Struct.* **2023**, *274*, 115007. [CrossRef]
16. Cai, K.; Yuan, H. Testing, numerical and analytical modelling of self-drilling screw connections between thin steel sheets in shear. *Thin Wall Struct.* **2023**, *182*, 110292. [CrossRef]
17. Zhou, X.; Shi, Y.; Zhou, T. Experimental study of the shear resistance of cold-formed steel stud walls. *Chin. Civil. Eng. J.* **2010**, *43*, 38–44. (In Chinese)
18. Yu, C. Shear resistance of cold-formed steel framed shear walls with 0.686 mm, 0.762 mm, and 0.838 mm steel sheet sheathing. *Eng. Struct.* **2010**, *32*, 1522–1529. [CrossRef]
19. Chen, W.; Ye, J.; Xu, Y. Shear experiments of load-bearing cold-formed thin-walled steel wall system lined with sandwich panels. *J. Build. Struc.* **2017**, *38*, 85–92. (In Chinese)
20. Zhang, W.; Mandavian, M.; Yu, C. Lateral strength and deflection of cold-formed steel shear walls using corrugated sheathing. *J. Constr. Steel Res.* **2018**, *148*, 399–408. [CrossRef]
21. Chao, S.; Lu, L.; Wu, H. Shearing Bearing Capacity of Screwed Connections of Thin Steel Sheets. *Int. J. Steel Struct.* **2019**, *19*, 577–590. [CrossRef]
22. *GB50018-2002*; Technical Code of Cold-Formed Thin-Walled Steel Structures. China Planning Publishing House: Beijing, China, 2010. (In Chinese)
23. Nie, J.; Hu, H.; Fan, J.; Tao, M. Experimental study on seismic behavior of high-strength concrete filled double-steel-plate composite walls. *J. Constr. Steel Res.* **2013**, *88*, 206–219. [CrossRef]
24. Zhao, Q.; Li, Y.; Tian, Y. Cyclic behavior of corrugated double-skin composite walls with different aspect ratios. *J. Struct. Eng. ASCE* **2020**, *146*, 04020214. [CrossRef]
25. *GB/T 50081-2019*; Standard for Test Methods of Concrete Physical and Mechanical and Mechanical Properties. China Architecture & Building Press: Beijing, China, 2019. (In Chinese)
26. *GB/T 228. 1-2010*; Metallic Materials: Tensile Testing: Part 1: Method of Test at Room Temperature. Standards Press of China: Beijing, China, 2021. (In Chinese)
27. *JGJ/T 101-2015*; Specification for Seismic Test of Buildings. China Architecture & Building Press: Beijing, China, 2015. (In Chinese)
28. Feng, P.; Qiang, H.; Ye, L. Discussion and definition on yield points of materials, members and structures. *Eng. Mech.* **2017**, *34*, 36–46. (In Chinese)
29. Park, R. State of the art report ductility evaluation from laboratory and analytical testing. In Proceedings of the Ninth World Conference on Earthquake Engineering, Tokyo/Kyoto, Japan, 2–9 August 1988; pp. 605–616.
30. *GB 50011-2010*; Code for Seismic Design of Buildings. China Construction Industry Press: Beijing, China, 2010. (In Chinese)
31. *ASCE/SEI 41-13*; Seismic Evaluation and Retrofit of Existing Buildings. American Society of Civil Engineers: Reston, VA, USA, 2017.
32. *AISI S213-07w/S1-09*; North American Standard for Cold Formed Steel Framing-Lateral Design with Supplement 1. American Iron and Steel Institute: Washington, DC, USA, 2012.
33. *AISI S400-15*; North American Standard for Seismic Design of Cold-Formed Steel Structural Systems. American Iron and Steel Institute: Washington, DC, USA, 2015.
34. Xie, Z.; Zhang, A.; Yan, W.; Zhang, Y.; Mu, T.; Yu, C. Study on shear performance and calculation method for self-pierce riveted joints in galvanized steel sheet. *Thin Wall Struct.* **2021**, *161*, 74–90. [CrossRef]
35. Yanagi, N.; Yu, C. Effective Strip Method for the Design of Cold-Formed Steel Framed Shear Wall with Steel Sheet Sheathing. *J. Struct. Eng.* **2014**, *140*, 04013101. [CrossRef]
36. *JGJ 138-2016*; Code for Design of Composite Structures. China Architecture & Building Press: Beijing, China, 2016. (In Chinese)

materials

MDPI

Article

The Influence of Mineral Additives on Aggregate Reactivity

Grzegorz Rogojsz and Tomasz Rudnicki *

Faculty of Civil Engineering and Geodesy, Military University of Technology, 2 Gen. Sylwestra Kaliskiego Str., 00-908 Warsaw, Poland; grzegorz.rogojsz@wat.edu.pl
* Correspondence: tomasz.rudnicki@wat.edu.pl

Abstract: In this article, the authors present the results of their research on assessing the effect of selected mineral additives on the alkaline reactivity of aggregates. The main objective of this research was to check whether the reactivity of aggregates that do not meet the standard requirements can be reduced. Due to the decreasing availability of crushed aggregates and the decreasing resources of sand used for cement concrete road surfaces, solutions should be sought that allow the use of lower-grade aggregates. Among the available mineral additives, dense microsilica, white microsilica, limestone flour, glass flour, basalt flour, and glass granulate were selected. Laboratory tests were carried out in accordance with the requirements for testing the alkaline reactivity of road aggregates in NaOH solution applicable in Poland. The tests included the use of mineral additives in the amounts of 10% and 20%. Based on the research conducted, it was observed that the most beneficial effect was obtained with the addition of white microsilica, for which a decrease in aggregate reactivity was observed by 76.7% for 10% of the additive and 95.8% for 20% of the content. The least beneficial effect, on the other hand, was the use of compacted microsilica, for which an increase in alkaline reactivity was observed by 9.3% for 10% of the additive and 20.9% for 20% of the additive. The research conducted shows that the alkaline reactivity of the aggregate can be reduced, due to which it is possible to use reactive aggregates for the construction of road surfaces made of cement concrete.

Keywords: mineral additives; alkaline reactivity; standard mortar

1. Introduction

Mineral additives, which are common by-products of various industries, can have a significant impact on the alkaline reactivity of concrete, especially in the context of the so-called alkaline–silica reaction (ASR). This reaction is a chemical process between alkaline ions (e.g., sodium and potassium ions) from cement and reactive silica species present in some aggregates [1–6]. The reaction results in the formation of products that swell in the presence of moisture, leading to serious damage to the concrete, such as micro-cracks, structural weakening, and loss of strength [7–12]. Fly ash and blast furnace slag have a very significant impact, reducing the risk of alkaline–silica reactions (ASRs) due to the reaction of alkaline ions from cement and reactive silica in the aggregate, causing swelling and damage to the concrete [13–15]. Mineral additives containing aluminum and silicon compounds (e.g., fly ash, blast furnace slag) can react with alkaline ions, binding them into insoluble phases. This leads to a decrease in the pH of the pore solution, which slows down the ASR. Mineral additives such as fly ash, metakaolin, colloidal silica, or granulated blast furnace slag can significantly reduce the alkaline reactivity of concrete [16]. An interesting phenomenon described in the literature is the dilution effect, i.e., mineral additives reduce the total clinker content in cement, which reduces the amount of alkali. As a result, concrete is less susceptible to alkali–silica reactions because the concentration of alkali responsible for initiating this reaction is reduced [17–20]. In addition, we dealt with a modification of the chemical composition of cement phases. Mineral additives can also modify the chemical composition of cement phases, limiting the formation of minerals with high alkaline reactivity. Another mechanism of action of mineral additives is to lower the pH of the pore solution. This prompted us to look for other available mineral additives to reduce the reactivity of mortars containing potentially reactive aggregates.

Analyzing the literature, we can find studies on the effect of fly ash in specific concrete applications [21] (in terms of strength and durability) and economic aspects (in terms of reuse of waste materials from industrial processes). A significant part of mineral additives also enters the pozzolanic reaction. In the case of fly ash and other pozzolanic additives, the pozzolanic reaction occurs, consisting of the binding of calcium hydroxide ($Ca(OH)_2$), one of the products of cement hydration, to insoluble silicon and calcium compounds [22–25]. This reaction reduces the availability of $Ca(OH)_2$, which indirectly reduces alkaline reactivity because the calcium ion stabilizes the alkaline–silica reaction.

Mineral additives also have a significant effect on the properties of the mortar, improving its durability, strength, and resistance to external factors. The main aspects of the action of these additives include, above all, increasing compressive strength. Additives such as colloidal silica, fly ash, or blast furnace slag enter into pozzolanic reactions, creating additional hydration products that improve the structure of the mortar. The influence of mineral additives also improves durability by sealing the mortar and increasing its resistance to moisture and aggressive ions (e.g., chlorides and sulfates). As a result, the mortar becomes more resistant to environmental factors and chemical degradation, which extends its service life. This proves that selected additives can have an impact on increasing compressive strength [26–32]. Another area of action of additives is the reduction in mortar and concrete shrinkage by reducing the amount of heat generated during cement hydration, which reduces the risk of plastic shrinkage and cracking. This also increases the adhesion and cohesion of the cement matrix [33–37].

Some additives, such as colloidal silica, improve the rheological properties of the mortar, which makes it easier to lay and increases adhesion to the substrate. Mortar with such additives is less susceptible to the segregation of components and is more homogeneous. Increased resistance to water and penetration of aggressive ions: mineral additives reduce the porosity of concrete, which limits the permeability of water and chemical compounds (e.g., chlorides and sulfates), which can accelerate the corrosion of reinforcement and damage to the concrete structure. The less permeable structure of concrete means that it is more resistant to corrosion and frost cracking. The positive effect of additives can also be observed through an increased resistance to high temperatures. Additives such as metakaolin can improve the resistance of mortar to high temperatures. Metakaolin is less susceptible to thermal degradation than standard clinker, which makes the mortar more resistant to fire and variable thermal conditions [38]. The last element of the impact of additives on mortar and cement concrete is undoubtedly a significant improvement in aesthetics: mineral additives affect the appearance of the mortar and can give it a more uniform and lighter color, which is desirable in prefabricated elements and in finishing works.

In summary, the use of appropriate mineral additives is one of the key methods of reducing the alkaline reactivity of concrete and improving its durability. Properly selected additives lower the pH of the pore solution, bind alkaline ions, reduce the number of capillary pores, and reduce the clinker content, which significantly reduces the risk of alkali–silica reactions. Additionally, they have a beneficial effect on the mortar in terms of strength, durability, aesthetics, and resistance to various external factors. The appropriate selection of mineral additives can, therefore, significantly improve the properties of the mortar, increasing its quality and suitability for various construction applications.

2. Materials and Methods

2.1. Materials

In order to assess the effect of mineral additives on the alkaline reactivity of the aggregate, river sand with a grain size of 0/2 mm was used. Since the natural grain size of the sand did not meet the requirements specified in [39], it was fractionated into individual fractions (Figure 1) and then mixed in accordance with the grain size requirements presented in Table 1.

Figure 1. Fractionated sand at 0/2 mm.

Table 1. Required sand grain size for alkali reactivity testing.

Fraction [mm]	Contents [%]
2–4	10
1–2	25
0.5–1	25
0.25–0.5	25
0.125–0.25	15

The cement class CEM I 42.5 R, hereinafter referred to as CEM I, was used to prepare the samples, with the parameters presented in Table 2.

The mineral additives used were compacted microsilica (MkZ) from Mikrosilika Trade in the form of silica dust generated in arc furnaces during the production of metallic silicon and ferrosilicon alloys, and white microsilica (MkB), which is a by-product of zirconium silicate production. Another additive was limestone flour (MW), which is used as a filler for mineral–asphalt mixtures created by drying and grinding limestone, the main component of which is calcium carbonate. The influence of glass waste generated on the basis of construction glass with a hardness of 6–7 on the Mosh scale in the form of glass flour (MS) and glass granulate (GS) was also examined. Another additive analyzed was basalt dust (PB), also called basalt flour, which is waste generated during the processing of aggregate for the production of mineral–asphalt mines. The chemical composition of the individual additives is presented in Table 3, and the additives are shown in Figures 2 and 3.

Table 2. Properties of cement CEM I 42.5 R.

Parameter	Unit	Value
Specific surface area	cm^2/g	4124
Start of setting time	Min	184
End of setting time	Min	242
Change in volume	Mm	1.0
Compressive strength		
After 2 days	MPa	30.1
After 28 days	MPa	60.2
Content of SO_3	%	2.95
Content of Cl	%	0.089
Insoluble residue	%	0.57
Loss of ignition	%	3.33

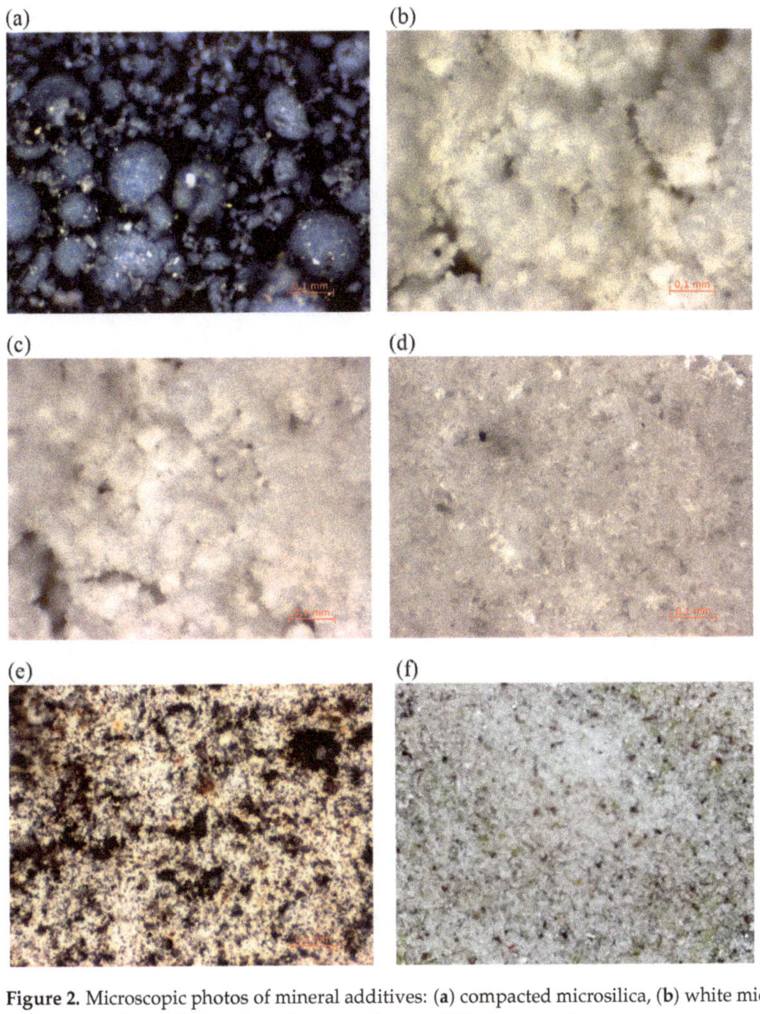

Figure 2. Microscopic photos of mineral additives: (**a**) compacted microsilica, (**b**) white microsilica, (**c**) limestone flour, (**d**) glass flour, (**e**) basalt dust, and (**f**) glass granulate.

Table 3. Chemical composition of the additives used.

Component	Unit	MkZ	MkB	MW	MS	PB	GS
SiO_2	%	>80.0	>94.0	3.5	>65.0	38.2	>65.0
CaO	%	<3.5	<1.0		>8.0	15.2	>8.0
ZrO_2	%		<4.0				
$CaCO_3$	%			93.0			
FeO_3	%		<1.0	0.3	<0.2	15.9	<0.2
MgO	%			0.7	<0.4	7.7	<0.4
SO_3	%	<4.0				0.2	
Na_2O	%	<8.0			>14.0	2.9	>14.0
Al_2O_3	%		<1.0		2.0	12.7	2.0
Cl^-	%	<1.8	<0.3			0.07	

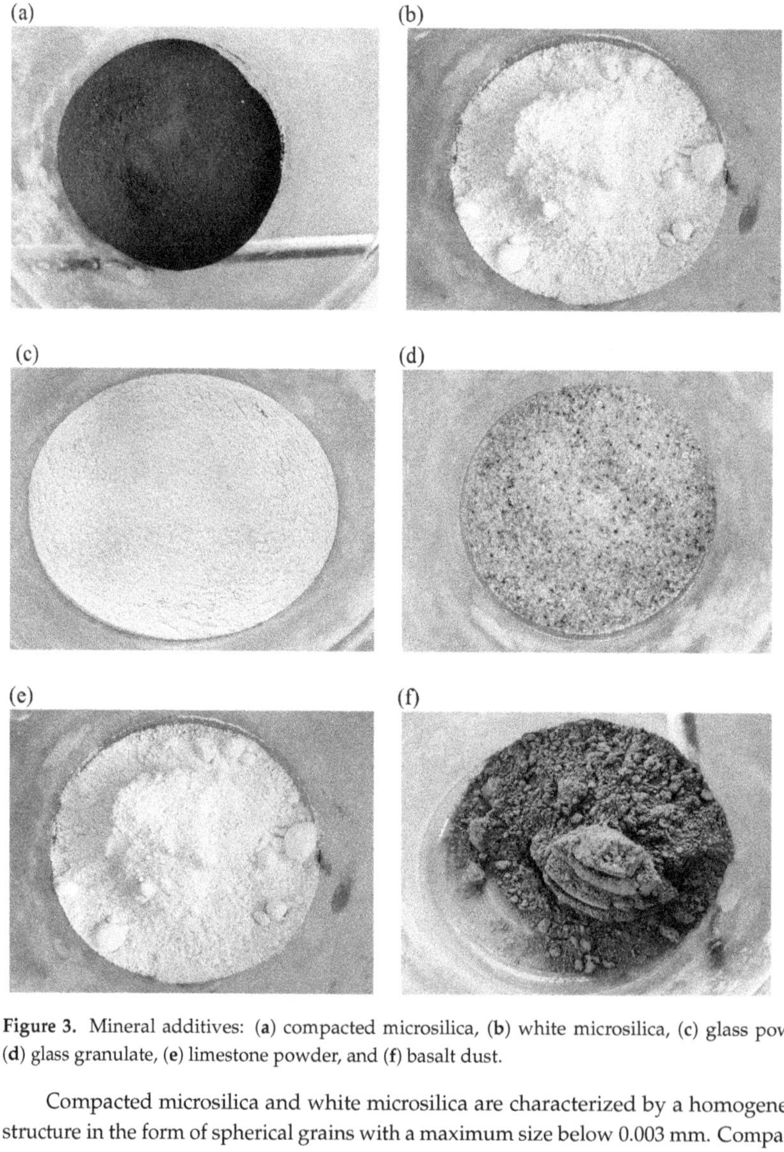

Figure 3. Mineral additives: (**a**) compacted microsilica, (**b**) white microsilica, (**c**) glass powder, (**d**) glass granulate, (**e**) limestone powder, and (**f**) basalt dust.

Compacted microsilica and white microsilica are characterized by a homogeneous structure in the form of spherical grains with a maximum size below 0.003 mm. Compacted microsilica in loose form creates clusters of grains of 0.1–0.2 mm in size, as visible in Figure 1, created by the agglomeration of small particles, which immediately disintegrate after contact with water. Limestone flour consists of non-homogeneous grains, mainly below 0.06 mm in size, which do not interconnect with each other, creating a loose material in the form of dust, as shown in Figure 2c. In contrast to limestone flour, glass flour is characterized by a very homogeneous structure, with grains below 0.06 mm in size (Figure 2d). Basalt dust, like limestone flour, is formed by non-homogeneous grains of crushed aggregate with a size below 0.08 mm (Figure 2e). Glass granulate is composed of irregular grains up to 0.5 mm in size, as shown in Figure 2f.

2.2. Methods

Samples for laboratory tests were prepared in accordance with the procedure described in the instructions [39] in the form of bars with dimensions of 25 mm × 25 mm × 285 mm, with 3 pieces for each additive. In the first stage of the tests, 10% of the cement content was replaced with a mineral additive, and in the second stage, 20% of the cement content was replaced with a mineral additive. In order to compare the effect of individual additives on the alkaline reactivity of the sand used in the tests, reference samples without mineral additives were also prepared. The composition of the individual mixes for stages 1 and 2 is presented in Table 4, and the description of the symbols of the test samples used is presented in Table 5.

Table 4. Recipes.

Research Stage	Aggregate Content [g]	Cement Content [g]	Contents of the Additive [g]	Water Content [g]
Control mix	792	352	0	165.44
Stage 1 10% additives	792	316.8	35.2	165.44
Stage 2 20% additions	792	281.6	70.4	165.44

Table 5. Symbols of research samples.

Mineral Supplement	Additive Content 10%	Additive Content 20%
No add-on	CEM100	
Compacted microsilica	MkZ10	MkZ20
White microsilica	MkB10	MkB20
Limestone flour	MW10	MW20
Glass flour	MS10	MS20
Basalt dust	PB10	PB20
Glass granulates	GS10	GS20

After preparation, the test samples were conditioned in a humidity chamber at a temperature of 20 ± 1 °C and a humidity of no less than 90% for a period of 24 ± 2 h. After this time, the samples were unmolded, and the length of the samples was measured, constituting a dimensional base with an accuracy of 0.001 mm. After the initial measurement, the test samples were placed in a container filled with distilled water and placed in a thermostatic chamber at a temperature of 80 ± 2 °C for a period of 24 h. After this time, a zero measurement was performed, and the test samples were placed in a container containing a 1 molar NaOH solution at a temperature of 80 ± 2 °C. The length of the samples was measured after 1, 7, 10, 14, and 16 days of storage in the NaOH solution at a temperature of 80 ± 2 °C. The measurement procedure used was based on the instruction

PB-1-18 [40] developed by the General Directorate for National Roads and Motorways in Poland. The evaluation of the effect of mineral additives on the reactivity of the aggregate was carried out based on the results of the average change in the length of the mortar samples after 14 days of storage in 1 molar NaOH solution, as shown in Table 6.

Table 6. Reactivity categories.

Aggregate Reactivity Category	Descriptive Term	14-day Sample Length Change [%]
R0	Non-reactive	≤0.15
R1	Moderately reactive	>0.15; ≤0.30
R2	Highly reactive	>0.30; ≤0.45
R3	Very strongly reactive	>0.45

3. Results

3.1. Determination of the Reactivity of the Mortar for Stage 1 (10% Addition)

The difference between the zero measurements of the sample length and the measurement of the length at each storage period was calculated with an accuracy of 0.001% for the effective distance between the ends of the plugs placed in the mortar. The change in the length of the sample tested was calculated using the following equation:

$$\text{Change in sample length [\%]} = 100 \times (L_n - L_0)/G \quad (1)$$

where

L_n—sample length after "n days" [mm], where n is the number of days from the zero measurement;
L_0—zero sample length [mm];
G—distance between the inner ends of the plugs in the mortar samples [mm] with an accuracy of 0.1 mm

The results of the percentage change in the length of the samples for Stage 1 are presented in Table 7, and the average values are presented in Figure 4.

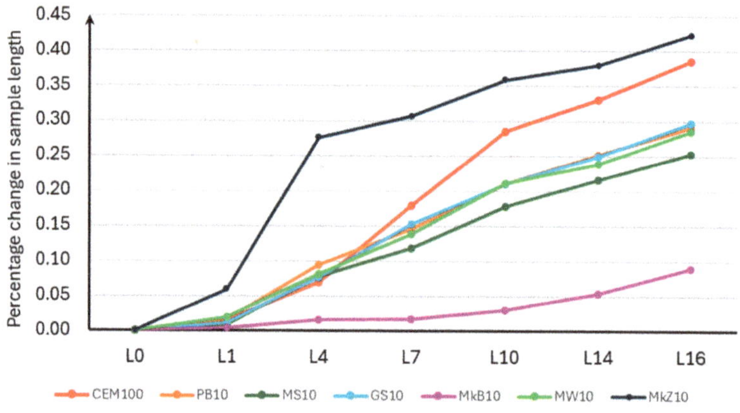

Figure 4. Average percentage change in length of samples for Stage 1.

Table 7. Percentage change in length of samples for Stage 1.

Sample	L1	L4	L7	L10	L14	L16
CEM100 P1	0.0139	0.1010	0.1847	0.2866	0.3290	0.3861
CEM100 P2	0.0154	0.0998	0.1733	0.2820	0.3340	0.3879
CEM100 P3	0.0143	0.0102	0.1798	0.2877	0.3301	0.3841
PB10 P1	0.0158	0.0939	0.1497	0.2162	0.2520	0.2939
PB10 P2	0.0158	0.0949	0.1475	0.2067	0.2513	0.2905
PB10 P3	0.0152	0.0944	0.1395	0.2094	0.2531	0.2919
MS10 P1	0.0046	0.0773	0.1211	0.1765	0.2134	0.2519
MS10 P2	0.0139	0.0789	0.1131	0.1801	0.2174	0.2544
MS10 P3	0.0099	0.0781	0.1205	0.1792	0.2184	0.2551
GS10 P1	0.0081	0.0737	0.1627	0.2099	0.2495	0.2959
GS10 P2	0.0138	0.0796	0.1419	0.2100	0.2492	0.2996
GS10 P3	0.0124	0.0762	0.1554	0.2105	0.2509	0.2967
MkB10 P1	0.0038	0.0162	0.0139	0.0289	0.0500	0.0851
MkB10 P2	0.0046	0.0158	0.0204	0.0324	0.0574	0.0944
MkB10 P3	0.0040	0.0166	0.0184	0.0304	0.0541	0.0911
MW10 P1	0.0197	0.0779	0.1384	0.2112	0.2386	0.2837
MW10 P2	0.0196	0.0834	0.1415	0.2126	0.2422	0.2888
MW10 P3	0.0184	0.0845	0.1378	0.2100	0.2377	0.2856
MkZ10 P1	0.0539	0.2674	0.3035	0.3551	0.3778	0.4216
MkZ10 P2	0.0639	0.2854	0.3100	0.3612	0.3809	0.4240
MkZ10 P3	0.0601	0.2771	0.3065	0.3593	0.3812	0.4221

Since the observed measurement error was less than 0.005% of the increase in THE sample length, it is not plotted on the graph below.

Analyzing the obtained results, it should be stated that mineral additives have a significant influence on the results of potential alkaline reactivity. The study analyzed the effectiveness of the influence of mineral additives on the change in the length of the tested samples. All analyzed additives, except for compacted microsilica, had a positive effect on the change in the length of the samples compared to the control mixture without additives. The most beneficial was the use of white microsilica, which reduced the percentage increase in the length of the samples by 76.6%. Therefore, the use of white microsilica (MkB10) reduced the alkaline reactivity category of the aggregate used from R2 to R0. The least beneficial effect was the use of compacted microsilica (MkZ10) because it increased the percentage increase in the length of the samples by 9.5% without affecting the change in the reactivity category of the aggregate. The remaining mineral additives in the form of limestone flour (MW10), glass flour (MS10), glass grit (GS10), and basalt dust (PB10) had a similar effect on the reduction in the percentage increase in the length of the samples, limiting their increase by 25.7%, 34.2%, 22.9%, and 24.3%, respectively, which allowed the reactivity class to be changed from R2 to R1.

3.2. Determination of Mortar Reactivity for Stage 2 (20% Addition)

As in the case of Stage 1, for samples with 20% mineral additives, the difference between the zero measurements of the sample length and the measurements of the length at each storage period was calculated to the nearest 0.001% of the effective distance between the ends of the plugs placed in the mortar, according to Equation (1). The results of the percentage change in the length of the samples for Stage 2 are presented in Table 8, and the average values are shown in Figure 5.

Table 8. Percentage change in length of samples for Stage 2.

Sample	L1	L4	L7	L10	L14	L16
CEM100 P1	0.0139	0.1010	0.1847	0.2866	0.3290	0.3861
CEM100 P2	0.0154	0.0998	0.1733	0.2820	0.3340	0.3879
CEM100 P3	0.0143	0.0102	0.1798	0.2877	0.3301	0.3841
PB20 P1	0.0088	0.0719	0.1100	0.1496	0.1888	0.2122
PB20 P2	0.0058	0.0642	0.1011	0.1457	0.1822	0.2064
PB20 P3	0.0074	0.0699	0.1085	0.1421	0.1723	0.2044
MS20 P1	0.0092	0.0488	0.0757	0.1096	0.1299	0.1703
MS20 P2	0.0100	0.0458	0.0673	0.0989	0.1208	0.1546
MS20 P3	0.0110	0.0399	0.0686	0.1160	0.1235	0.1599
GS20 P1	0.0119	0.0651	0.0924	0.1414	0.1645	0.2011
GS20 P2	0.0062	0.0620	0.0905	0.1432	0.1644	0.2002
GS20 P3	0.0091	0.0701	0.0899	0.1451	0.1654	0.2023
MkB20 P1	0.0004	0.0027	-0.0058	0.0058	0.0069	0.0166
MkB20 P2	0.0008	0.0015	-0.0042	0.0027	0.0046	0.0154
MkB20 P3	0.0005	0.0021	-0.0056	0.0047	0.0051	0.0171
MW20 P1	0.0100	0.0758	0.1340	0.1848	0.2136	0.2514
MW20 P2	0.0108	0.0819	0.1416	0.1946	0.2254	0.2573
MW20 P3	0.0116	0.0901	0.1399	0.1908	0.2213	0.2605
MkZ20 P1	0.0169	0.1047	0.2341	0.2868	0.3465	0.4224
MkZ20 P2	0.0196	0.0943	0.3290	0.3979	0.4518	0.5149
MkZ20 P3	0.0184	0.1005	0.2667	0.3658	0.4298	0.4654

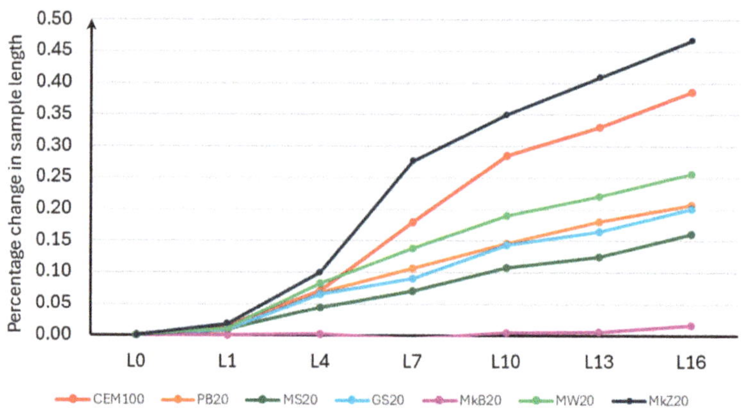

Figure 5. Average percentage change in length of samples for Stage 2.

Since the observed measurement error was less than 0.005% of the increase in the sample length, it is not plotted on the graph below.

Analyzing the presented results, it should be noted that the 20% content of mineral additives also had a significant influence on aggregate alkaline reactivity. Similarly to the previous case, the most beneficial influence on limiting the percentage increase in the sample length was that of white microsilica (MkB20). Increasing its share in the mixture allowed us to practically eliminate the phenomenon of a sample length increase. Limiting the percentage increase in the sample length by 95.7% also allowed us, in this case, to change the alkaline reactivity class from R2 to R0. The use of compacted microsilica caused a percentage increase in the sample length of 21.1% to the value of 0.4094%. This change did not adversely affect the decrease in the alkaline reactivity category of the aggregate. However, it should be noted that in the case of testing the length increase after 16 days, it was already 0.4679%, which allowed us to classify the aggregate as very strongly reactive with category R3. For the 20% additive content, in the case of the length increase test after 14 days, the use of glass flour (MS20) also allowed us to reduce the reactivity category from R2 to R0, and the limitation of the percentage increase in the sample length was 67.7%; however, considering the results after 16 days, the limitation of the percentage increase in the sample length was 58.1%, which allowed us to change the aggregate reactivity class from R2 to R1. Basalt dust (PB20) and glass granulate (GS20) have almost the same effect on the limitation of the percentage length of the sample by 46.2% and 47.9%, respectively. Also, in this case, we dealt with a change in the alkaline reactivity category of the aggregate from R2 to R1. The mineral additive in the form of limestone flour limits the percentage increase in the length of the samples to the lowest extent, only by 33.6%, which allowed us to change the alkaline reactivity category of the aggregate used from R2 to R1.

Figure 6 presents the results of the average percentage change in the length of the samples for each mineral additive, comparing the 10% and 20% contents of additives.

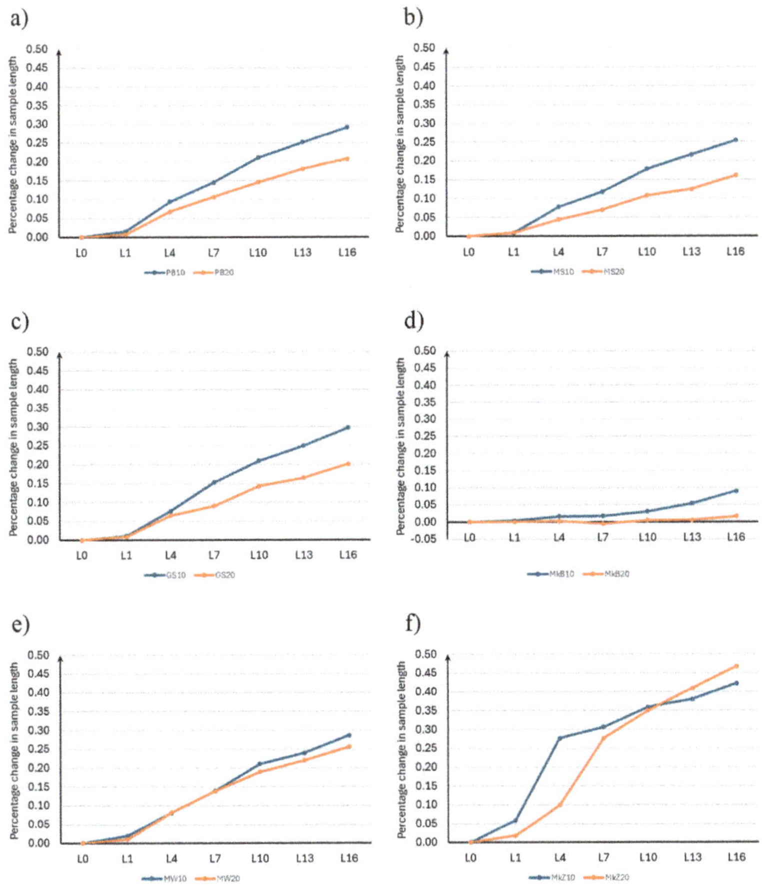

Figure 6. Summary of the average percentage change in sample length for Stages 1 and 2: (**a**) basalt dust, (**b**) glass powder, (**c**) glass granulate, (**d**) white microsilica, (**e**) limestone powder, and (**f**) compacted microsilica.

Comparing the use of 10% and 20% mineral additives, it should be noted that in the case of basalt dust, glass flour, glass grit, and white microsilica, increasing the content of the additive had a significant beneficial effect on limiting the percentage increase in the sample length, which increased by about 0.1%. In the case of limestone flour, doubling its share in the mixture did not cause significant differences in the final results of the percentage increase in the length of the samples, which was only 0.03%. Increasing the content of the additive in the form of compacted microsilica in the initial phase of the study had a more beneficial effect on the increase in the length of the samples; however, after the 10th day of conditioning, greater increases in the length of the samples occurred than in the case of 10% compacted microsilica additive. The difference in the obtained results was similar to that for limestone flour and was less than 0.05%.

The unfavorable effect of compacted microsilica on the tested samples was also observed in the form of cracks and scratches appearing on the samples, as shown in Figure 7.

Figure 7. Cracks in samples with 20% addition of compacted microsilica.

4. Discussion

As mentioned at the beginning of this article, the course of the alkaline reactivity phenomenon is influenced by many factors, and three basic alkaline reactions can be distinguished:

- The reaction between amorphous silica (a component of aggregates) and alkalis:

$$SiO_2 + 2NaOH \rightarrow Na_2SiO_3 + H_2O$$

- The reaction between alkalis and carbonates (mainly in the case of aggregates from carbonate rocks—limestones and dolomites):

$$CaMg(CO_3)_2 + {}_2NaOH^- \rightarrow Mg(OH)_2 + CaCO_3 + Na_2CO_3$$

$$Na_2CO_3 + Ca(OH)_2 \rightarrow 2NaOH + CaCO_3$$

- The reaction between alkali and poorly crystallized forms of silicates.

As a result of the alkali–aggregate reaction, gel shells are formed around the aggregate grain, which swells under the influence of water, causing stress in the concrete. These stresses can cause scratches or cracks.

Analyzing the reactions presented above, it can be concluded that mixtures with an increased content of silica or calcium carbonate should be more susceptible to alkaline reactivity. Among the analyzed mineral additives, compacted microsilica and white microsilica are characterized by a SiO_2 content above 80% and 95%, respectively.

In the case of the compacted microsilica additive at the amount of both 10% and 20%, an increase in alkaline reactivity was obtained, which was observed at 9.3% and 20.9%, respectively. The authors of the work [41] using microsilica at the amount of 10% and 20% achieved a reduction in sample expansion by about 50%; however, this is mainly due to the use of slag as an additive at the amount of 60% and 70%. For the additive in the form of white microsilica, which has the highest SiO_2 content, a decrease in alkaline reactivity was observed, amounting to 76.7% for the content of 10% white microsilica and 95.8% for the content of 20%.

Also, in the case of the mineral additive in the form of limestone flour, which is characterized by the highest $CaCO_3$ content of over 93%, a decrease in alkaline reactivity was observed for both 10% and 20% of the additive, which was obtained at 25.8% and 33.7%, respectively, and was also confirmed in the experimental studies carried out by the authors of these works [42,43].

The total content of alkalis in concrete also affects alkaline reactivity. It should, therefore, be assumed that mortars with a higher Na_2Oeq content should be characterized by a higher alkaline reactivity. In the experimental studies conducted, the highest Na_2O content, exceeding 14%, was characterized by glass flour and glass granulates. However, for both additives, a decrease in the alkaline reactivity of the aggregate was obtained. For the content of additives at the amount of 10%, it was 34.4% for glass flour and 26.1% for glass grit. On the other hand, for the content of additives at the amount of 20%, the decrease in alkaline reactivity was 58.2% for glass flour and 47.9% for glass grit. It can, therefore, be stated that the increase in the alkali content in the mixture contributed to the decrease in the alkaline reactivity of the aggregate. According to the authors of the work [16], the additives that affect the decrease in the alkaline reactivity of aggregates are mainly zeolites, fly ash, or lithium compounds.

5. Conclusions

Based on the tests conducted involving the introduction of mineral additives at the amount of 10% and 20% to the standard mortar as a cement substitute, the following conclusions can be drawn:

(1) Compacted microsilica, which is a material commonly used in UHPC concretes, increases the alkaline reactivity of aggregates. In the discussed tests, the reactivity was 9.3% for the 10% additive content and 20.9% for the 20% additive content, respectively.

It should, therefore, be stated that compacted microsilica is not suitable for reducing the alkaline reactivity of aggregates intended for road surfaces.

(2) The addition of white microsilica at the amounts of 10% and 20% caused a decrease in the alkaline reactivity of the aggregate by 76.7% and 95.8%, respectively, which contributed to reducing the aggregate reactivity class from R2 to R0. This is very beneficial because reactive aggregates can be used to build road surfaces made of cement concrete. This directly contributes to the possibility of using aggregates for road construction, which until now has not met the standard requirements. Nevertheless, there is a risk of increased shrinkage of the concrete mix, which will also constitute a further stage of the authors' research.

(3) A very positive effect can be observed after adding limestone flour, which is characterized by the highest $CaCO_3$ content of over 93%, and an additive in the form of glass flour and glass granulate, characterized by an increased Na_2Oeq content. The research conducted shows that the use of the above additives also has a beneficial effect on reducing alkaline reactivity and aggregates from carbonate rocks and will not necessarily enter into a reaction between alkalis and carbonates. It is, therefore, possible to use limestone aggregates mainly in the form of limestone flour as a substitute for cement in order to reduce the alkaline reactivity of the aggregate.

(4) The results obtained in the research constitute an important direction in the scope of the possibilities of using mineral additives to minimize the alkaline reactivity of aggregates, and additionally, as was confirmed in the results presented in the work [32], by using mineral additives, it is possible to reduce the cement content in the mortar, which also has a positive effect due to environmental aspects and reducing carbon footprint.

(5) In Poland, over 70% of sand mines are characterized by alkaline reactivity at the level of R1 or R2; therefore, due to the use of the above-mentioned additives, this level can be significantly reduced, and the amount of sand used in the construction of roads and bridges can be increased.

Author Contributions: Conceptualization, T.R. and G.R.; methodology, T.R. and G.R.; software, T.R. and G.R.; validation, T.R. and G.R.; formal analysis, T.R. and G.R.; investigation, T.R. and G.R.; writing—original draft preparation, T.R. and G.R.; writing—review and editing, T.R. and G.R.; visualization, T.R. and G.R.; supervision, T.R. and G.R. All authors have read and agreed to the published version of the manuscript.

Funding: This research received no external funding.

Institutional Review Board Statement: Not applicable.

Informed Consent Statement: Not applicable.

Data Availability Statement: The original contributions presented in this study are included in the article. Further inquiries can be directed to the corresponding author.

Conflicts of Interest: The authors declare no conflict of interest.

References

1. Segura, I.P.; Ranjbar, N.; Damø, A.J.; Jensen, L.S.; Canut, M.; Jensen, P.A. A review: Alkali-activated cement and concrete production technologies available in the industry. *Heliyon* **2023**, *9*, e15718. [CrossRef] [PubMed]
2. West, G. *Alkali-Aggregate Reaction in Concrete Roads and Bridges*; Thomas Telford: London, UK, 1996.
3. Figueira, R.; Sousa, R.; Coelho, L.; Azenha, M.; de Almeida, J.; Jorge, P.; da Silva, C.J.R. Alkali-silica reaction in concrete: Mechanisms, mitigation and test methods. *Constr. Build. Mater.* **2019**, *222*, 903–931. [CrossRef]
4. Davies, G.; Oberholster, R. Alkali-silica reaction products and their development. *Cem. Concr. Res.* **1988**, *18*, 621–635. [CrossRef]

5. Jóźwiak-Niedźwiedzka, D.; Gibas, K.; Glinicki, M.A. Petrographic identification of reactive minerals in domestic aggregates and their classification according to RILEM and ASTM recommendations. *Roads Bridges Drog. Mostly* **2017**, *16*, 223–239. [CrossRef]
6. Jóźwiak-Niedźwiedzka, D.; Antolik, A.; Dziedzic, K.; Lisowski, P. Potencjalna reaktywność alkaliczna piasku z krajowych złóż. *Roads Bridges Drog. Mostly* **2022**, *21*, 253–271. [CrossRef]
7. Bavasso, I.; Costa, U.; Mangialardi, T.; Paolini, A.E. Assessment of Alkali–Silica Reactivity of Aggregates by Concrete Expansion Tests in Alkaline Solutions at 38 °C. *Materials* **2020**, *13*, 288. [CrossRef]
8. Naziemiec, Z. Alkali-silica reactivity of selected domestic fine aggregates. *Roads Bridges Drog. Mostly* **2018**, *17*, 271–283. [CrossRef]
9. Moranville-Regourd, M. Modelling of expansion induced by ASR—New approaches. *Cem. Concr. Compos.* **1997**, *19*, 415–425. [CrossRef]
10. Samarakoon, M.; Ranjith, P.; Rathnaweera, T.; Perera, M. Recent advances in alkaline cement binders: A review. *J. Clean. Prod.* **2019**, *227*, 70–87. [CrossRef]
11. Ahmed, T.; Burley, E.; Rigden, S.; Abu-Tair, A.I. The effect of alkali reactivity on the mechanical properties of concrete. *Constr. Build. Mater.* **2003**, *17*, 123–144. [CrossRef]
12. Marzouk, H.; Langdon, S. The effect of alkali-aggregate reactivity on the mechanical properties of high and normal strength concrete. *Cem. Concr. Compos.* **2003**, *25*, 549–556. [CrossRef]
13. Thomas, M. The effect of supplementary cementing materials on alkali-silica reaction: A review. *Cem. Concr. Res.* **2011**, *41*, 1224–1231. [CrossRef]
14. Williamson, T.; Juenger, M.C. The role of activating solution concentration on alkali–silica reaction in alkali-activated fly ash concrete. *Cem. Concr. Res.* **2016**, *83*, 124–130. [CrossRef]
15. Way, S.; Shayan, A. Study of some synthetically prepared hydrous alkali calcium silicates. *Cem. Concr. Res.* **1992**, *22*, 915–926. [CrossRef]
16. Jackiewicz-Rek, W. Reaktywność alkaliczna kruszywa. *Constr. Mater.* **2015**, *1*, 198–200. [CrossRef]
17. Hasdemir, S.; Tuğrul, A.; Yilmaz, M. Evaluation of alkali reactivity of natural sands. *Constr. Build. Mater.* **2012**, *29*, 378–385. [CrossRef]
18. Owsiak, Z.; Czapik, P.; Zapała-Sławeta, J. Methods of Mitigating Alkali Reactivity of Gravel Aggregate. *Struct. Environ.* **2022**, *14*, 102–109. [CrossRef]
19. Gogte, B. An evaluation of some common Indian rocks with special reference to alkali-aggregate reactions. *Eng. Geol.* **1973**, *7*, 135–153. [CrossRef]
20. Jóźwiak-Niedźwiedzka, D.; Antolik, A.; Dziedzic, K.; Glinicki, M.A.; Gibas, K. Resistance of selected aggregates from igneous rocks to alkali-silica reaction: Verification. *Roads Bridges Drog. Mostly* **2019**, *18*, 67–83. [CrossRef]
21. Longarini, N.; Crespi, P.; Zucca, M.; Giordano, N.; Silvestro, G. The advantages of fly ash use in concrete structures. *Inżynieria Miner.* **2014**, *15*, 141–145.
22. Coo, M.; Pheeraphan, T. Effect of sand, fly ash and limestone powder on preplaced aggregate concrete mechanical properties and reinforced beam shear capacity. *Constr. Build. Mater.* **2016**, *120*, 581–592. [CrossRef]
23. Coo, M.; Pheeraphan, T. Effect of sand, fly ash, and coarse aggregate gradation on preplaced aggregate concrete studied through factorial design. *Constr. Build. Mater.* **2015**, *93*, 812–821. [CrossRef]
24. Nayak, D.K.; Abhilash, P.; Singh, R.; Kumar, R.; Kumar, V. Fly ash for sustainable construction: A review of fly ash concrete and its beneficial use case studies. *Clean. Mater.* **2022**, *6*, 100143. [CrossRef]
25. Szcześniak, A.; Zychowicz, J.; Stolarski, A. Influence of Fly Ash Additive on the Properties of Concrete with Slag Cement. *Materials* **2020**, *13*, 3265. [CrossRef]
26. Faysal, R.M.; Maslehuddin, M.; Shameem, M.; Ahmad, S.; Adekunle, S.K. Effect of mineral additives and two-stage mixing on the performance of recycled aggregate concrete. *J. Mater. Cycles Waste Manag.* **2020**, *22*, 1587–1601. [CrossRef]
27. Mazilu, C.; Deju, R.; Georgescu, D.P.; Apostu, A.; Barbu, A. Effects of Micro- and Nanosilica on the Mechanical and Microstructural Characteristics of Some Special Mortars Made with Recycled Concrete Aggregates. *Materials* **2024**, *17*, 2791. [CrossRef]
28. Khan, M.I.; Abbas, Y.M.; Fares, G.; Alqahtani, F.K. Flowability and Strength Characteristics of Binary Cementitious Systems Containing Silica Fume, Fly Ash, Metakaolin, and Glass Cullet Powder. *Materials* **2023**, *16*, 6436. [CrossRef] [PubMed]
29. Gao, S.; Ji, Y.; Qin, Z.; Zhang, H.; Xing, F.; Liu, A. A comprehensive analysis of pore structures and performances of mineral admixtures modified recycled aggregate concrete based on experiment and theory. *Constr. Build. Mater.* **2022**, *358*, 129451. [CrossRef]
30. Bentz, D.P.; Ferraris, C.F.; Jones, S.Z.; Lootens, D.; Zunino, F. Limestone and silica powder replacements for cement: Early-age performance. *Cem. Concr. Compos.* **2017**, *78*, 43–56. [CrossRef]
31. Perraki, T.; Kontori, E.; Tsivilis, S.; Kakali, G. The effect of zeolite on the properties and hydration of blended cements. *Cem. Concr. Compos.* **2010**, *32*, 128–133. [CrossRef]
32. Rogojsz, G.; Rudnicki, T. Influence of Mineral Additives on Strength Properties of Standard Mortar. *Materials* **2024**, *17*, 4158. [CrossRef]

33. Lee, K.; Lee, H.; Lee, S.; Kim, G. Autogenous shrinkage of concrete containing granulated blast-furnace slag. *Cem. Concr. Res.* **2006**, *36*, 1279–1285. [CrossRef]
34. Bouasker, M.; Mounanga, P.; Turcry, P.; Loukili, A.; Khelidj, A. Chemical shrinkage of cement pastes and mortars at very early age: Effect of limestone filler and granular inclusions. *Cem. Concr. Compos.* **2008**, *30*, 13–22. [CrossRef]
35. Itim, A.; Ezziane, K.; Kadri, E.-H. Compressive strength and shrinkage of mortar containing various amounts of mineral additions. *Constr. Build. Mater.* **2011**, *25*, 3603–3609. [CrossRef]
36. Ezziane, K.; Bougara, A.; Kadri, A.; Khelafi, H.; Kadri, E. Compressive strength of mortar containing natural pozzolan under various curing temperature. *Cem. Concr. Compos.* **2007**, *29*, 587–593. [CrossRef]
37. Chindaprasirt, P.; Homwuttiwong, S.; Sirivivatnanon, V. Influence of fly ash fineness on strength, drying shrinkage and sulfate resistance of blended cement mortar. *Cem. Concr. Res.* **2004**, *34*, 1087–1092. [CrossRef]
38. Cassagnabère, F.; Mouret, M.; Escadeillas, G. Early hydration of clinker–slag–metakaolin combination in steam curing conditions, relation with mechanical properties. *Cem. Concr. Res.* **2009**, *39*, 1164–1173. [CrossRef]
39. *PN-EN 196-1:2016-07*; Cement Testing Method—Part 1: Determinaation of Strength. Slovenian Institute for Standardization: Ljubljana, Slovenia, 2016.
40. General Directorate for National Roads and Motorways. *Technical Guidelines for the Classification of Domestic Aggregates and Prevention of Alkaline Reaction in Concrete Used in Road Surfaces and Road Engineering Structures*; General Directorate for National Roads and Motorways: Warsaw, Poland, 2019.
41. Hussain, S.E. Effect of microsilica and blast furnace slag on pore solution composition and alkali-silica reaction. *Cem. Concr. Compos.* **1991**, *13*, 219–225. [CrossRef]
42. Wang, H.; Wu, D.; Mei, Z. Effect of fly ash and limestone powder on inhibiting alkali aggregate reaction of concrete. *Constr. Build. Mater.* **2019**, *210*, 620–626. [CrossRef]
43. Rashad, A.M. Effect of limestone powder on the properties of alkali-activated materials—A critical overview. *Constr. Build. Mater.* **2022**, *356*, 129188. [CrossRef]

materials

MDPI

Article

Design of Self-Compacting Concrete with Reduced Cement Content by Aggregate Packing Method

Tomasz Rudnicki

Faculty of Civil Engineering and Geodesy, Military University of Technology, 2 Kaliskiego St., 00-908 Warsaw, Poland; tomasz.rudnicki@wat.edu.pl

Abstract: This article presents the procedure for designing self-compacting cement concrete characterized by minimal free space and a maximally compacted mineral skeleton. Such a designed mix allows for lower cement consumption and an increased amount of mineral additives. The paper presents a broad analysis of the influence of different aggregate proportions (36 recipes) and verification of the properties of the concrete mix using CEM I 42.5 R cements and fly ash. As a result of the appropriately compacted mineral skeleton, only 17% free space was obtained, which will allow the amount of cement to be reduced from 500 kg/m^3 to 350 kg/m^3 while fully maintaining the properties of the mix and hardened concrete. After 90 days of curing, SCC concrete was characterized by a compressive strength above 68 MPa and a small 2.1% decrease in compressive strength after 100 freeze–thaw cycles.

Keywords: SCC; concrete durability; fly ash

1. Introduction

Self-compacting concrete (SCC) is an innovative type of special concrete. It is characterized by specific rheological properties of a fluid concrete mix due to the precise selection of ingredients and their mutual proportions [1]. These properties constitute the criterion for identifying a self-compacting mix. A mix with a fluid consistency is able to tightly fill the formwork and envelop and fill dense reinforcement while maintaining a uniform composition and minimal porosity, and what is more, it compacts and deaerates under the influence of its own weight [2,3]. The fluid consistency of the mix is achieved thanks to a large amount of binder and the addition of an appropriate superplasticizer based on polycarboxylates and polyethers in an increased amount (2–4%). Their presence allows us to obtain a uniform structure, enabling the efficient use of cement and additives. Low porosity is achieved by reducing the amount of water added to the concrete and the presence of fine-grained mineral additives (fine fractions tightly fill the air gaps between cement grains). Formwork filling occurs without segregation of components and bleeding (leakage of water from the mixture). After laying, it self-levels, which results in an even and smooth surface. In addition, the process of mechanical vibration and noise, which is a negative factor for health, is eliminated. The lack of mechanical compaction results in a reduction in the thickness of the contact zone, which is usually characterized by much greater porosity than deeper layers of grout [4]. Self-compacting concrete is characterized by a reduced number of air pores by up to 25%. The phenomenon of spontaneous deaeration of the mixture is also considered to be one of the basic rheological properties. The need for layered dosing of the mixture disappears, which, together with the high efficiency of self-compacting concrete, results in a reduction in work time [5] and a reduction in labor input by up to 30%.

This concrete undoubtedly improves the technological process, because the mixture does not change its consistency during transport and laying, for up to 3 h. It becomes possible to concrete structures with complex, atypical shapes and high-quality surfaces. The structure of the concrete is homogeneous, which results from the continuous process of laying the mixture and eliminating differences in the degree of compaction of individual layers. The exceptional durability and strength of hardened concrete means that self-compacting concretes are usually also high-quality concretes. When it comes to the technical parameters of SCC concrete, a certain analogy can be seen to ordinary concretes. The main, leading value defining the suitability of concrete for a specific application is compressive strength. It influences other technical features, which in the case of compatibility of the compressive strength of these two types of concrete are also similar. The strength range of self-compacting concretes oscillates around 30–100 MPa, while obtaining concretes of lower classes is often problematic. It turns out that, for concretes of classes lower than C30/37, hardened self-compacting concrete after 28 days is characterized by strength 40–80% higher than originally assumed. This is undoubtedly influenced by the high cement content and the presence of mineral additives, sealing the structure of concrete, affecting the increase in strength and durability. One feature that highlights ambiguities is the resistance of concrete to creep. On the one hand, a larger amount of grout contributes to the increase in creep, and on the other hand, lower porosity limits this phenomenon and concrete is more durable. This issue still requires numerous studies and analyses, as the obtained results indicate that the increase in creep is closely related to the high cement content in the mixture. The situation is similar with shrinkage; it was found that the standard procedures have absolutely no practical application in the case of SCC concretes. A larger amount of grout promotes the phenomenon of shrinkage, while a tight structure and a low w/c ratio limit it [3,4]. Numerous studies show that the issue of adhesion of self-compacting concrete to reinforcement is also more favorable than in the case of ordinary concretes [1–5]. This is mainly due to the homogeneous structure of the self-compacting mixture and the lack of differences in the degree of its compaction [4–6]. Lower porosity in the contact zone of the grout with the reinforcement causes tight and precise covering of the bars, for both ordinary and prestressing tendons. Greater strength, uniformity, and tightness of concrete does not always have a 100% positive connotation—in the event of fire, it causes more spalling than in the case of ordinary concrete, and additionally, flaking of the surface layer. One method may be the use of propylene fibers, which, by melting in fire conditions, increase the number of pores in the concrete and loosen its structure, increasing permeability and allowing water vapor to escape. When it comes to greater durability of concrete, it results mainly from the tight structure and low content of micropores [7,8]. In this matter, the issue of the appropriate degree of aeration of the mixture becomes important, which is influenced by both air-entraining admixtures and superplasticizer. This requires strict control of their effect on the properties of the mixture with simultaneous verification of the effects of their action and mutual cooperation. The small number of air pores in SCC concretes and the presence of mineral additives, together with a low w/c ratio, additionally seal and strengthen the microstructure. The phenomenon of absorption and diffusion of gases and liquids occurs much slower in SCC concrete. The carbonation process, which deprives concrete of protective properties towards steel and is initiated by the presence of carbon dioxide, is not more aggressive than in the case of ordinary concretes due to the low value of the diffusion coefficient. Reduced diffusion of chloride ions ensures greater durability and protection of reinforcing steel. The main research objective presented in the article is to present the procedure for designing self-compacting concretes by properly selecting the proportions of aggregates (sand and gravel) in order to obtain the minimum free space (vacuum) [2,5]. Concrete designed in this way will be characterized by a minimum amount of cement and water needed to fill the free spaces. The excess binder necessary for the flow of the concrete mix was obtained by increasing the amount of mineral additives, such as limestone flour or fly ash [7]. A new element in the context of design methods is the presented simplified procedure for designing a mineral skeleton characterized by minimal free space. Additionally, a blocking factor for the mineral aggregate mix was introduced into the calculations [8–10].

2. Materials and Methods

The research work used:

- CEM I 42.5 R cement in accordance with PN-EN 197-1 [11];
- Natural aggregate of fractions 2/8 mm and 8/16 mm, with a density of 2.65 g/cm^3 and water absorption of 1.60% and natural sand of fraction 0/2 density of 2.65 g/cm^3;
- In order to increase the viscosity of the concrete mix and reduce the amount of cement, a mineral additive in the form of fly ash with a density of 2.50 g/cm^3 was used;
- A third-generation high-range/strong water-reducing admixture SP and VMA viscosity regulators complying with NP EN 934-1 [12] and PN-EN 934-2 [13] (amodified polycarboxylic high-range water-reducing admixture in liquid form with a density of 1.02 to 1.07 g/cm^3)—tap water in accordance with PN-EN 1008 [14].

2.1. Methods

The concept behind the method of blocking aggregate in concrete is as follows. The appropriate selection of fractions and determination of the proportions of coarse aggregate and sand is the key to designing a tight crushed pile [15–17]. Such a designed mineral mixture provides the desired consistency and workability, while the resulting minimum free space allows the amount of cement to be limited to the minimum amount necessary to obtain compressive strength. From the experience of the authors of the blocking method [6,15,18], coarse aggregates should be avoided because they are characterized by a larger specific surface area compared to natural gravel aggregates. Excessive filling of the mixture with grout, including both filling empty spaces and surrounding the aggregate surface with grout, can result in increased water demand.

Elimination of the phenomenon of blocking coarse aggregate grains between reinforcing bars and ensuring their correct encapsulation is only possible in the case of correct encapsulation of grains with grout while maintaining appropriate distances between them, which translates into determining the optimal content in the mixture. Van B.K. and S. Tangtermsirikul (Bangkok 1995) [18], in their research on the blocking criterion, considered the influence of both the shape and the size of the grains on the blocking coefficient, taking into consideration natural river aggregate of 0–8 mm and crushed limestone of 8–16 mm. Based on the graph, a formula was developed to determine the possibility of aggregate grains getting blocked in the mixture:

$$\sum_{i=1}^{n}\left(\frac{n_{ai}}{n_{abi}}\right)=\sum_{i=1}^{n}\left[\frac{\left(\frac{V_{ai}}{V_t}\right)}{\left(\frac{V_{abi}}{V_t}\right)}\right]=\sum_{i=1}^{n}\left(\frac{V_{ai}}{V_{abi}}\right)=1 \quad (1)$$

where:

V_{ai}—volume of a given aggregate fraction;
V_{abi}—blocking index for a given aggregate fraction (value ≤ 1).

Moreover, the use of crushed aggregates is an important issue. They are characterized by a larger specific surface area and, at the same time, a larger friction surface, which requires the use of a larger content of cement paste [19,20]. Assuming a constant amount of cement in the case of natural and crushed aggregate, increasing the amount of paste involves the need to use a large amount of filler (in extreme cases, its amount in concrete on crushed aggregate can be over 230 kg/m^3).

The voidness criterion determines the optimal content of coarse aggregate (gravel 8–16 mm), or more precisely, its ratio to the total aggregate in the *Nga* mixture, at which the content of voids will be the smallest. The Nga (content of voids in the coarse aggregate in relation to the total aggregate) coefficient also determines the required amount of cement paste [21–25]. The correct design of the mixture composition is realized in the case of the amount of paste greater than the volume of voids in the crushed pile—only then will we obtain the required consistency of the mixture due to fluidity, and consequently the free filling of the formwork with the mixture. Based on analyses and experience, correlations were developed between the minimum required amount of paste in relation to the share of coarse aggregate content. They show that the smallest value of the required amount of paste (in L/m^3) can be used in the case of 40–60% coarse aggregate content. The most unfavorable cases occur with a complete lack of 8–16 mm gravel (at the same time, obtaining a self-compacting mixture without the presence of this fraction is often impossible) or with excessive overfilling of the crushed pile with coarse aggregate (when the Nga ratio changes by +/− 0.7, even to a value of 1). The void criterion is usually of secondary importance, but it plays an important role in concreting with a narrow stream, in the case of, for example, building walls.

2.2. Functional Design Stages

When analyzing the mechanism of self-compacting concrete, it can be assumed, in accordance with the assumptions of the physicochemical mechanics of dispersion systems, that it is a dispersion system consisting of a dispersed phase (mineral skeleton above 0.25 mm) and a liquid phase (water, cement, filler, superplasticizer, and fine sand < 0.25 mm) [9,15]. The assumptions of the functional method for designing self-compacting concrete are based on Professor Paszkowski's method, the so-called double-envelopment method, i.e., it was assumed that the coarse aggregate grains are covered with a layer of cement mortar, while the sand grains are covered with cement paste. It was assumed that the functional method consists of 4 stages, each of which is responsible for the selection of a different component, and the final verification of properties takes place at the stage of trial mixing [9]. The stages of the functional method are presented in Table 1:

Table 1. Design stages for SCC [9].

Stage	Description	Stage Objective
Stage I	Determination of w/c, type of cement, type of mineral additives and chemical admixture	Achieving the project requirements contained in the technical specifications
Stage II	Selection of grain size of the mineral mixture by the blocking method	Obtaining a mineral mixture with minimal free space
Stage III	Calculation of the specific surface area of the aggregate pile and assumption of the thickness of the aggregate grain covering in order to separate the grains by the film distance, *b*	Determining the minimum amount of grout that will allow the concrete mix to flow without blocking the aggregate grains
Stage IV	Preparation of a test batch, testing of the concrete mix and hardened concrete	Verification of SCC properties (L-box, V-funnel, Abrams inverted cone, strength and durability tests)

2.3. Designing the Mineral Composition of Concrete

The first stage of design consisted of selecting the components of the concrete mix in order to meet the requirements included in the technical design in terms of compressive strength and exposure class. In the second stage, an analysis was carried out in terms of determining the proportions of mineral aggregate fractions 0/2, 2/8, and 8/16 using the blocking method in order to obtain a mineral mix characterized by minimal free space, i.e., the best packing. The proportions of the 36 analyzed mixes are included in Table 2:

Table 2. Analyzed proportions of the mineral mixture (%).

	Recipes	1	2	3	4	5	6
A	Gravel 8/16	50	50	50	50	50	50
	Gravel 2/8	50	40	30	20	10	0
	Sand 0/2	0	10	20	30	40	50
B	Gravel 8/16	40	40	40	40	40	40
	Gravel 2/8	50	40	30	20	10	0
	Sand 0/2	10	20	30	40	50	60
C	Gravel 8/16	30	30	30	30	30	30
	Gravel 2/8	50	40	30	20	10	0
	Sand 0/2	20	30	40	50	60	70
D	Gravel 8/16	20	20	20	20	20	20
	Gravel 2/8	60	50	40	30	20	10
	Sand 0/2	20	30	40	50	60	70
E	Gravel 8/16	10	10	10	10	10	10
	Gravel 2/8	60	50	40	30	20	10
	Sand 0/2	30	40	50	60	70	80
F	Gravel 8/16	0	0	0	0	0	0
	Gravel 2/8	70	60	50	40	30	20
	Sand 0/2	30	40	50	60	70	80

The third stage of design consists of calculating the specific surface area of the aggregate pile and assuming the thickness of the aggregate grain cover in order to separate the grains by the film distance, *b*. The aim of this stage is to determine the minimum amount of grout that will allow the concrete mix to flow without blocking the aggregate grains [9]. Calculation of the minimum amount of cement grout, Z_{cmin} that will allow the concrete to flow based on the determined free space in the mineral mix and the specific surface area of the aggregate pile was performed according to Formula (2):

$$Z_{c_{min}} = Z_1 + Z_2 \tag{2}$$

The minimum amount of grout filling the free space between aggregate grains is calculated using Formula (3) and the minimum amount of grout causing the aggregate grains to be separated by a distance, *b*, is calculated using Formula (4) and is designated Z_2:

$$Z_1 = \frac{P_{min} \cdot \rho_{wz}}{\rho_{0\ max}} \tag{3}$$

where Z_1 is the minimum amount of grout filling the free space in the mineral mixture, g_z/g_k; P_{min}—is the minimum free space in the mixture determined experimentally, %; and ρ_{wz} is the density of the cement paste, g/cm^3.

$$Z_2 = P_w \cdot b \cdot \rho_{wz} \tag{4}$$

where Z_2 is the minimum amount of grout causing overfilling of free spaces in the mineral mixture, g_z/g_k; P_w is the specific aggregate surface area, cm^2/g; and b is the thickness of the film on the aggregate, cm.

Finally, the requirement for the minimum amount of cement paste [9] enabling the concrete to flow is determined according to Formula (5):

$$Z_{cmin} = \left(\frac{P_{min} \cdot \rho_{wz}}{\rho_{0\ max}} + P_w \cdot b \cdot \rho_{wz} \right) \tag{5}$$

The minimum amount of cement paste, determined according to the assumptions of this method, should allow the desired fluidity of the designed concrete. The final amount of paste (liquid phase) was determined using Formula (6):

$$Z = Z_{cmin} \cdot K \tag{6}$$

where Z is the final amount of grout in the concrete mix, kg/m^3, and K is the amount of fine and coarse aggregate, kg/m^3.

According to Formula (7), for absolute volume:

$$\frac{Z}{\rho_{wz}} + \frac{K}{\rho_k} = 1000 \tag{7}$$

where ρ_k is the aggregate density, g/cm^3.

The final amount of aggregate in the concrete mix can be calculated using Formula (8):

$$K = \frac{1000}{\frac{Z_{cmin}}{\rho_{wz}} + \frac{1}{\rho_k}} \tag{8}$$

In order to check the correctness of the calculations of all the components of the concrete mix, the so-called absolute volume equation (or tightness condition) should be used, which has the following form [26]:

$$\frac{c}{\rho_c} + \frac{k}{\rho_k} + \frac{p}{\rho_p} + \frac{d}{\rho_d} + \frac{w}{\rho_w} + \frac{m}{\rho_m} = 1000 \tag{9}$$

where c is the amount of cement, kg/m^3; k is the amount of coarse aggregate, kg/m^3; p is the amount of sand, kg/m^3; d is the amount of the chemical admixture (the superplasticizer), kg/m^3; w is the amount of water, kg/m^3; ρ_c is the density of the cement, kg/dm^3; ρ_k is the density of the aggregate, kg/dm^3; ρ_p is the density of the sand, kg/dm^3; ρ_d is the density of the chemical admixture (superplasticizer), kg/dm^3; ρ_w is the density of water, kg/dm^3; m is the amount of limestone dust, kg/m^3; and ρ_m is the density of the limestone dust, kg/dm^3.

The last stage of the design consists of making a trial batch and testing the concrete mix and hardened concrete to verify the SCC properties (L-box, V-funnel, inverted Abrams cone). As part of the work, compressive strength tests and determination of concrete durability were performed.

3. Results

In the first stage, analyzing the project requirements, it was determined that the SCC concrete mix should be characterized by strength class C30/37, consistency class SF2, VS1, VF1 and frost resistance F100. In the second stage, a mineral mix was designed using the blocking method and the results are included in Tables 1 and 2. Further in the text, Table 3 presents sample calculations for one of the regulations, SCC C2. The aim of this stage was to obtain a mix characterized by minimum free space and maximum density. A summary of the obtained results is presented in Table 4.

In the third stage, the specific surface area was calculated for 36 different aggregate mixes with variable proportions of the content of individual fractions, which allows for the analysis of their impact on the remaining parameters in the design. It was assumed that each of the aggregates must be covered with a paste with a film thickness, b, equal to 0.1 mm for aggregates from 2 to 31.5 mm and 0.02 mm for aggregates below 2 mm.

At this stage, the minimum amount of cement paste, Z_{zmin} Z_{cmin}, was determined, which will allow filling the free spaces, defined as Z_1 and Z_2, which is responsible for moving the aggregate grains apart by a distance, *b*, without blocking the grains during the flow of the concrete mix. A summary of the obtained results of the minimum amount of cement paste is presented in Table 5. The next step in the design is to determine the amount of aggregate using Formula (8) for the total volume of all components and to determine the amount of filler in the form of fly ash.

In order to verify the values of individual components adopted at the design stage, included in Tables 6–8, the final design stage was carried out, i.e., preparation of a batch in the laboratory and determination of the properties of the concrete mix. In terms of the concrete mix, the ability of the mix to flow was determined as the degree of fluidity of the concrete mix using the cone flow method. In parallel, the time t_{500} was determined using this method, measured from the moment the cone was lifted until the mixture flowed out to a diameter of 500 mm (mix viscosity). The time of flow of the concrete mix from the V-funnel was also measured [27–29]. The measurement of viscosity using the V-funnel method allowed for the assessment of the viscosity and ability of the self-compacting mix to fill the mold. Both tests described above are helpful in confirming the homogeneity of self-compacting concrete for different proportions of aggregate and the amount of ash added. Flowability was also determined, i.e., the ability of the concrete mixture to flow, without losing its homogeneity or blocking, through limited spaces and narrow gaps, such as densely reinforced areas in the L-box device. The obtained results of the properties of self-compacting concrete mixtures are presented in Table 9 and in Table 10 are presented the results of the strength properties and durability of concrete. Below in Table 3 are sample calculations for one of the regulations, SCC C2:

Table 3. Recipe ingredients for SCC C2.

Designing Self-Compacting Concrete by the Aggregate Locking Method				
Recipe Composition SCC C2		**[%]**	**[kg/m^3]**	**Density [kg/m^3]**
Aggregate				
30%	Gravel 8/16 mm	23.68	576	2.65
40%	Gravel 2/8 mm	31.57	768	2.65
30%	Sand 0/2 mm	23.68	576	2.64
Filer	Fly ash FA	0.00	0	2.66
Cement	CEM I 42.5 R	6.47	350	3.10
Water		2.91	157	1.00
Admixture	ViscoCrete	0.22	5.24	1.05
	-	100.0	2431	

Table 4. Calculations of aggregate grain size and volume in m^3.

C = 34 mm									
			Recipe Ingredients in [%]			Volume [L/m³]			Sum:
Sieve [mm]	Daf	C/Daf	Sand 0/2	Gravel 2/8	Gravel 8/16	Sand 0/2	Gravel 2/8	Gravel 8/16	
31.500	-	-	0.00	0.00	0.00	0.00	0.00	0.00	0.00
16.000	27.625	1.231	0.00	0.00	0.00	0.00	0.00	0.00	0.00
8.000	14.000	2.429	0.00	1.00	71.00	0.00	2.90	154.23	157.13
4.000	7.000	4.857	0.00	43.00	26.00	0.00	124.55	56.48	181.03
2.000	3.500	9.714	0.00	46.50	3.00	0.00	134.68	6.52	141.20
1.000	1.750	19.429	9.00	9.50	0.00	19.62	27.55	0.00	47.14
0.500	0.875	38.857	36.00	0.00	0.00	78.50	0.00	0.00	78.50
0.250	0.438	77.714	24.00	0.00	0.00	52.33	0.00	0.00	52.33
0.125	0.219	155.429	23.50	0.00	0.00	51.24	0.00	0.00	51.24
0.075	0.113	302.222	2.50	0.00	0.00	5.24	0.00	0.00	5.45
0.000	0.056	604.444	5.00	0.00	0.00	10.90	0.00	0.00	10.90

For the designed aggregate proportions of 30% 8/16 gravel, 40% 2/8 gravel and 30% 0/2 mm sand, the following good grain size distribution curve was obtained, as shown in Figure 1:

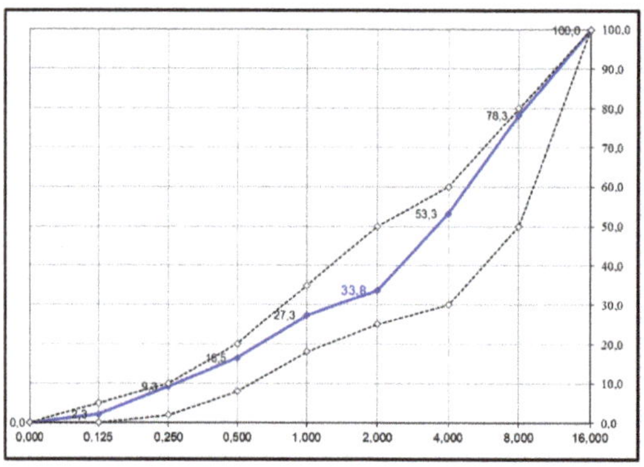

Figure 1. Grain size diagram of the designed mineral mixture.

Values q_{0max} and P_{min}, which are necessary to determine the minimum amount of grout, were determined experimentally. The following values of the maximum bulk density, q_{0max}, and free space in the designed mix, P_{min}, for case C2 were obtained:

$$q_{0\ max} = 2.098\ \text{g/cm}^3$$

$$P_{min} = 20.83\% = 0.2083\ \text{g/cm}^3$$

which is why:

$$Z_1 = \frac{0.2083 \cdot 1.81}{2.098} = 0.1797\ g_z/g_k$$

Assuming the density of the cement paste is equal to 1.81 g/cm^3.

Table 5. Calculation of the blocking index, BL, for a given aggregate composition, C2.

Blocking for a Given Type of Aggregate						
Sieve	Daf	C/Daf	for Natural Aggregate	for Crushed Aggregate	Sum Volume	Blocking
[mm]			[L/m^3]	[L/m^3]	[L/m^3]	
31.500	-	-	-	-		-
16.000	27.625	1.231	80	60		0.000
8.000	14.000	2.429	508	408		0.326
4.000	7.000	4.857	622	522		0.254
2.000	3.500	9.714	724	670		0.138
1.000	1.750	19.429	840	840		0.050
0.500	0.875	38.857	840	840		0.108
0.250	0.438	77.714	840	840		0.080
0.125	0.219	155.429	840	840		0.087
0.075	0.113	302.222	840	840		0.014
0.000	0.056	604.444	840	840		0.028
						1.088

Based on the calculated specific surface area, the demand for grout, Z_2, is determined according to the calculations according to the adopted calculation model:

$$Z_2 = 0.1932\ g_z/g_k$$

Table 6. Determination of the demand for Z_2 grout based on the determined specific surface area and the thickness of the grout film for recipe SCC C2.

SCC C2				
Calculated Specific Surface Area P_w	Wrap Radius b	$P_w \cdot b$	Grout Density	Demand for Leaven Z_2
[cm^2/g]	[cm]	[cm^3/g]	[g/m^3]	[g_z/g_k]
0.00	0.01	0.0000	1.81	0.0000
0.00	0.01	0.0000	1.81	0.0000
0.54	0.01	0.0054	1.81	0.0098
1.25	0.01	0.0125	1.81	0.0226
1.95	0.01	0.0195	1.81	0.0353
1.30	0.002	0.0026	1.81	0.0047
4.32	0.002	0.0086	1.81	0.0156
5.76	0.002	0.0115	1.81	0.0209
11.28	0.002	0.0226	1.81	0.0408
2.40	0.002	0.0048	1.81	0.0087
9.60	0.002	0.0192	1.81	0.0348
38.40		0.1067	1.81	0.1932

The calculated amount of grout needed to fill the empty spaces in the mixture is given by $Z_1 = 0.1797\ g_z/g_k$, while the amount of leaven, Z_2, due to the specific surface area and film thickness b for this composition is 0.1932 g_z/g_k. The blocking factor for this case is 1.09. In summary, the total slurry requirement (liquid phase) for the SCC C2 recipe is:

$$Z_{cmin} = 0.1797 + 0.1932 = 0.3729 \ g_z/g_k$$

With the above values at our disposal, we calculate the amount of aggregate needed:

$$K = \frac{1000}{\frac{0.3729}{1.81} + \frac{1}{2.6470}} = 1712.89 \text{ kg/m}^3$$

Therefore, the final amount of cement paste (liquid phase) is as follows:

$$Z = 0.3729 \cdot 1712.89 = 638.74 \text{ kg/m}^3$$

The amount of grout (cement + water) is known and assumed in the recipe; it is as follows: cement 350 kg and water 157.5 kg, i.e., together, 507.5 kg/m^3. We have obtained the amount of necessary grout of 638.74 kg/m^3. This means that the remaining amount (131.24 kg/m^3) must be supplemented with a filler, e.g., fly ash, so that the concrete can flow freely (is self-compacting) by spreading the aggregate with the grout. The amount of superplasticizer added, 1.5%, to the cement mass is 5.25 kg/m^3.

Final summary of all grout quantity results, Z_1, Z_2, and Z_{cmin}, for each of the 36 analyzed compositions are presented in Table 7:

Table 7. Calculation of the minimum amount of grout and the aggregate blocking index.

Recipe	Gravel 8/16 [%]	Gravel 2/8 [%]	Sand 0/2 [%]	q_{0max} [g/cm^3]	P_{min} [%]	q_{wz} [g/cm^3]	Z_1 [g_z/g_k]	P_w [g/cm^2]	Z_2 [g_z/g_k]	Z_{cmin} [kg/m^3]	Blocking Indicator
A1	50	50	0	1.648	37.81	1.81	0.4153	6.1	0.0957	0.511	1.218
A2	50	40	10	1.724	34.94	1.81	0.3668	16.5	0.1236	0.490	1.197
A3	50	30	20	1.789	32.49	1.81	0.3287	26.9	0.1515	0.480	1.177
A4	50	20	30	1.855	30.00	1.81	0.2927	37.3	0.1793	0.472	1.157
A5	50	10	40	1.962	25.96	1.81	0.2395	47.8	0.2072	0.447	1.137
A6	50	0	50	2.078	21.58	1.81	0.1880	58.2	0.2351	0.423	1.117
B1	40	50	10	2.172	18.04	1.81	0.1503	17.0	0.1305	0.281	1.169
B2	40	40	20	2.186	17.51	1.81	0.1450	27.4	0.1584	0.303	1.149
B3	40	30	30	2.196	17.13	1.81	0.1412	37.9	0.1863	0.327	1.128
B4	40	20	40	2.199	17.02	1.81	0.1401	48.3	0.2141	0.354	1.108
B5	40	10	50	2.205	16.79	1.81	0.1378	58.7	0.242	0.380	1.088
B6	40	0	60	2.214	16.45	1.81	0.1345	69.2	0.2699	0.404	1.068
C1	30	50	20	2.065	22.08	1.81	0.1935	28.0	0.1653	0.359	1.12
C2	30	40	30	2.098	20.83	1.81	0.1797	38.4	0.1932	0.373	1.1
C3	30	30	40	2.135	19.43	1.81	0.1647	48.8	0.2211	0.386	1.08
C4	30	20	50	2.186	17.51	1.81	0.1450	59.3	0.2489	0.394	1.059
C5	30	10	60	2.199	17.02	1.81	0.1401	69.7	0.2768	0.417	1.039
C6	30	0	70	2.211	16.57	1.81	0.1356	80.1	0.3047	0.440	1.019
D1	20	60	20	2.096	20.91	1.81	0.1806	28.5	0.1723	0.353	1.091
D2	20	50	30	2.119	20.04	1.81	0.1712	38.9	0.2001	0.371	1.071
D3	20	40	40	2.127	19.74	1.81	0.1680	49.4	0.228	0.396	1.051
D4	20	30	50	2.145	19.06	1.81	0.1608	59.8	0.2559	0.417	1.031
D5	20	20	60	2.167	18.23	1.81	0.1523	70.2	0.2837	0.436	1.011
D6	20	10	70	2.185	17.55	1.81	0.1454	80.6	0.3116	0.457	0.991
E1	10	60	30	2.115	20.19	1.81	0.1728	39.5	0.2071	0.380	1.043
E2	10	50	40	2.128	19.70	1.81	0.1676	49.9	0.2349	0.402	1.022
E3	10	40	50	2.142	19.17	1.81	0.1620	60.3	0.2628	0.425	1.002
E4	10	30	60	2.158	18.57	1.81	0.1558	70.8	0.2907	0.446	0.982
E5	10	20	70	2.162	18.42	1.81	0.1542	81.2	0.3185	0.473	0.962
E6	10	10	80	2.181	17.70	1.81	0.1469	91.6	0.3464	0.493	0.942
F1	0	70	30	2.189	17.40	1.81	0.1439	40.0	0.2214	0.358	1.014
F2	0	60	40	2.207	16.72	1.81	0.1371	50.4	0.2419	0.379	0.994
F3	0	50	50	2.244	15.32	1.81	0.1236	60.9	0.2697	0.393	0.974
F4	0	40	60	2.265	14.53	1.81	0.1161	71.3	0.2976	0.414	0.954
F5	0	30	70	2.279	14.00	1.81	0.1112	81.7	0.3255	0.437	0.933
F6	0	20	80	2.286	13.74	1.81	0.1088	92.1	0.3533	0.462	0.913

Knowing the minimum amount of grout, Z_{cmin}, the remaining components of the concrete mix can be determined from Formulas (5)–(8). The final compositions of the recipes, together with the determination of the minimum amount of fly ash (FA) [29], are presented in Table 8. The calculation of the amount of FA to meet the demand for the Z paste was made assuming that we have a constant amount of cement, C 350 kg/m^3 and w/c 0.45.

Table 8. Composition calculations for the analyzed recipes differing in the amount of individual aggregate fractions and the demand for fly ash FA.

Recipe	Gravel 8/16 [%]	Gravel 2/8 [%]	Sand 0/2 [%]	P_{min} [%]	P_w [g/cm^2]	Blocking Indicator	Aggregate [kg/m^3]	Cement Grout [kg/m^3]	Cement [kg/m^3]	Water [kg/m^3]	FA [kg/m^3]
A1	50	50	0	37.81	6.1	1.218	1514	773.8	350	157.5	266.3
A2	50	40	10	34.94	16.5	1.197	1541	755.6	350	157.5	248.1
A3	50	30	20	32.49	26.9	1.177	1554	746.4	350	157.5	238.9
A4	50	20	30	30.00	37.3	1.157	1565	738.9	350	157.5	231.4
A5	50	10	40	25.96	47.8	1.137	1600	714.9	350	157.5	207.4
A6	50	0	50	21.58	58.2	1.117	1634	691.5	350	157.5	184.0
B1	40	50	10	18.04	17.0	1.169	1875	526.7	350	157.5	19.2
B2	40	40	20	17.51	27.4	1.149	1833	556.0	350	157.5	48.5
B3	40	30	30	17.13	37.9	1.128	1789	585.8	350	157.5	78.3
B4	40	20	40	17.02	48.3	1.108	1743	617.3	350	157.5	109.8
B5	40	10	50	16.79	58.7	1.088	1701	646.0	350	157.5	138.5
B6	40	0	60	16.45	69.2	1.068	1663	672.3	350	157.5	164.8
C1	30	50	20	22.08	28.0	1.12	1735	622.6	350	157.5	115.1
C2	30	40	30	20.83	38.4	1.1	1712	638.4	350	157.5	130.9
C3	30	30	40	19.43	48.8	1.08	1691	652.6	350	157.5	145.1
C4	30	20	50	17.51	59.3	1.059	1679	661.2	350	157.5	153.7
C5	30	10	60	17.02	69.7	1.039	1644	685.2	350	157.5	177.7
C6	30	0	70	16.57	80.1	1.019	1609	708.7	350	157.5	201.2
D1	20	60	20	20.91	28.5	1.091	1745	615.8	350	157.5	108.3
D2	20	50	30	20.04	38.9	1.071	1715	636.6	350	157.5	129.1
D3	20	40	40	19.74	49.4	1.051	1675	663.5	350	157.5	156.0
D4	20	30	50	19.06	59.8	1.031	1644	685.1	350	157.5	177.6
D5	20	20	60	18.23	70.2	1.011	1616	704.4	350	157.5	196.9
D6	20	10	70	17.55	80.6	0.991	1586	724.7	350	157.5	217.2
E1	10	60	30	20.19	39.5	1.043	1701	646.1	350	157.5	138.6
E2	10	50	40	19.70	49.9	1.022	1665	670.3	350	157.5	162.8
E3	10	40	50	19.17	60.3	1.002	1632	693.2	350	157.5	185.7
E4	10	30	60	18.57	70.8	0.982	1601	714.6	350	157.5	207.1
E5	10	20	70	18.42	81.2	0.962	1564	739.5	350	157.5	232.0
E6	10	10	80	17.70	91.6	0.942	1537	758.2	350	157.5	250.7
F1	0	70	30	17.40	40.0	1.014	1737	621.5	350	157.5	114.0
F2	0	60	40	16.72	50.4	0.994	1702	645.2	350	157.5	137.7
F3	0	50	50	15.32	60.9	0.974	1680	660.6	350	157.5	153.1
F4	0	40	60	14.53	71.3	0.954	1648	682.0	350	157.5	174.5
F5	0	30	70	14.00	81.7	0.933	1615	705.1	350	157.5	197.6
F6	0	20	80	13.74	92.1	0.913	1579	729.6	350	157.5	222.1

The last stage of the design is the experimental verification of the properties of the concrete mix and the strength and durability characteristics of the cement concrete. Analyzing the results of the free space size, blocking index, and specific surface area of the designed aggregate and the resulting demand for the amount of cement paste, the following formula compositions were selected for further tests: B2, B3, B4, B5, C2, C3, C4, C5, D3, and D4. During the preparation of laboratory batches, the properties of the concrete mix were determined in terms of the ability of the mix to flow. The fluidity of the concrete mix was determined according to the cone flow method, in addition to the flow time, t_{500}, measured from the moment of lifting the cone to the moment of the mix flowing to a diameter of 500 mm [29–33]. The time of the concrete mix flowing out of the V-funnel was also measured, as well as the ability of the concrete mix to flow without losing its homogeneity or blocking in the L-box device. The obtained results of the properties of self-compacting concrete mixes are presented in Table 9.

Table 9. Properties of the analyzed self-compacting concrete mixtures.

Recipe	Gravel 8/16 [%]	Gravel 2/8 [%]	Sand 0/2 [%]	Cement [kg/m³]	Water [kg/m³]	FA [kg/m³]	SP [kg/m³]	SF [mm]	VS t_{500} [s]	VF [s]	L-Box >0.80
B2	40	40	20	350	157.5	48.5	4.90	490	4.4	12.1	0.77
B3	40	30	30	350	157.5	78.3	5.25	540	4.0	9.3	0.79
B4	40	20	40	350	157.5	109.8	5.25	630	3.3	10.1	0.81
B5	40	10	50	350	157.5	138.5	4.90	640	3.1	8.4	0.84
C2	30	40	30	350	157.5	130.9	5.60	650	2.6	7.5	0.83
C3	30	30	40	350	157.5	145.1	5.60	670	1.8	6.6	0.85
C4	30	20	50	350	157.5	153.7	5.95	710	1.4	6.9	0.86
C5	30	10	60	350	157.5	177.7	5.60	730	1.5	5.9	0.87
D3	20	40	40	350	157.5	156.0	5.60	750	1.6	5.5	0.86
D4	20	30	50	350	157.5	177.6	5,25	730	1.3	5.7	0.88

The last stage of experimental verification of the design of SCC concrete with reduced cement content was a series of strength tests and durability tests for the selected recipes. The results of compressive strength performed after 7, 28, and 90 days are presented in Figure 2 below:

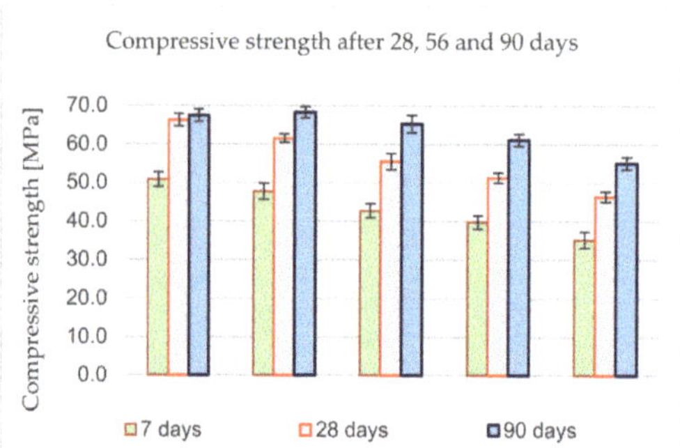

Figure 2. Compressive strength results after 7, 28, and 90 days.

The full results of determining the compressive strength, frost resistance, F100, and the air entrainment structure of concrete are presented in Table 10:

Table 10. Determination of compressive strength, frost resistance, F100, and air entrainment structure of SCC concrete.

Recipe	Gravel 8/16	Gravel 2/8	Sand 0/2	Cement	Water	FA	Compressive Strength			F100	A_{300}	L
	[%]	[%]	[%]	[kg/m³]	[kg/m³]	[kg/m³]	[MPa]	[MPa]	[MPa]	[MPa]	[%]	[mm]
							After 7 Days	After 28 Days	After 90 Days			
B2	40	40	20	350	157.5	48.5	40.5	51.4	60.2	−31	1.96	0.27
B3	40	30	30	350	157.5	78.3	42.1	57.2	63.5	−1.9	2.05	0.24
B4	40	20	40	350	157.5	109.8	40.6	60.3	68.3	−3.8	1.91	0.22
B5	40	10	50	350	157.5	138.5	43.3	55.7	64.2	−7.7	2.56	0.17
C2	30	40	30	350	157.5	130.9	47.7	61.6	68.3	0.8	2.51	0.20
C3	30	30	40	350	157.5	145.1	50.8	60.3	67.4	−2.1	2.90	0.15
C4	30	20	50	350	157.5	153.7	48.4	61.4	69.8	0.9	3.31	0.09
C5	30	10	60	350	157.5	177.7	41.8	63.7	72.5	−4.1	2.29	0.16
D3	20	40	40	350	157.5	156.0	44.5	58.9	68.6	−3.0	2.36	0.21
D4	20	30	50	350	157.5	177.6	42.5	56.4	66.7	−8.9	1.88	0.26

4. Discussion

The presented research work presented a procedure for designing self-compacting concretes using a reduced amount of cement by designing a composition of the aggregate mixture that will be characterized by the smallest vacuum. Although the design of self-compacting concrete is a much more complicated process and requires increased control and knowledge of innovative concretes, it allows obtaining a mixture without the need for vibration. Hardened SCC concrete surpasses ordinary concretes in its properties and technical parameters; it provides new possibilities and a spectrum of applications and could revolutionize the approach to design and execution in construction. Despite the lack of unambiguous design methods in Poland, it is worth using the available achievements of European countries, where self-compacting concrete technology is widely used. The emerging methods are based on recommendations regarding SCC and are still being developed; an example is the blocking method, according to which the composition of the concrete mixture was designed. For 36 designed mineral aggregate formulas, the grain size distribution curve, specific surface area, minimum value of the demand for grout, and the blocking index of the aggregate composition were calculated. A large number of the considered mixture compositions allowed for a broad analysis of the obtained results and the drawing of conclusions. It should be stated that the obtained results confirmed compliance with the theoretical assumptions regarding the composition of SCC mixtures. From among the analyzed cases, several compositions were selected that were optimal in terms of free spaces and the blocking index. Of course, this is not the only correct composition of the mixture, but when selecting it all parameters were taken into account and an attempt was made to select the most advantageous variant in terms of technology and economy. Further experimental research and continuous expansion of knowledge in the field of SCC concretes is extremely important and will certainly contribute in the future to the free use of self-compacting concrete technology, which is increasingly commonly used in special and architectural concretes in Poland. In further studies, the research will be extended to include studies involving crushed aggregates (basalt, granite) and other additives, such as blast furnace slag or limestone flour.

5. Conclusions

Based on the calculations, tests, and analyses performed for 36 different formulations, the following conclusions can be drawn:

1. The increase in the specific surface area of the aggregate causes an increase in the demand for cement paste, which significantly affects the increase in the amount of cement.
2. The greatest demand for paste is characteristic for mixtures with a high sand fraction content, above 40%, which confirms the theoretical assumption of the blocking method, while mineral mixtures with the largest specific surface area are characterized by the lowest blocking index.
3. In most cases, the grain size curve did not fit into the limit curves recommended for road and bridge concretes; therefore, new limit curves should be determined for SCC. The grain size curve for cases from group F did not fit into the limit curves, which indicates that obtaining self-compacting concrete without a gravel fraction of 8/16 is very difficult and does not guarantee obtaining the assumed strength parameters,
4. The analyzed recipes, in which the share of coarse aggregate was 60–70% and sand was 30–40%, were characterized by the lowest demand for grout and were within the limit curves. This was confirmed in the SCC tests.

5. After analyzing 36 different mineral compositions of self-compacting concretes for further analysis, the smallest free space and the lowest blocking index were found in formulas from groups B2–B4, C2–C5, and D3–D4.
6. The use of this method allows for a significant reduction in the content of voids and the amount of cement in the SCC mixture without a significant decrease in the flow properties and durability characteristics of concrete over time.

Funding: This research received no external funding.

Institutional Review Board Statement: Not applicable.

Informed Consent Statement: Not applicable.

Data Availability Statement: The original contributions presented in this study are included in the article. Further inquiries can be directed to the corresponding author.

Conflicts of Interest: The author declares no conflict of interest.

References

1. Okamura, H. Self-compacting high performance concrete. *Concr. Int.* **1997**, *19*, 50–54.
2. Okamura, H.; Maekawa, K.; Ozawa, K. *High Performance Concrete*; Gihoudou Pub.: Tokyo, Japan, 1993. (In Japanese)
3. Henry, G.R. ACI defines high performance concrete. *Concr. Int.* **1999**, *21*, 56–57.
4. Aitcin, P.C. *High Performance Concrete*; E&FN SPON: London, UK, 1998.
5. Okamura, H. *Self-Compacting High Performance Concrete*; Social System Institute: Tokyo, Japan, 1999. (In Japanese)
6. Bartos, P.J.M.; Grauers, M. Self-compacting concrete. *Concrete* **1999**, *33*, 9–14.
7. Chai, H.W. Design and Testing of Self-Compacting Concrete. Ph.D. Thesis, Department of Civil and Environmental Engineering, University College London, London, UK, 1998.
8. Ding, X.; Zhao, M.; Qiu, X.; Wang, Y.; Ru, Y. The Optimization of Mix Proportion Design for SCC: Experimental Study and Grey Relational Analysis. *Materials* **2022**, *15*, 1305. [CrossRef] [PubMed]
9. Rudnicki, T. Functional Method of Designing Self-Compacting Concrete. *Materials* **2021**, *14*, 267. [CrossRef]
10. de Larrard, F. *Concrete Mixture Proportioning—A Scientific Approach*; Mindess, S., Bentur, A., Eds.; Modern Concrete Technology Series No. 7; E&FN SPON: London, UK, 1999; p. 421.
11. *PN-EN 197-1:2012*; Cement—Part 1: Composition, Specifications and Conformity Criteria for Common Cements. PKN: Warsaw, Poland, 2012.
12. *PN-EN 934-1:2009*; Admixtures for Concrete, Mortar and Grout—Part 1: Common Requirements. PKN: Warsaw, Poland, 2009.
13. *EN 934-2:2010*; Administrative for Concrete, Mortar and Grout—Part 2: Concrete Admixtures—Definitions, Requirements, Conformity, Marking and Labelling. PKN: Warsaw, Poland, 2010.
14. *PN-EN 1008:2004*; Mixing Water for Concrete—Specification for Sampling, Testing and Assessing the Suitability of Water, Including Water Recovered from Processes in the Concrete Industry, as Mixing Water for Concrete. PKN: Warsaw, Poland, 2004.
15. Rudnicki, T. The method of aggregate skeleton in self-compacting concrete designing with segment regression. *Cem. Wapno Beton* **2016**, *21*, 10–19.
16. Ding, X.; Geng, H.; Shi, K.; Song, L.; Li, S.; Liu, G. Study on Adaptability of Test Methods for Workability of Fresh Self-Compacting SFRC. *Materials* **2021**, *14*, 5312. [CrossRef] [PubMed]
17. Billberg, P. Fine mortar rheology in mix design of SCC. In *First International RILEM Symposium on Self-Compacting Concrete*; RILEM Publications, S.A.R.L.: Stockholm, Sweden, 1999; pp. 47–58.
18. Tangtermsirikul, S.; Van, B.K. Blocking criteria for aggregate phase of self-compacting high-performance concrete. In Proceedings of the Regional Symposium on Infrastructure Development in Civil Engineering, Bangkok, Thailand, 19–20 December 1995; pp. 58–69.
19. Billberg, P. *Self-Compacting Concrete for Civil Engineering Structures—The Swedish Experience*; CBI Rapport 2; Swedish Cement and Concrete Research Institute: Stockholm, Sweden, 1999.
20. Montgomery, D.; Bui, V.K. Rapid testing method for segregation resistance of self-compacting concrete. *Cem. Concr. Res.* **2002**, *32*, 1489–1496.
21. Su, N.; Hsu, K.-C.; Chai, H.-W. A simple mix design method for self-compacting concrete. *Cem. Res.* **2001**, *31*, 1799–1807. [CrossRef]
22. Ouchi, M. State-of-the-art Report: Self-Compatibility Evaluation for Mix-proportioning and Inspection. In Proceedings of the International Workshop on Self-Compacting Concrete, Kochi University of Technology, Kami, Japan, 23–26 August 1998.

23. Persson, B. *Mix Proportions and Strength of SCC for Production of High Strength Poles, Piles and Pillars;* Münchener Baustoffseminar Selbstverdichtender Beton; Technical University: München, Germany, 2001; pp. 31–39.
24. Petersson, O.; Billberg, P.; Van, B.K. *A Model for Self-Compacting Concrete, Production Methods and Workability of Concrete;* E&FN Spon: London, UK, 1996; pp. 483–492.
25. Van, B.K.; Montgomery, D. Mixture proportioning method for SCC HPC with minimum pasta volume. In Proceedings of the 1st International RILEM Symposium on SCC, Stockholm, Sweden, 13–14 September 1999; pp. 373–384.
26. Neville, A.M. *Właściwości Betonu, Book V Edycja;* Polski Cement: Kraków, Poland, 2012.
27. Yang, Z.; Liu, S.; Yu, L.; Xu, L. A Comprehensive Study on the Hardening Features and Performance of Self-Compacting Concrete with High-Volume Fly Ash and Slag. *Materials* **2021**, *14*, 4286. [CrossRef]
28. Japanese Ready-Mixed Concrete Association. *Manual of Producing High Fluidity (Self-Compacting) Concrete;* Japanese Ready-Mixed Concrete Association: Tokyo, Japan, 1998. (In Japanese)
29. Bouzoubaa, N.; Lachemi, M. Self-compacting concrete incorporating high volumes of class F fly ash Preliminary results. *Cem. Concr. Res.* **2001**, *31*, 413–420. [CrossRef]
30. Glinicki, M.A.; Gołaszewski, J.; Cygan, G. Formwork Pressure of a Heavyweight Self-Compacting Concrete Mix. *Materials* **2021**, *14*, 1549. [CrossRef] [PubMed]
31. Devore, J.L. *Probability and Statistics for Engineering and the Science*, 7th ed.; Thomson Brooks/Cole: Pacific Grove, CA, USA, 2008; pp. 513–514.
32. Chiara, F.F.; Lynn, B.; Celik, O.; Joseph, D. Durable Bridges and Transportation Structures. In Proceedings of the International Symposium on High Performance Concrete, Orlando, FL, USA, 25–27 September 2000.
33. Horszczaruk, E.; Brzozowski, P.; Adamczewski, G.; Rudnicki, T. Influence of Hydrostatic Pressure on Compressive Strength of Self-Consolidating Concrete. *J. Civ. Eng. Archit.* **2013**, *12*, 1549–1555.

materials

MDPI

Article

Preparation and Performance Study of Novel Foam Vegetation Concrete

Teng Zhang *, Tianbin Li, Hua Xu, Mengyun Wang and Lingling Lu

State Key Laboratory of Geohazard Prevention and Geoenvironment Protection, Chengdu University of Technology, Chengdu 610059, China
* Correspondence: zhangteng@cdut.edu.cn

Abstract: Vegetation concrete is one of the most widely used substrates in ecological slope protection, but its practical application often limits the growth and nutrient uptake of plant roots due to consolidation problems, which affects the effectiveness of slope protection. This paper proposed the use of a plant protein foaming agent as a porous modifier to create a porous, lightweight treatment for vegetation concrete. Physical performance tests, direct shear tests, plant growth tests, and scanning electron microscopy experiments were conducted to compare and analyze the physical, mechanical, microscopic characteristics, and phyto-capabilities of differently treated vegetation concrete. The results showed that the higher the foam content, the more significant the porous and lightweight properties of the vegetation concrete. When the foam volume was 50%, the porosity increased by 106.05% compared to the untreated sample, while the volume weight decreased by 20.53%. The shear strength, cohesion, and internal friction angle of vegetation concrete all showed a decreasing trend with increasing foaming agent content. *Festuca arundinacea* grew best under the 30% foaming agent treatment, with germinative energy, germinative percentage, plant height, root length, and underground biomass increasing by 6.31%, 13.22%, 8.57%, 18.71%, and 34.62%, respectively, compared to the untreated sample. The scanning electron microscope observation showed that the pore structure of vegetation concrete was optimized after foam incorporation. Adding plant protein foaming agents to modify the pore structure of vegetation concrete is appropriate, with an optimal foam volume ratio of 20–30%. This study provides new insights and references for slope ecological restoration engineering.

Keywords: vegetation concrete; foaming agent; physical and mechanical properties; plant growth; slope protection

1. Introduction

Large-scale infrastructure and urbanization projects have resulted in numerous exposed slopes [1]. These slopes, when subjected to natural factors such as wind, sun exposure, and rainfall, are prone to geological disasters like collapses, landslides, and mudflows [2–4], posing severe threats to human life safety. Additionally, these events lead to a reduction in biodiversity, damage to landscape features, and create visual pollution due to their stark contrast with the surrounding natural scenery [5,6]. Traditional masonry measures, such as shotcrete, retaining walls, and frame beams, often sacrifice ecological functions and landscape structures without guaranteeing long-term protection [7–9]. With the global increase in environmental concerns, ecological protection of slopes has become increasingly important.

Vegetation concrete is a composite material that combines concrete with plant growth technologies [10]. It is made from planting soil, cement, aggregates, additives, fertilizers, water, and mixed green plant seeds [11]. This mixture is sprayed onto slopes using special equipment and is widely used for the comprehensive management and ecological restoration of exposed slopes, especially rocky ones. Vegetation concrete uses cement as a binder, and combined with common support materials such as anchors, dowels, and protective nets, it can stabilize the slope. Once the seeds germinate and grow into turf, the above-ground part of the vegetation can intercept rainfall, reduce splash erosion, and inhibit runoff. The roots form a dense reinforcement and anchorage system that further enhances slope stability [12,13]. However, in practical applications of slope ecological protection, it has been observed that the addition of cement can lead to consolidation of the vegetation concrete (Figure 1). This consolidation reduces the porosity and aeration of the vegetation concrete, adversely affecting root growth and nutrient uptake of the slope vegetation, ultimately hindering normal plant growth and compromising the effectiveness of ecological slope protection [14,15].

Figure 1. Vegetation concrete: (**a**) Vegetation concrete, (**b**) Consolidation of Vegetation concrete (Magnified view).

The consolidated vegetation concrete displays a reduction in porosity [16]. As the spatial environment for the extension of vegetation roots, vegetation concrete pores affect the growth of plant roots, determine the length and diameter of roots directly, and play an important role in slope protection [17,18]. Many scholars have improved the pore structure of vegetation concrete by adding plant fibers or residues. For example, Liu et al. [19] enhanced the pore structure of ecological slope protection substrates by adding organic materials such as rice husks, sawdust, corn, and distiller grains. Qiu et al. [20] increased the porosity by incorporating straw into the ecological slope protection substrate. These studies have improved the porosity of vegetation concrete; however, the plant fiber and plant residue can lead to discontinuous cracks in the vegetation concrete. Liu et al. [21] found that biochar could effectively increase the porosity of vegetation concrete. Currently, most methods to improve the porosity of vegetation concrete involve adding solid materials, which can increase the overall porosity. However, excessive amounts or uneven mixing can lead to agglomeration or the formation of cracks.

Lightweight foam concrete is a typical porous material characterized by low cost, lightweight, convenient construction, and green, low-carbon properties. It has been widely used in civil and industrial construction projects [22,23]. Foam concrete is made by incorporating foam from a foaming agent into concrete. The foaming agent is the primary substance that forms pores and can be divided into chemical and physical foaming agents [24,25]. Chemical foaming agents produce bubbles through chemical reactions, generating a large amount of gas but are challenging to control during gas production [26], and some foaming agents can produce harmful gases [27]. Physical foaming agents generate bubbles through mechanical action, mainly including rosin-based, synthetic surfactants, protein-based, and composite types [28], with more foam stability. Foaming agents are also used in soil improvement. For example, Vázquez et al. [29] and León et al. [30] used sugar foam as a soil amendment for planting vegetation, promoting crop growth; Chan et al. [31] found that Hypucem foam could improve soil physical properties, aiding healthy plant growth; Gao et al. [32] incorporated foaming agents into municipal sludge, inventing an artificial planting soil with a rich pore structure; Wang et al. [33] applied foaming agents in the formulation of plant growth soil, preparing improved soil that promoted vegetation growth. Currently, foaming agents are primarily used in the preparation of lightweight foam concrete and soil improvement, with fewer applications in the field of slope ecological protection.

To effectively reduce the consolidation of vegetation concrete and promote the growth of slope-protecting plants, this study attempts to apply a foaming agent as a modifier to vegetation concrete. Specifically, foam generated by the foaming agent is used to improve the pore structure of vegetation concrete, forming a lightweight, porous foamed vegetation concrete. Through experiments measuring volume weight and porosity, direct shear tests, plant growth tests, and SEM (Scanning Electron Microscope) analysis, the effects of incorporating foaming agents on the physical properties, mechanical properties, plant growth performance, and microstructure of vegetation concrete were systematically studied. Additionally, the mechanisms by which foaming agents influence the performance of vegetation concrete and slope ecological protection were also explored. This research provides new insights and references for slope ecological restoration projects.

2. Materials and Methods

2.1. Materials

This study was conducted based on the vegetation concrete developed by our research group [11,34]. The primary materials involved included planting soil, cement, aggregates, green additives, ecological magnetic slow-release fertilizers, plant fibers, and water. The planting soil is red-layer clay from the Sichuan Basin in China, sourced locally. It was cleaned of debris such as stones and dead branches, air-dried, and then sifted through a 2 mm sieve for use, with its basic physical properties shown in Table 1. The cement used is P·O 42.5R ordinary Portland cement produced in Sichuan. The aggregate consists of gravel and sand with a mass ratio of 2:1. A single-size grade is used, with the gravel particle size being less than 12 mm and the sand particle size being less than 4.75 mm. Green additives mainly include superabsorbent polymer (PAM), carboxymethylcellulose (CMC) as a binder, and plant residue as a pH regulator.

Table 1. Basic physical properties of soil samples.

Specific Gravity *Gs*	Dry Density (g/cm^3)	Void Ratio	Liquid Limit (%)	Plastic Limit (%)	Permeability Coefficient *k* (cm/s)
2.68	1.59	0.82	49	23	2.7×10^{-6}

2.2. Experimental Design

Protein foaming agents exhibit rapid foaming speed, uniform bubble size, and remarkably stable foam with superior foaming performance [35]. These agents can be classified into plant protein foaming agents and animal protein foaming agents [25]. The study used a plant protein foaming agent, which is characterized by its resistance to external factors such as temperature and pH, widespread production and applications, low cost, and environmental friendliness. The foaming agent was diluted with water at a ratio of 1:60 and then stirred using a high-speed mixer to produce foam, as shown in Figure 2. Six cases were established for the amount of foam added, corresponding to foam volume percentages of 0%, 10%, 20%, 30%, 40%, and 50% relative to the volume of vegetation concrete. The mix proportions of the vegetation concrete components are shown in Table 2.

Figure 2. Foam made by high-speed mixer: (**a**) High-speed mixer, (**b**) Foam.

Table 2. Vegetation concrete proportioning.

Soil (kg/m³)	Aggregate (kg/m³)	Cement (kg/m³)	Green Additive (kg/m³)			Ecological Magnetic Slow-Release Fertilizer (kg/m³)	Plant Fibers (kg/m³)	Water (kg/m³)
			PAM	CMC	pH Regulator			
1500	400	70	4.5	5.5	8	15	15	500

2.3. Sample Preparation

According to the designed mix ratio, the components of vegetation concrete were weighed and mixed. First, powdered admixtures were thoroughly blended, followed by the addition of water to create a mixed slurry. The foaming agent was diluted with water and stirred to produce foam, which was then added to the mixed slurry according to the mix ratio and stirred until uniform. Subsequently, aggregates and plant fibers, which are larger-particle admixtures, were added and mixed evenly. The process of preparing the sample is shown in Figure 3. Specimens were prepared using a 60 cm^3 cutting ring (diameter 61.8 mm $\times$ height 20 mm), wrapped in cling film immediately after preparation, and placed in a standard curing room (humidity above 95%, temperature 20 $\pm$ 2 °C) for curing. For the planting test, pots with a diameter of 15 cm and a height of 11.5 cm were used, with 50 *Festuca arundinacea* seeds sown in each pot. Watering was performed appropriately according to weather conditions to ensure consistent watering amounts per pot, and the germination and seedling growth of *Festuca arundinacea* seeds were monitored.

Figure 3. Sample preparation: (a) Add powdered components and stir; (b) Add water and stir; (c) Add foam. (d) Stirring, (e) Add larger granular components and stir; (f) Foam vegetation concrete.

2.4. Experimental Methods

2.4.1. Porosity

Three replicates were randomly selected from each treatment group, as described above. The porosity of vegetation concrete was measured using the mass method [36]. Porosity was calculated based on the apparent volume of the vegetation concrete specimen, the mass of the specimen when saturated in water, and the dry mass of the specimen, and the calculation formula is shown in Formula (1):

$$P = \left(1 - \frac{M_1 - M_0}{V}\right) \times 100\% \quad (1)$$

In the formula, P is the porosity of the sample (%); V is the apparent volume of the specimen (cm^3); M_0 is the mass of the saturated specimen in water (g); M1 is the mass of the dried specimen (g).

2.4.2. Volume Weight

Volume weight was measured using the drying and weighing method, and the calculation formula is shown in Formula (2):

$$\rho = \frac{M - M'}{V_1} \quad (2)$$

In the formula, ρ is the bulk weight of the test sample (g/cm^3), M is the mass of the cutting ring plus the dried soil sample (g), M' is the mass of the cutting ring (g), and V_1 is the volume of the cutting ring (cm^3).

2.4.3. Mechanical Properties

The direct shear test was conducted using a DZJ-4 multifunction direct shear apparatus, applying vertical pressures of 50 kPa, 100 kPa, 150 kPa, and 200 kPa, respectively. The shearing speed was controlled at 0.8 mm/min, and the test stopped when the displacement reached 10 mm. The peak shear stress from the test was used as the shear strength value. A graph of shear strength versus vertical pressure was plotted, with the angle of the line representing the internal friction angle and the intercept on the vertical axis representing the cohesion.

2.4.4. Planting Performance

The germinative energy and germinative percentage of *Festuca arundinacea* were assessed at 7 and 10 days post-sowing, employing the method of dividing the total number of germinated seeds. Subsequently, plant height and underground biomass were measured 28 days after emergence. For each treatment, a random selection of 10 plants was made for measurement using a ruler. The underground portion was passed through a sieve and washed with water to determine the length of the primary root. Following this, the root was subjected to degreening in an oven at 105 °C for one hour before being dried at 80 °C until reaching constant mass; this dry weight represented the underground biomass.

2.4.5. SEM

SEM testing was conducted using a FEI Nano SEM450 scanning electron microscope (FEI, Hillsboro, OR, USA) Specimens were cut into samples measuring 2 mm × 2 mm × 2 mm, and the surfaces of the specimens were coated with gold before observing the microstructure.

3. Results

3.1. Volume Weight and Porosity

Figure 4 shows the visual representation of vegetation concrete samples with different foam volumes after 28 days of curing. The samples with 0%, 10%, and 20% foam volume, shown in Figure 4a, Figure 4b, and Figure 4c, respectively, exhibit no significant changes in appearance. In contrast, the 30%, 40%, and 50% foamed specimens, shown in Figure 4d, Figure 4e, and Figure 4f, respectively, display a significant increase in surface porosity, with the 50% foamed specimen being the loosest. As the foam volume increases, the vegetation concrete becomes increasingly loose and porous. During sample preparation, it was found that adding foam caused the mixture to expand in volume and increase in porosity (Figure 3b,d). In addition, mixing became easier after the addition of foam.

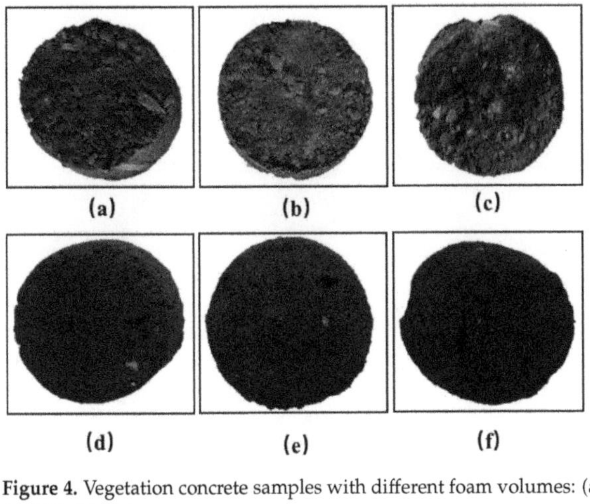

Figure 4. Vegetation concrete samples with different foam volumes: (**a**) 0%, (**b**) 10%, (**c**) 20%, (**d**) 30%, (**e**) 40%, (**f**) 50%.

The results of volume weight and porosity measured after 28 days of curing for vegetation concrete with different treatments are shown in Figure 5. It can be seen that with the increase in foam, the volume weight of vegetation concrete significantly decreased, while the porosity significantly increased. As a result, the characteristics of vegetation concrete, which are its porosity and lightweight nature, become increasingly evident. The volume weight of vegetation concrete with 50% foam volume was reduced by 20.53% compared to that without foam, while the porosity increased by 106.05%. When the foam volume was 40% and 50%, the difference in volume weight and porosity of vegetation concrete was only 0.04 g/cm^3 and 5.21%, respectively, showing a tendency towards stability. The incorporation of foam cannot infinitely affect the volume weight and porosity of vegetation concrete; there is a threshold. Within a reasonable range, the porous and lightweight properties of vegetation concrete can reduce slope load, increase ventilation, improve drainage, and benefit slope stability, and lightweight vegetation concrete is more straightforward to construct and transport, helping to improve construction efficiency and reduce construction costs [37,38].

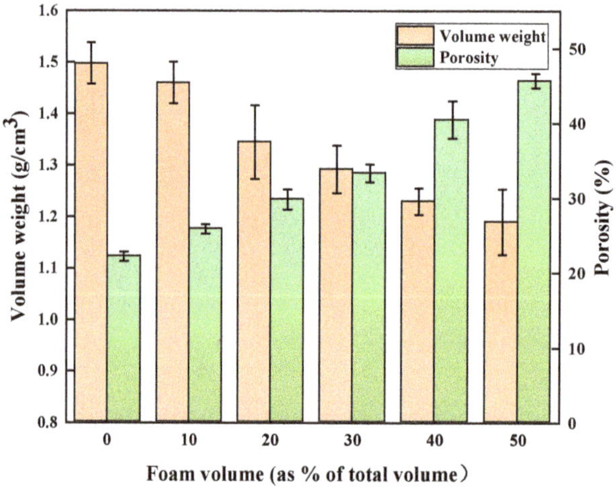

Figure 5. Change in volume weight and porosity of vegetation concrete with added foam volume.

3.2. Shear Strength

Slope superficial layer failures are mostly shear failures [39]. The shear strength of vegetation concrete determines the shear stress and bearing capacity of slope vegetation concrete, which is one of the important mechanical properties affecting the stability of slope ecological protection [40]. The shear strength of vegetation concrete changes with age. Vegetation concrete with different treatments was selected for curing 3 d, 7 d, 14 d, and 28 d, and its shear strength under a vertical pressure of 200 kPa was measured. The results are shown in Figure 6.

Figure 6 shows the shear strength of vegetation concrete at different curing ages. It can be seen that with the extension of the curing age, the shear strength of vegetation concrete with different foam volumes continuously increases, with a rapid increase in the early stage, and the growth rate slows down from 14 days to 28 days. Using vegetation concrete with 30% foam volume as an example, the shear strength at 7 days increased by 24 kPa compared to 3 days, an increase of 17.02%. The shear strength at 14 days increased by 16 kPa compared to 7 days, an increase of 9.70%. The shear strength at 28 days increased by 6 kPa compared to 14 days, an increase of 3.31%. Within the studied range of foam volume, as the foam volume increases, the shear strength of vegetation concrete at the same curing age shows a downward trend, with the decreasing rate gradually increasing. At 7 days, when the foam volume ratio is 10%, 20%, 30%, 40%, and 50%, the shear strength of vegetation concrete is 178 kPa, 173 kPa, 165 kPa, 153 kPa, and 136 kPa, respectively, which is a decrease of 1.66%, 4.42%, 8.84%, 15.47%, and 24.86% compared to vegetation concrete without foam additives.

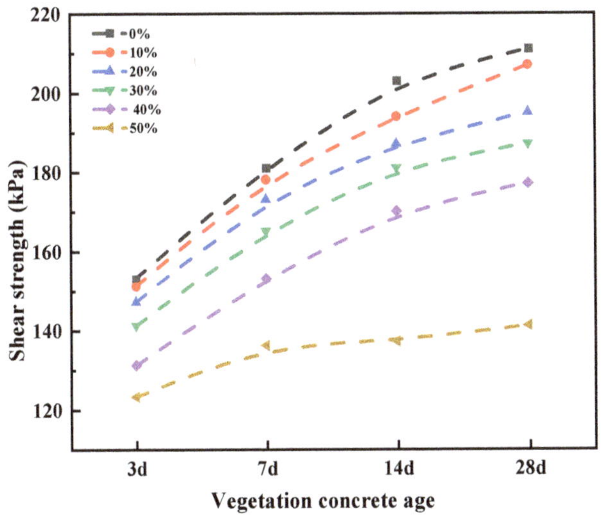

Figure 6. Shear strength of vegetation concrete at different ages.

Draw the relationship curve between the shear strength of vegetation concrete at 28 days under different treatments and vertical pressure and obtain its cohesion and internal friction angle, as shown in Figure 7. As the foam volume increases, the cohesion and internal friction angle of vegetation concrete decrease significantly, with the maximum decrease in cohesion being 17 kPa and the maximum decrease in internal friction angle being 6.8°.

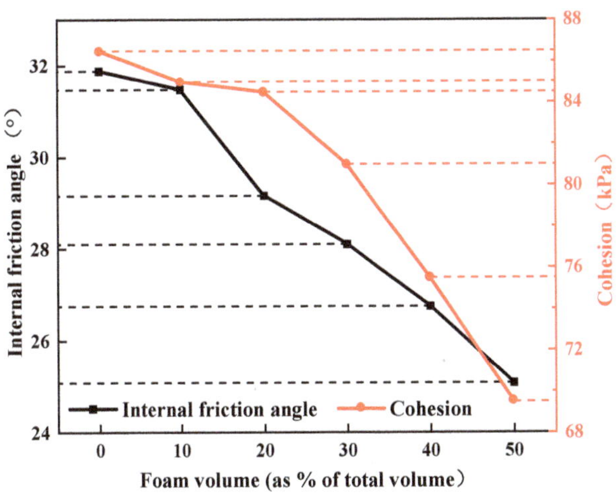

Figure 7. Influence of foam volume on cohesion and angle of internal friction of vegetated concrete.

3.3. Planting Properties of Festuca Arundinacea

Festuca arundinacea is a common grass species in the slope areas of Southwest China, characterized by its perennial growth habit, well-developed root system, rapid growth rate, and strong adaptability. The growth conditions of *Festuca arundinacea* on vegetation concrete with different treatments are shown in Figure 8.

Figure 8. The growth conditions of *Festuca arundinacea* on vegetation concrete with different treatments: (**a**) 0%, (**b**) 10%, (**c**) 20%, (**d**) 30%, (**e**) 40%, (**f**) 50%.

Figure 9 illustrates the effect of foaming agent dosage on the biomass of *Festuca arundinacea*. The figure demonstrates that germinative energy (Figure 9a), germinative percentage (Figure 9b), plant height (Figure 9c), root length (Figure 9d), and underground biomass (Figure 9e) all show a trend of increasing before decreasing as the foaming volumes increase. At foam volumes of 10%, 20%, and 30%, all growth indicators for *Festuca arundinacea* were higher than those of the blank control group. At a foam volume of 40%, the plant height was lower than that of the blank control group, while other indicators were higher or equal to the blank control group. At a foam volume of 50%, all growth indicators for *Festuca arundinacea* were lower than those of the blank control group. In the 30% foam treatment, *Festuca arundinacea* grew best, with germinative energy, germinative percentage, plant height, root length, and underground biomass increasing by 6.31%, 13.22%, 8.57%, 18.71%, and 34.62%, respectively, compared to the treatment without foaming agent.

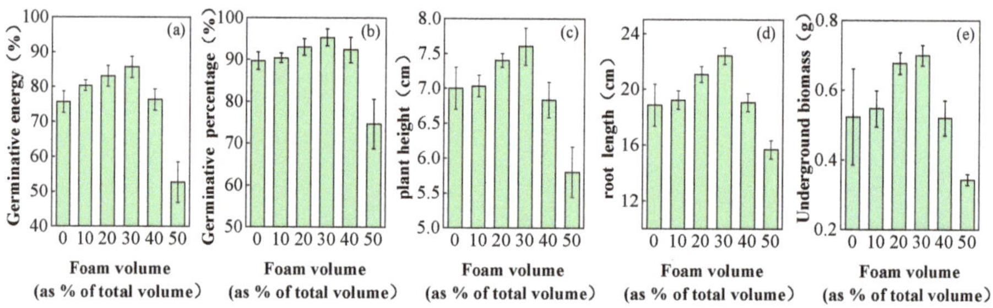

Figure 9. Influence of foam volume on the biomass of *Festuca arundinacea*.

3.4. Microstructure

To further investigate the changes in the internal structure of vegetation concrete upon the addition of a foaming agent, scanning electron microscopy (SEM) tests were conducted on samples that had been cured for 28 days. The findings are presented in Figures 10 and 11. Specifically, Figure 10 displays the SEM images of vegetation concrete with varying foaming agent concentrations, magnified to 2000×. Additionally, Figure 11 illustrates the SEM images of vegetation concrete with 0% and 50% foam volume content, magnified to 5000×.

From Figure 10, it is evident that the microstructures of the vegetation concrete vary based on the foaming agent dosage. The unfoamed sample (Figure 10a) exhibits a dense structure with larger particles and fewer pores. With the introduction of the foaming agent, both the number of pores and the fineness of the particles in the vegetation concrete increase. As the foaming agent content rises, the pore structure of the vegetation concrete becomes finer, and the distribution and proportion of pores become more uniform. Figure 11a illustrates that without any foam, the vegetation concrete has fewer internal pores, and some microcracks are present. However, when the foam volume content reaches 50% (Figure 11b), there is a significant increase in the number of pores, and the internal structure undergoes substantial changes.

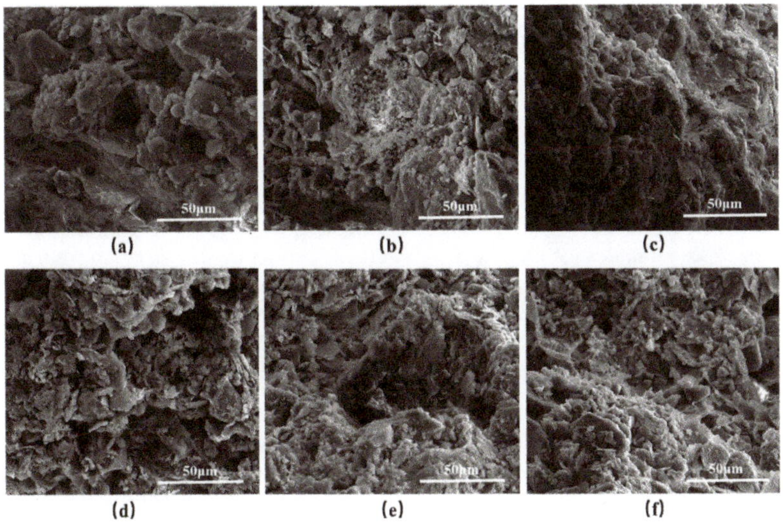

Figure 10. SEM images of vegetation concrete samples at ×2000 magnification: (**a**) 0%, (**b**) 10%, (**c**) 20%, (**d**) 30%, (**e**) 40%, (**f**) 50%.

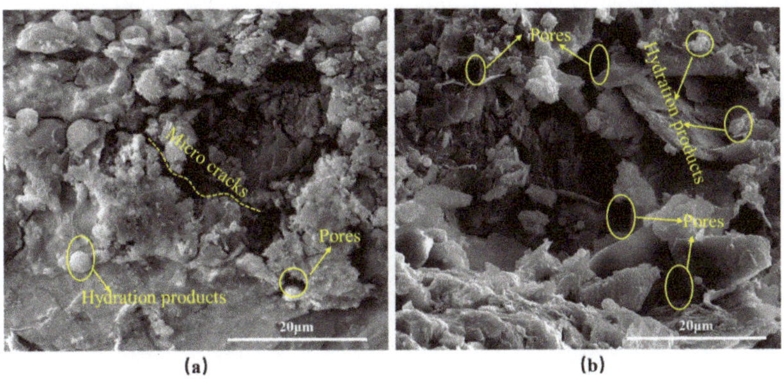

Figure 11. SEM images of vegetation concrete with 0% and 50% foam volume at ×5000 magnification: (**a**) 0%, (**b**) 50%.

4. Discussion

4.1. The Effect of Foaming Agents on the Physical Properties and Plant Growth Performance of Vegetation Concrete

The incorporation of foam not only creates pores within the vegetation concrete, thereby increasing its porosity but also reduces the volume of the mixture per unit of space, which consequently lowers the volume weight of the vegetation concrete. Consequently, the resulting vegetation concrete exhibits a lightweight and porous characteristic. Volume weight and porosity are important parameters for evaluating soil structure [41], representing the physical properties of the soil. High volume weight and low porosity indicate compacted soil with poor structural integrity. Conversely, a low volume weight and high porosity indicate good soil structure [42]. A reduction in volume weight and an increase in porosity generally signify improved vegetation concrete structure. This is corroborated by Figures 10 and 11, which illustrate that the foaming agent significantly enhances the pore structure within the vegetation concrete, leading to a more homogeneous distribution of pores throughout the cement-soil matrix.

The germinative energy, germinative percentage, plant height, root length, and underground biomass of *Festuca arundinacea* exhibit a trend of initial increase followed by a decrease with the increasing dosage of foaming agents, reaching a maximum at 30% foam volume. Beyond this point, the growth-promoting effect on *Festuca arundinacea* diminishes, demonstrating that while the addition of foam promotes growth, more is not necessarily better. The findings of Wang et al. [43] suggest that soil pores are of significant importance in the context of soil water retention and fertility maintenance. The capacity of soil pores to retain water, nutrients, and air directly impacts the soil's ability to retain water, nutrients, and air, which in turn affects soil fertility and the ecological status of vegetation [44,45]. An appropriate foam dosage leads to a favorable pore structure in vegetation concrete, which is conducive to water retention and air circulation, providing the necessary conditions of moisture and oxygen for the germination of *Festuca arundinacea* seeds and the growth of their root systems. However, exceeding the optimal dosage leads to increased porosity, weakening the water and nutrient retention capabilities of vegetation concrete and inhibiting seed germination and seedling growth.

Vegetation plays a crucial role in slope ecological protection. It intercepts rainfall and slows down water flow through its aboveground parts, protecting the soil surface from wind and rain erosion. Additionally, vegetation stabilizes the soil through its roots, strengthens soil structure, and improves soil chemical properties [46]. For *Festuca arundinacea*, high germination percentage and energy are beneficial for protecting slope surfaces at an early stage. An increase in plant height can significantly reduce surface runoff and soil erosion [47]. An increase in underground biomass enlarges the contact area between roots and soil, enhancing the friction between them. This enhances the interaction between roots and soil, increasing the shear strength of the root-soil composite [48,49]. Overall, an appropriate foam dosage is conducive to improving slope stability.

4.2. The Impact of Foaming Agents on the Strength of Vegetation Concrete

After treatment with cement, the planting soil in vegetation concrete transitions from contact bonding to cementation bonding [50], thereby forming a cement-soil framework structure composed of soil particles, cement particles, hydration products, and crystalline products [51]. This enhances the structural strength of the soil. However, the introduction of foam increases inner porosity, reducing the cementation effect and affecting the mechanical properties of vegetation concrete itself. This observation is consistent with the study by Ma et al. [52]. They noted that the strength of soil-based foamed concrete primarily depends on the pore structure of the matrix and the bonding interface. When the foam content increases, it enlarges the inner pores, significantly reducing the compressive strength of concrete.

The cohesion of vegetation concrete primarily depends on the crystallization and cementation between particles. As foam volume increases, the cement-soil particle framework fills more pores, reducing the cementation force and making the overall structure prone to damage, leading to decreased cohesion. The internal friction angle is related to the sliding of soil particles and their rearrangement. With increasing foam volume, more pores replace the connections between particles, reducing the direct contact area between particles. Consequently, the internal friction angle also decreases.

Vegetation concrete must meet specific shear strength requirements to serve its purpose in slope protection. Huang et al. [53] conducted research indicating that to prevent shear slippage between vegetation concrete and the slope surface, the cohesion required to maintain self-stability of vegetation concrete on a 70° slope should be no less than 15 kPa, and the internal friction angle is generally maintained at around 30°. In this study, all sample cohesions were greater than 15 kPa, except for samples with 40% and 50% foam volume, where the rest had internal friction angles ranging between 28° and 32°, meeting slope protection requirements.

4.3. Suitable Foam Volumes

The introduction of foam into vegetation concrete enhances its lightweight and porous characteristics, but it also leads to a decrease in strength, resulting in both positive and negative effects. The positive effects of foam include increased porosity and reduced volume weight, while the negative effect is a reduction in strength. By carefully selecting the appropriate foam volume, it is possible to maximize the benefits of the foaming agent while keeping the drawbacks within acceptable limits. A foam volume ratio of 20% to 30% is considered optimal for achieving an excellent pore structure and suitable porosity in vegetation concrete, which promotes seed germination and plant growth while ensuring that shear strength and other performance indicators meet slope protection standards.

Additionally, when the incorporation volume ratio of foam is between 20% and 30%, the reduction in volume weight ranges from 10.2% to 13.7%. This translates to a savings of approximately 205 to 276 kg/m^3 in raw materials. Consequently, there is a decrease in the use of sand, gravel aggregates, and cement, which contributes positively to environmental protection and promotes the efficient use of global resources.

5. Conclusions

This paper demonstrates the use of plant protein foaming agents to create lightweight vegetation concrete, enhance its pore structure, reduce consolidation, and promote the germination and growth of *Festuca arundinacea* seeds. A lightweight porous vegetation concrete suitable for ecological slope protection was developed, leading to the following conclusions:

(1) Incorporating foam made from plant protein foaming agents into vegetation concrete as a modifier effectively increased the porosity and reduced the volume weight. The higher the dosage was, the more significant the lightweight and porous properties of the vegetation concrete became. When the foam volume was 50%, the porosity increased by 106.05% compared to the untreated sample, while the volume weight decreased by 20.53%. The germinative energy, germinative percentage, plant height, root length, and underground biomass of *Festuca arundinacea* exhibited a trend of initial increase followed by a decrease. *Festuca arundinacea* grew best in the 30% foaming agent treatment, with all phytoindicators higher than in the no foaming agent treatment.

(2) As the foam dosage increased, the shear strength of vegetation concrete gradually decreased. Additionally, both cohesion and internal friction angles exhibited a downward trend. When the foam content is $\leq$30%, it can meet the requirements for slope protection. However, at the same foam dosage, the shear strength of vegetation concrete increased progressively with the curing age. Scanning electron microscopy observations revealed that after incorporating plant protein foaming agent foam, the pore structure of vegetation concrete became finer, and the proportion of pore structures became more uniform, significantly improving the pore structure. The optimum volume ratio for incorporation of plant protein foaming agent foam is 20–30%, which ensures that the mechanical properties meet the requirements for slope stability while maximally improving the physical properties and plant growth environment of vegetation concrete.

(3) This research provides a new technical approach to ecological slope protection. Optimized lightweight porous vegetation concrete can reduce slope loading, improve soil aeration and drainage, reduce consolidation, and promote the growth of slope vegetation, which helps to improve slope stability and reduce construction costs. Our study focused primarily on examining how a plant protein-based composite foaming agent influences the performance characteristics of vegetation concrete; validation was conducted mainly in controlled laboratory settings. To advance the field, future research should investigate the adaptability of this newly developed foamed vegetation concrete to different environmental conditions and assess its long-term durability.

Author Contributions: Conceptualization, methodology, formal analysis, investigation, validation, writing—original draft preparation, T.Z.; supervision, writing—review and editing, T.L. and H.X.; funding acquisition, H.X.; Investigation, M.W.; Data curation, L.L. All authors have read and agreed to the published version of the manuscript.

Funding: This research was funded by Key Science and Technology Projects of the Chinese Ministry of Transport (2022-MS5-127).

Institutional Review Board Statement: Not applicable.

Informed Consent Statement: Not applicable.

Data Availability Statement: Date are contained within the article.

Conflicts of Interest: The authors declare no conflicts of interest.

References

1. Bao, X.; Liao, W.; Dong, Z.; Wang, S.; Tang, W. Development of Vegetation-Pervious Concrete in Grid Beam System for Soil Slope Protection. *Materials* **2017**, *10*, 96. [CrossRef] [PubMed]
2. Xu, Y.; Su, C.; Huang, Z.; Yang, C.; Yang, Y. Research on the Protection of Expansive Soil Slopes under Heavy Rainfall by Anchor-Reinforced Vegetation Systems. *Geom. Geomembr.* **2022**, *50*, 1147–1158. [CrossRef]
3. Huang, Z.; Peng, Z.; Jiao, W.; Liu, Y.; Xu, Y.; Ma, S. Field Study on Vegetation Eco-Protection Technology for Red Sandstone Fill Slope against Water Damage. *Case Stud. Const. Mater.* **2024**, *20*, e03311. [CrossRef]
4. Mickovski, S.B.; Gonzalez-Ollauri, A.; Sorolla, A.; Löchner, A.; Emmanuel, R. A Case History of Co-Design and Co-Deployment of a Nature-Based Solution (NbS) against Erosion and Slope Instability. *Ecol. Eng.* **2024**, *209*, 107406. [CrossRef]
5. Li, Y.H.; Zhao, J.Z.; Zhang, W.H.; Guo, R.; Nan, G.J.; Li, Y.C. Research Status and Development Trend of Ecological Restoration Technology for High and Steep Rock Slope in Open-Pit Mine. *J. Hebei GEO Univ.* **2021**, *44*, 82–86.
6. Tomassi, A.; Falegnami, A.; Meleo, L.; Romano, E. The GreenSCENT Competence Frameworks. In *The European Green Deal in Education*; Routledge: London, UK, 2024.
7. Wang, J.-P.; Bai, M.-H.; Tan, Y.-R.; Ge, S.-Q.; Gao, X.-G.; Dadda, A.; Shen, J.-Y.; Zhang, J. Effect of Vegetation on Unsaturated Soil Hydraulic and the Slope Stability under Rainfall. *Rhizosphere* **2024**, *31*, 100933. [CrossRef]
8. Bai, Y.; Liu, J.; Xiao, H.; Song, Z.; Ma, K.; Deng, Y. Soil Stabilization Using Synthetic Polymer for Soil Slope Ecological Protection. *Eng. Geol.* **2023**, *321*, 107155. [CrossRef]
9. Yao, D.; Qian, G.P.; Yao, J.L.; Liu, J.W.; Yu, X.L. Polymer Curing Agent in Ecological Protection Design Weak Rock Slope Engineering Application. *J. Perform. Constr. Facil.* **2020**, *34*, 04020034. [CrossRef]
10. Liu, C.; Xia, Y.; Chen, J.; Huang, K.; Wang, J.; Wang, C.; Huang, Z.; Wang, X.; Rao, C.; Shi, M. Research and Application Progress of Vegetation Porous Concrete. *Materials* **2023**, *16*, 7039. [CrossRef]
11. Xu, H.; Li, T.B.; Chen, J.N.; Liu, C.N.; Zhou, X.H.; Xia, L. Characteristics and Applications of Ecological Soil Substrate for Rocky Slope Vegetation in Cold and High-Altitude Areas. *Sci. Total Environ.* **2017**, *609*, 446–455. [CrossRef]
12. Yang, L.X.; Li, S.C.; Sun, H.L.; Ye, F.F.; Liu, W.; Luo, S. Polyacrylamide Molecular Formulation Effects on Erosion Control of Disturbed Soil on Steep Rocky Slopes. *Can. J. Soil Sci.* **2011**, *91*, 917–924.
13. Alam, S.; Manzur, T.; Borquist, E.; Williams, J.; Matthews, J.C. In-situ Assessment of Soil-Root Bonding Strength to Aid in Preventing Soil Erosion. *Soil Till. Res.* **2021**, *213*, 105140. [CrossRef]
14. Chang, J.; Im, G.C.; Cho, G.C. Introduction of Microbial Biopolymers in Soil Treatment for Future Environmentally-Friendly and Sustainable Geotechnical Engineering. *Sustainability* **2016**, *8*, 251. [CrossRef]
15. Chang, M.; Lee, A.T.; Tran, S.; Lee, Y.M.; Kwon, J.; Im, G.C.; Cho, G.C. Review on Biopolymer-Based Soil Treatment (BPST) Technology in Geotechnical Engineering Practices. *Transp. Geotech.* **2020**, *24*, 100385. [CrossRef]
16. Finch, H.J.S.; Samuel, A.M.; Lane, G.P.F. Soils and Soil Management. In *Lockhart & Wiseman's Crop Husbandry Including Grassland*, 9th ed.; Woodhead Publishing Series in Food Science, Technology and Nutrition; Finch, H.J.S., Samuel, A.M., Lane, G.P.F., Eds.; Woodhead Publishing: Cambridge, UK, 2014; pp. 37–62.
17. Zhang, J.; Ran, Y.G.; Ma, D.H.; Chen, L.; Wu, Y.B.; Huang, P. Study on the Growth Dynamics of Cynodon dactylon Roots and Their Impact on Soil Pore Evolution. *Acta Pedol. Sin.* **2024**, *61*, 1–18.
18. Tian, L.; Pang, Z.M.; Quan, H.Z.; Wang, J.; Li, Z.X.; Wang, H.J. Study on the Physical Properties and Vegetation Adaptability of Plant-Growing Porous Concrete. *J. Chin. Ceram. Soc.* **2016**, *35*, 3381–3386.
19. Liu, D.X.; Li, S.L.; Xu, W.N.; Cheng, Z.L. Selection Tests for Type and Ratio of Organic Matter in Vegetation Concrete. *Adv. Sci. Technol. Water Resour.* **2012**, *32*, 37–40+54.
20. Qiu, C.; Han, X.Z.; Chen, X.; Lu, X.; Yan, J.; Feng, Y.; Gan, J.; Zou, W.; Liu, G. Effects of Organic Amendment Depths on Black Soil Pore Structure Using CT Scanning Technology. *Trans. Chin. Soc. Agric. Eng.* **2021**, *37*, 98–107.
21. Liu, D.; Gao, X.; Xu, Y.; Yang, Y.; Chen, J.; Ding, Y.; Xia, D.; Xiao, H.; Xu, W. Influence of Biochar Addition Amount on Physicochemical Properties of Vegetation Concrete and Biomass of Cynodon dactylon. *J. Basic Sci. Eng.* **2021**, *29*, 1–14.

22. Guo, M.; He, Y.; Zhi, X. Experimental Research on Dynamic Mechanical Properties of High-Density Foamed Concrete. *Materials* **2024**, *17*, 4781. [CrossRef]
23. Shang, X.; Qu, N.; Li, J. Development and Functional Characteristics of Novel Foam Concrete. *Constr. Build. Mater.* **2022**, *324*, 126666. [CrossRef]
24. Yan, Z.; He, Y. *High-Performance Foam Concrete Insulation Products: Practical Technology*, 1st ed.; China Building Materials Industry Press: Beijing, China, 2015; pp. 1–4.
25. Gu, Y.; Wang, Y.; Wang, X. Research Progress of Foam Concrete in Different Processes. *J. Build. Mater.* **2013**, *16*, 148–151.
26. Liu, X.; Jiao, S.; Wang, Z. A Review on Foaming Agents and Foam Concrete. *Value Eng.* **2017**, *28*, 236.
27. Rong, H.; Zhang, J.; Zhang, L.; Zhang, Y.; Xu, R.; Liu, Z. Preparation and Performance of Bio-Based Foaming Agents for Foam Concrete. *Bull. Chin. Ceram. Soc.* **2020**, *39*, 90–95.
28. Yan, Z.; He, Y. *Foam Concrete Foaming Agent Production and Application Technology*, 1st ed.; China Building Materials Industry Press: Beijing, China, 2021; pp. 1–16.
29. Vázquez, E.; Teutscherová, N.; Almorox, J.; Navas, M.; Espejo, R.; Benito, M. Seasonal Variation of Microbial Activity as Affected by Tillage Practice and Sugar Beet Foam Amendment under Mediterranean Climate. *Appl. Soil Ecol.* **2017**, *117–118*, 70–80. [CrossRef]
30. León, P.; Espejo, R.; Gómez-Paccard, C.; Hontoria, C.; Mariscal, I.; Renella, G.; Benito, M. No-Tillage and Sugar Beet Foam Amendment Enhanced Microbial Activity of Degraded Acidic Soils in South West Spain. *Appl. Soil Ecol.* **2017**, *109*, 69–74. [CrossRef]
31. Chan, C.-L.; Joyce, D.C. Effects of Urea Formaldehyde Foam Soil Amendment on Growth and Response to Transient Water Deficit Stress of Potted Flindersia schottiana Saplings. *Sci. Horticult.* **2007**, *114*, 112–120. [CrossRef]
32. Gao, H.; Zhang, B.; Zhou, K.; Cheng, F. Preparation Method for Municipal Sludge-Based Artificial Planting Soil. Chinese Patent Application 202110677081.9, 3 September 2021.
33. Wang, Y.; Dong, B.; Xing, F.; Hong, S.; Fang, G.; Zhang, Y. A Method for Preparing Landscape Planting Soil Particles and Methods for Using the Same. Chinese Patent Application 202111191835.6, 18 January 2022.
34. *DB50/T 10006-2023 DB51/T 10006-2023*; Technical Guidelines for Ecological Protection of Highway Slopes Using Mesh Anchorage and Spraying Vegetation Concrete. Chongqing Municipal Market Supervision Administration, Sichuan Provincial Market Supervision Administration: Chongqing, China, 2023.
35. Zhao, Y.; Wu, Z. Effect of Foaming Agent Content on Properties of Foamed Cement Banking. *J. Pingdingshan Univ.* **2016**, *31*, 38–41.
36. Kim, H.H.; Park, C.G. Plant Growth and Water Purification of Porous Vegetation Concrete Formed of Blast Furnace Slag, Natural Jute Fiber, and Styrene Butadiene Latex. *Sustainability* **2016**, *8*, 386. [CrossRef]
37. Qu, H.L.; Wang, C.X.; Huang, X.L.; Ding, Y.; Huang, X. Seismic Performance of Substrate for Vegetation Concrete from Large-Scale Shaking Table Test. *Shock Vib.* **2020**, *2020*, 6670726. [CrossRef]
38. Liu, D.; Liu, D.; Gao, J.; Yang, Y.; Ding, Y.; Guo, C.; Zhang, X.; Xia, Z.; Xu, W. Influence of Addition of Two Typical Activated Carbons on Fertility Properties and Mechanical Strength of Vegetation Concrete under Freeze-Thaw Conditions. *Sci. Total Environ.* **2022**, *838*, 156446. [CrossRef] [PubMed]
39. Wang, Y.; Liu, X.; Zhang, Z.; Ma, D.; Cui, Y. Experimental Research on Influence of Root Content on Strength of Undisturbed and Remolded Grassroots-Reinforced Soil. *China J. Geotech. Eng.* **2015**, *37*, 1405–1410.
40. Jia, S. Reinforcement Mechanism and Mechanical Properties of Lime-Cement Soil. Master's Thesis, Inner Mongolia Agricultural University, Huhhot, China, 2011.
41. Yu, B.; Zhao, L.P.; Gao, J.L.; Wang, X.C.; Li, Y.; Liu, X.Y.; Zhang, X.Y.; Wang, H.J.; Li, S.Y.; Wang, X.M. Study on Response and Mechanism of Soil Porosity to Freezing in High-Yield Farmland in Songliao Plain. *J. Soil Water Conserv.* **2012**, *24*, 80–85.
42. Yang, W.; Lan, H.; Li, M.; Meng, C. Predicting Bulk Density and Porosity of Soil Using Image Processing and Support Vector Regression. *Trans. Chin. Soc. Agric. Eng.* **2021**, *37*, 144–151.
43. Wang, Y.; Su, Z.; Zhou, M. Characteristics and Influence of Topsoil Porosity in the Northern Agro-Pastoral Ecotone. *Pratacultural Sci.* **2020**, *37*, 1249–1258.
44. Li, H.; Fan, S.; Zhang, G.; Zhang, S.; Zhou, Z. Characteristics of Soil Water Holding and Soil Porosity under Different Tree Species after Conversion of Cropland to Forest in the Loess Hilly Region. *Bull. Soil Water Conserv.* **2010**, *30*, 27–30.
45. Pei, Z.; Wu, Y.; Wang, Q.; Zhong, Z.; Ren, M.; Wei, C.; Lu, J.; Wang, W. Correlations between Soil Porosity Related Parameters and Other Soil Parameters in Songnen Plain, Northeastern China. *Res. Soil Water Conserv.* **2016**, *23*, 134–138.
46. Zuazo, V.H.D.; Pleguezuelo, C.R.R. Soil Erosion and Runoff Prevention by Plant Covers: A Review. In *Sustainable Agriculture*; Lichtfouse, E., Navarrete, M., Debaeke, P., Véronique, S., Alberola, C., Eds.; Springer: Dordrecht, The Netherlands, 2009; pp. 785–811.
47. Lann, T.; Bao, H.; Lan, H.; Zheng, H.; Yan, C.; Peng, J. Hydro-mechanical Effects of Vegetation on Slope Stability: A Review. *Sci. Total Environ.* **2024**, *926*, 171691. [CrossRef]
48. Xu, H.; Wang, X.Y.; Liu, C.N.; Chen, J.N.; Zhang, C. A 3D Root System Morphological and Mechanical Model Based on L-Systems and Its Application to Estimate the Shear Strength of Root-Soil Composites. *Soil Tillage Res.* **2021**, *212*, 105074. [CrossRef]
49. Xu, H.; Yuan, H.L.; Wang, X.Y.; Wang, D.; Chen, J.X.; Rong, C.Q. Influences of Morphology and Hierarchy of Roots on Mechanical Characteristics of Root-Soil Composites. *Chin. J. Geotech. Eng.* **2022**, *44*, 926–935.

50. Wang, Q.; Chen, H.E.; Cai, K.Y. Quantitative Evaluation of Microstructure Characteristics of Cement-Soil. *Rock Soil Mech.* **2003**, *24* (Suppl. S1), 12–16.
51. Xiong, J.; Li, Y.; Hu, S.; Liao, C.; Zhou, M.; Wang, J.; Luo, S. Influence of Different Parent Soils on the Microstructure and Strength of Cemented Soil. *Bull. Chin. Ceram. Soc.* **2024**, *43*, 1832–1839.
52. Ma, C.; Chen, B. Properties of a Foamed Concrete with Soil as Filler. *Constr. Build. Mater.* **2015**, *76*, 61–69.
53. Huang, X. Effect of Herb Roots on Shearing Strength in Vegetation-Growing Concrete Matrix. Master's Thesis, China Three Gorges University, Yichang, China, 2011.

materials

MDPI

Article

New Discovery of Natural Zeolite-Rich Tuff on the Northern Margin of the Los Frailes Caldera: A Study to Determine Its Performance as a Supplementary Cementitious Material

Jorge L. Costafreda [1,*], Domingo A. Martín [1,2], Miguel A. Sanjuán [3] and Jorge L. Costafreda-Velázquez [4]

1 Escuela Técnica Superior de Ingenieros de Minas y Energía, Universidad Politécnica de Madrid, C/Ríos Rosas, 21, 28003 Madrid, Spain; domingoalfonso.martin@upm.es
2 Laboratorio Oficial Para Ensayos de Materiales de Construcción (LOEMCO), C/Eric Kandell, 1, 28906 Getafe, Spain
3 Department of Science and Technology of Building Materials, Civil Engineering School, Technical University of Madrid, 28040 Madrid, Spain; masanjuan@ieca.es
4 Department of Constructions, University of Holguín, Avenida XX Aniversario, Vía Guardalavaca, Piedra Blanca, Holguín 80100, Cuba; costafreda1992@gmail.com
* Correspondence: jorgeluis.costafreda@upm.es

Abstract: The release of Neogene volcanism in the southeastern part of the Iberian Peninsula produced a series of volcanic structures in the form of stratovolcanoes and calderas; however, other materials also accumulated such as large amounts of pyroclastic materials such as cinerites, ashes, and lapilli, which were later altered to form deposits of zeolites and bentonites. This work has focused on an area located on the northern flank of the San José-Los Escullos zeolite deposit, the only one of its kind with industrial capacity in Spain. The main objective of this research is to characterize the zeolite (SZ) of this new area from the mineral, chemical, and technical points of view and establish its possible use as a natural pozzolan. In the first stage, a study of the mineralogical and chemical composition of the selected samples was carried out using X-ray diffraction (XRD), scanning electron microscopy (SEM), X-ray fluorescence (XRF), and thermogravimetric analysis (TGA); in the second stage, chemical-qualitative and pozzolanicity technical tests were carried out at 8 and 15 days. In addition, a chemical analysis was performed using XRF on the specimens of mortars made with a standardized mixture of Portland cement (PC: 75%) and natural zeolite (SZ: 25%) at the ages of 7, 28, and 90 days. The results of the mineralogical analyses indicated that the samples are made up mainly of mordenite and subordinately by smectite, plagioclase, quartz, halloysite, illite, and muscovite. Qualitative chemical assays indicated a high percentage of reactive silica and reactive CaO and also negligible contents of insoluble residues. The results of the pozzolanicity test indicate that all the samples analyzed behave like natural pozzolans of good quality, increasing their pozzolanic reactivity from 8 to 15 days of testing. Chemical analyses of PC/SZ composite mortar specimens showed how a significant part of SiO_2 and Al_2O_3 are released by zeolite while it absorbs a large part of the SO_3 contained in the cement. The results presented in this research could be of great practical and scientific importance as they indicate the continuation of zeolitic mineralization beyond the limits of the San José-Los Escullos deposit, which would result in an increase in geological reserves and the extension of the useful life of the deposit, which is of vital importance to the local mining industry.

Keywords: mordenite; smectite; SiO_2 reactive; CaO reactive; cement; mortars; pozzolanicity; mechanical strength

1. Introduction

There are many studies in Spain where natural zeolites are mentioned in relation to genetic and spatial bentonite formations of volcanic origin [1–5]. Authors such as Mattei et al. [6], De la Fuente and Linares [7], and López-Ruiz and Cebriá [8] explain in detail the formation processes of bentonites and highlight the affinity between clay minerals, such as smectites and mordenite, assuring that their origin is typically hydrothermal and related to the volcanism of southeastern Spain. Other researchers [9,10] mention the association of zeolite with bentonite formations located in known deposits in southeastern Spain, such as Morrón de Mateo, Los Trancos, and the proximal surroundings of San José and Los Escullos. De La Villa et al. [11] state that these zeolites may have been formed during the alkaline reaction processes of bentonite. Deon et al. [12] detected zeolite associated with layers of illite and smectite in their investigations in the Rodalquilar caldera in the southeast of the Iberian Peninsula. In the report of the Geological Map of Spain [13], zeolite is referred to as a residual material of the bentonitization process, along with other minerals such as smectite (montmorillonite variety), quartz, plagioclase, calcite, amphibole, mica, and tridymite. Despite the above, the most detailed and specific work on the study of zeolite began in 2005 [14–16], with important contributions not only in mineralogical characterization but also in heat treatment to improve its technological properties, which was also stated by Suárez et al. [17]. However, despite the fact that zeolitic mineralization is practically ubiquitous in the volcanic environment of the Neogene, some authors have confirmed that the highest concentration of zeolite of the mordenite variety occurs in Los Frailes caldera [18], where the only industrial zeolite deposit in Spain is found. In fact, Martín et al. [19] have proven by electrical resistivity tomography (ERT) that the thickness of the mineralized horizon of this deposit exceeds 40 m. The most recent applications of zeolite from this region have been aimed at the chemical stabilization of wastewater from mining and metallurgical processes where there is an abundant presence of heavy metals [20]. Uses in the manufacture of waste-based compost for soil improvement are also reported due to its effectiveness in retaining metals and ammonium [21]. Domene et al. [22] have effectively restored vegetation in semi-desert areas of Almeria (Spain) through the standardized use of organic compost prepared with a mixture of zeolite and greenhouse waste from that region. Possibly the most widespread application of this zeolite in recent times is as a highly reactive pozzolan for the improvement of cements, mortars, and concretes [23,24]. This is due to the need to control and reduce CO_2 emissions into the atmosphere. In this sense, Byung-Wan et al. [25] designed zeolite mortars using an alkaline activator (NaOH), obtaining a mechanical strength of 43.5 MPa at 7 days. This is essentially due to the mineralogical composition, texture, structure, physical, and mechanical properties of the zeolites [26]. Iswarya and Beulah M [27] carried out an extensive review on the specific use of zeolite in the manufacture of high-strength concretes. The use of zeolite in the manufacture of high-strength concrete is a common practice today, resulting in lighter structures that are resistant to sulphate attack, seawater, and frost [28–30]. Finally, Boháč et al. [31] and Vejmelková et al. [32] have shown that in zeolite-containing cementitious pastes, an acceleration of hydration occurs much faster than the time at which the peak of the exothermic flow occurs.

The main objective of this work is the mineral, chemical, and technical characterization of the natural zeolite-rich tuff from new evidence discovered in an unexplored area within the Los Frailes caldera and to prove that it is sufficiently pozzolanic in capacity to be used effectively as a reactive aggregate in mortars and concretes. The second objective is to demonstrate that these materials have properties similar to those of the main deposit known as San José-Los Escullos, located 655 m to the south (Figure 1), which would have a positive impact on the expansion of the mining area and an increase in geological reserves.

Figure 1. Location of the research area.

2. Materials and Methods

2.1. Materials

In this research, 6 natural zeolite-rich tuff samples weighing 20 kg each were used. The specimens are made of a zeolite tuff of pale green, light gray, and white. It is usually compact, although in parts it is friable and has a low density. It forms extensive outcrops embedded in andesitic and dacitic rocks. All the samples were carefully crushed and sifted to the particle size fraction of 63 μm. In addition, a Type I Portland cement was used, the chemical composition of which is shown in Table 1.

Table 1. Chemical composition of Portland cement as determined by XRF.

	% Oxides Weight												
Materials	SiO_2	CaO	Fe_2O_3	Al_2O_3	Na_2O	SO_3	MgO	K_2O	TiO_2	P_2O_5	MnO	PPC	Total
PC [1]	17.45	64.04	3.35	5.59	0.091	4	0.641	1.37	0.326	0.072	0.094	2.43	99.454

[1] Portland cement.

In the dosage of the mortar, normalized sand (NS) was used; this sand is made of quartz of rounded grains, where the SiO_2 content is 98%. The granulometric distribution of normalized sand (NS) is listed in Table 2.

Table 2. Granulometric distribution of the normalized sand (NS) used in this research.

Square Mesh Dimensions (mm)	2.0	1.60	1.00	0.5	0.16	0.08
Residue Retained on Sieves (%)	0.00	7 ± 5	33 ± 5	67 ± 5	87 ± 5	99 ± 1

Table 3 shows details of the proportions of materials used in both the mixed mortar specimens (PC/SZ) and the reference specimen (PCSR).

Table 3. Formulation of Portland cement (PC:75%)/natural zeolite-rich tuff (SZ:25%) mix ratios in mortar specimens.

Sample	Mortar Components	Formulation (g)	Test Age (Days)
PC/SZ-01 [1] PC/SZ-02 PC/SZ-03 PC/SZ-04 PC/SZ-05 PC/SZ-06	Portland cement Natural zeolite-rich tuff Normalized sand Distilled water	Portland cement: 375 g Natural zeolite-rich tuff: 125 g Distilled water: 225 g Normalized sand: 1350 g	7/28/90
PCSR [2]	Portland cement Normalized sand Distilled water	Portland cement: 450 g Distilled water: 225 g Normalized sand: 1350 g	7/28/90

[1] PC/SZ-01 to 06: mortar specimens prepared with Portland cement and natural zeolite-rich tuff; [2] PCSR: mortar specimens manufactured with Portland cement only as a reference element.

2.2. Methods

A study of the mineralogical phases in the samples was carried out by X-ray diffraction (XRD) with the help of the Rigaku Miniflex-600 diffractometer of the Escuela Técnica Superior de Ingenieros de Minas y Energía (Universidad Politécnica de Madrid, Madrid, Spain). The study of the phases was performed in an interval of 4° to 60°, with a step of 0.01° and at every 5° intervals. A voltage of 40 kV and a current of 15 mA were applied.

A scanning electron microscopy (SEM) study was carried out to determine the morphological properties of the minerals in the samples, such as the crystalline texture, species, and size as well as surface characteristics and the presence of pores, cavities, and channels. A Hitachi S–570 electron microscope was used from the Centralized Laboratory of the Escuela Técnica Superior de Ingenieros de Minas y Energía (Universidad Politécnica de Madrid, Madrid, Spain). The microscope is equipped with a Kevex-1728 analyzer, a BIORAD Polaron, a power supply for evaporation, and a Polaron SEM coating system. It has a resolution of 3.5 nm and an amplification of 200×10^3.

The chemical composition of the samples in their natural state was determined by X-ray fluorescence (XRF). Compounds such as SiO_2, Al_2O_3, CaO, Na_2O, K_2O, MgO, Fe_2O_3, TiO_2, and MnO were found. In addition, loss on ignition (LOI), Si/Al ratio, and Si/(Al + Fe) ratio were determined. This test was also carried out on mixed mortar specimens (cement/natural zeolite-rich tuff) hardened at different ages (7, 28 and 90 days) to determine the behavior of compounds over time. A Philips WDXRF spectrometer (PW1404) was used. The intensity of the radiation ranged from 10 to 100 kV. This equipment belongs to the Universidad Politécnica de Madrid (Madrid, Spain).

Thermogravimetric analysis was carried out to study the thermal behavior of the samples and establish the main thermal events that take place until their definitive collapse. The samples were analyzed by (TGA) with air atmosphere and at 20 °C/min. The heating range was 25 °C to 950 °C. This analysis was carried out in the Laboratorio de Tamices Moleculares of the Instituto de Catálisis y Petroleoquímica of the Centro de Investigaciones Científicas (CSIC) of Madrid (Madrid, Spain).

The pozzolan quality of the zeolite samples was determined by a qualitative-technological chemical analysis, following the specifications of the Standard UNE–EN 196-2-2014 [33]. The main objective of this test is to determine the percentages of reactive SiO_2 in relation to the total SiO_2 present in the sample capable of reacting with Portland cement in the hydraulic reaction process. Other compounds such as total-reactive CaO, MgO, Al_2O_3, Fe_2O_3, and SO_3 were determined. The proportion of insoluble residue (IR) and loss on ignition (LOI) of each sample were calculated. This study was carried out at the Laboratorio Oficial de Ensayos de Materiales de Construcción (LOEMCO), Getafe (Spain).

The pozzolanic reactivity of the samples was calculated by chemical analysis of pozzolanicity. This test takes into account the reaction capacity of zeolite with the $Ca(OH)_2$

present in a solution together with a certain amount of Portland cement. In total, 75% Portland cement and 25% zeolite were mixed. The assay considers the variation in the concentration of hydroxyl ions and calcium ions in solution over a period of 8 and 15 days [34].

A mechanical compressive strength test was carried out on specimens made with a mixed mixture of Portland cement and natural zeolite-rich tuff. A type 1 Portland cement, class 52.5 R, was chosen, whose characteristics are detailed in the Standard UNE–EN 197-1:2011 [35]. Standardized sand was used as a fine aggregate to produce the mixed mortar specimens hardened to 7, 28, and 90 days in a cement to natural zeolite-rich tuff ratio of 75:25% [36]. Distilled water was used in the dosing of the paste. The proportion of materials in the cement/natural zeolite-rich tuff mortars was as follows: cement 375 g, zeolite: 125 g, fine aggregate (standard sand Type CEN EN 196-1 [36]): 1350 g, and distilled water: 225 g. In the reference mortars (Portland cement only), the formulation was PC: 450 g, fine aggregate: 1350 g, and distilled water: 225 g. The mortar specimens were placed in a container with water in a humid chamber at a temperature of 20 °C ± 1 °C.

3. Results and Discussion

3.1. X-ray Diffraction (XRD)

The results of XRD analyses revealed a major presence of a crystalline phase consisting of mordenite. The secondary phases were smectite (montmorillonite), illite, halloysite, quartz, plagioclase, and muscovite (Figure 2).

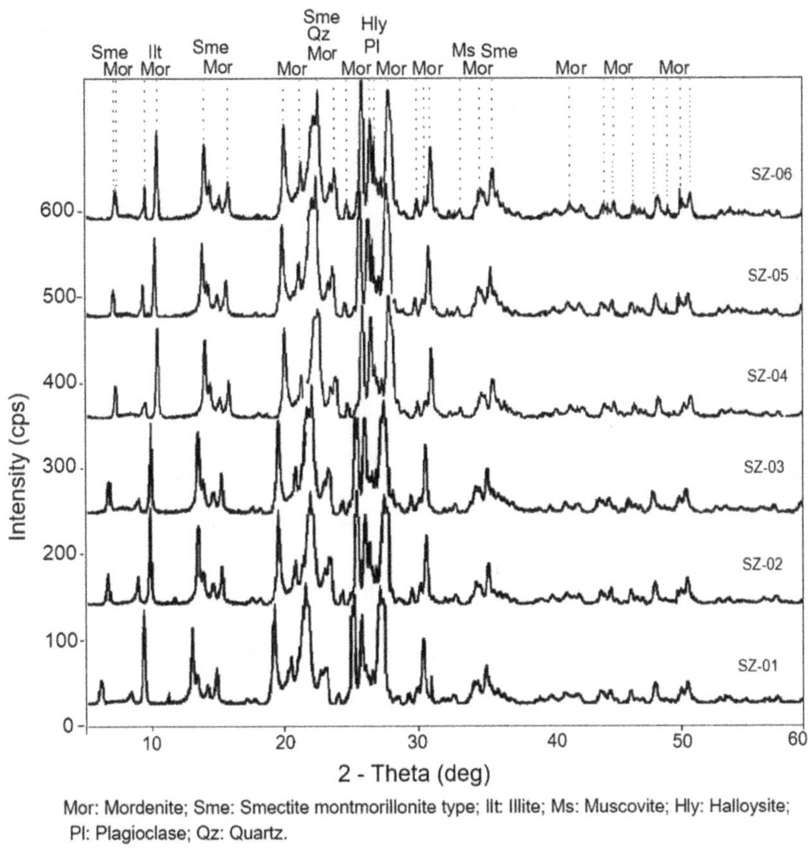

Figure 2. X-ray diffraction patterns of the samples of natural zeolite-rich tuff.

The morphology and distribution of the various peaks of mordenite indicate a highly developed degree of crystallinity in relation to the other phases.

The mentioned mineral phases have also been found in the main deposit (San José-Los Escullos) by other researchers [18], who have established that the content of mordenite in the natural zeolite-rich tuff formations of the deposit ranges from 47 to 95%. However, they describe a wider variety of mineralogical phases not found in the study area, such as orthoclase, sanidine, calcite, hematite, illite, chlorite, and pyrite. These same phases have been located at 40 m depth in the Los Frailes caldera through the study of drill cores and geophysical work by Martín et al. [19]. On the other hand, Stamatakis et al. [1] have found opal and cristobalite, which they attribute to the action of residual fluids enriched in silica.

The newly discovered area could be considered an extension of the zeolitic mineralization towards the northern flank of the deposit.

3.2. Scanning Electron Microscopy

The results obtained by SEM showed a major presence of mordenite crystals in relation to smectite and halloysite (Figure 3a–f). From a morphological and textural point of view, mordenite exhibits markedly idiomorphic textures. Its habits are preferably acicular, fibrous, rhombohedral-hexagonal, and sometimes bacillary. The crystals crisscross each other to form very compact radial aggregates. The spatial arrangement of aggregates often forms cavities and intergranular voids.

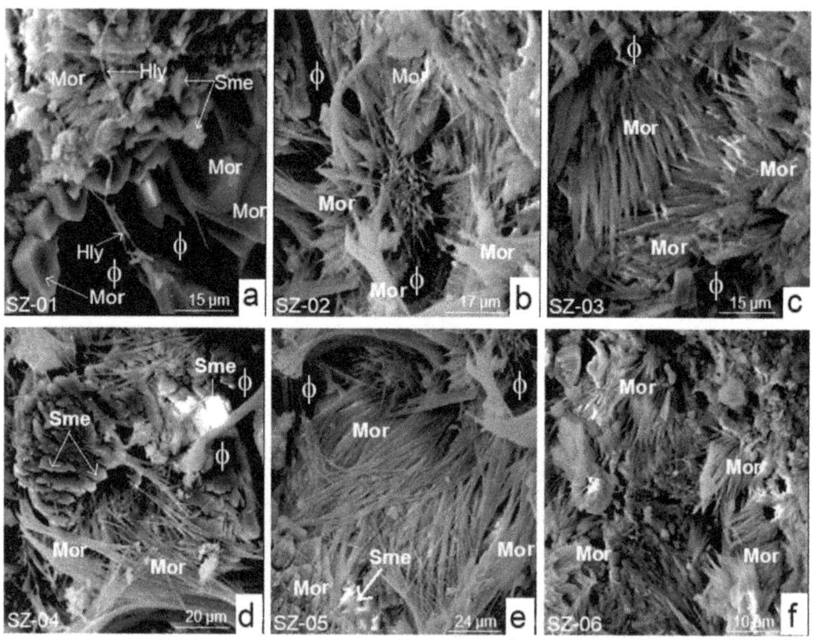

Figure 3. SEM micrographs (**a**–**f**) of the analyzed samples.

The crystals of mordenite are frequently intergrown with smectite; both species are formed from the alteration of amorphous materials [18,19]. According to Stamatakis et al. [1] the mordenite that appears in the main deposit in the form of fibrous crystals has formed at the expense of glassy matter of volcanic origin.

3.3. X-ray Fluorescence (XRF)

The data obtained from the XRF analysis showed high SiO_2 and Al_2O_3 contents (Table 4). In addition, the contents of the alkaline compounds (Na_2O and K_2O) were significantly higher than those of the alkaline–earth compounds (CaO and MgO). The contents of SO_3 were practically negligible. The calculation of the Si/Al ratio showed values in the range of 4.0 to 5.2, which indicates that the samples studied have pozzolanic properties [37,38]. The Si/(Al + Fe) ratio ranged from 3.6 to 3.9. Loss on ignition varied from 10.7% to 11.9%. The ratio between SiO_2 and the alkaline and alkaline–earth compounds was high.

Table 4. Chemical composition of XRF samples.

Samples	% Oxides Weight												
	S_iO_2	Al_2O_3	CaO	Na_2O	K_2O	MgO	Fe_2O_3	TiO_2	MnO	SO_3	LOI	Si/Al	Si/ (Al + Fe)
SZ-01	65.65	15.63	1.02	2.61	4.33	2.04	1.77	0.12	0.091	0.14	11.5	4.2	3.8
SZ-02	64.48	15.22	1.47	3.13	3.61	2.01	1.81	0.11	0.115	0.08	10.7	5.2	3.8
SZ-03	64.91	16.43	1.30	2.81	3.14	2.24	1.79	0.10	0.112	0.10	10.9	4.0	3.6
SZ-04	65.14	15.13	1.08	2.52	3.31	2.19	1.75	0.13	0.083	0.06	11.9	4.3	3.9
SZ-05	64.47	16.21	1.35	3.03	3.18	2.11	1.80	0.10	0.072	0.11	11.1	4.0	3.6
SZ-06	64.93	16.30	1.31	2.55	3.22	2.14	1.73	0.14	0.051	0.10	11.3	4.0	3.6

It is inferred that the low CaO and MgO contents in the samples are due to the fact that Ca^{2+} and Mg^{2+} cations are removed during hydrothermal processes [39], which confirms the hydrothermal genesis of zeolite deposits in southeastern Spain [40]. However, for the same reason, the contents of SiO_2, Na_2O, and K_2O tend to increase [18].

3.4. Thermogravimetric Analysis

The behavior of the TGA curves showed a simple decomposition process of the samples, with three well-defined thermogravimetric events (Figure 4). The first event occurred from relatively low temperatures (0–39 °C) to 250 °C, with a mass loss equivalent to 8.5% and 7.6%. In this interval, all the samples experienced rapid moisture loss and surface dehydration with the expulsion of the gases contained in the pores. The second event was recorded in the thermal range of 250 °C to 750 °C, with a mass loss of 2.3% to 1.2%. In this interval, the process of dehydration of the samples continued, possibly due to the loss of intrareticular or zeolitic water from mordenite and smectite; however, there were still no signs of structural collapse. In this thermal range, the dehydroxylation of the smectites occurred by the disintegration of the group $(OH)^-$. The third and last thermal event occurred between 750 °C and 1200 °C, the fundamental feature being the stabilization of the TG curve. At this stage, the structural rearrangement of the samples occurred without processes of mass loss or gain.

Research carried out at the San José-Los Escullos deposit established that the zeolite from that deposit behaves similarly to those in this work when subjected to thermal analysis, the most notable feature being the simple decomposition in three consecutive stages [41]. However, the percentage of mass loss of zeolite from the main deposit in the first thermal event was much lower, which can be interpreted as the zeolites being of a higher purity in this research.

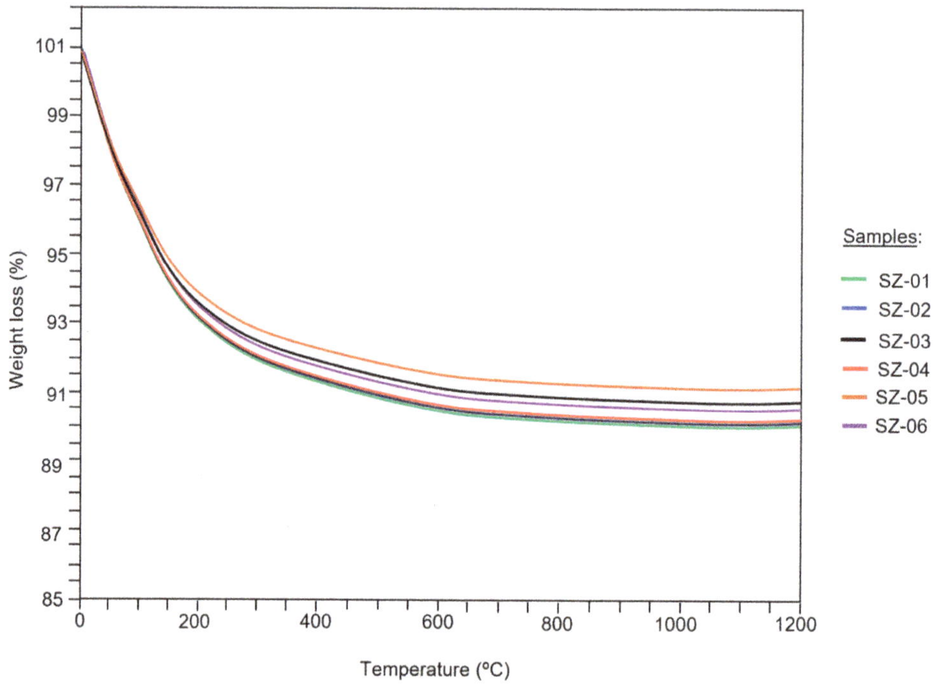

Figure 4. Behavior of the TGA curves in the samples studied.

3.5. Chemical Analysis of Technical Quality

The results obtained through the chemical-technological test showed that the samples behave like natural pozzolans capable of reacting hydraulically with Portland cement (Table 5). As far as could be seen, virtually all the SiO_2 was reactive in all the cases. For example, in sample SZ-01, the total SiO_2 calculated was 69.25%, of which 68.22% was the part that reacted to the cement; that is, more than 98% of the original SiO_2 in this sample was reactive. This was also proven in the remaining samples. A similar fact has been verified with reactive CaO in relation to total CaO. According to established standards, a sample is chemically suitable from a qualitative-technological point of view if its SiO_2 content > 25%, Al_2O_3 < 16%, MgO < 5%, SO_3 < 4%, and insoluble residue (I.R.) < 3% [35]. The values calculated and shown in Table 5 reflect this approach.

Table 5. Qualitative-technological chemical composition of the samples.

Compounds (%)	Samples					
	SZ-01	**SZ-02**	**SZ-03**	**SZ-04**	**SZ-05**	**SZ-06**
Total SiO_2	69.25	68.80	68.51	68.74	68.07	68.53
Reactive SiO_2	68.22	67.77	67.48	67.71	67.04	67.50
Total CaO	1.22	1.31	1.29	1.12	1.33	1.28
Reactive CaO	0.62	0.71	0.69	0.52	0.73	0.68
Al_2O_3	15.33	14.92	16.13	14.53	15.61	15.70
MgO	1.54	1.51	1.74	1.69	1.61	1.64
Fe_2O_3	1.36	1.40	1.38	1.34	1.39	1.32
SO_3	0.09	0.11	0.08	0.10	0.13	0.09
I.R. [1]	1.03	1.13	2.01	1.90	1.78	1.83
SiO_2/(CaO + MgO)	25.0	24.4	22.6	24.5	23.1	23.5

[1] I.R.: insoluble residue.

Presa et al. [41] have calculated reactive SiO_2 contents for zeolite from the San José-Los Escullos deposit in the range of 58.68% to 63.16%, which are comparatively lower than those of the samples analyzed in this work. These authors also calculated between 14.08% and 19.87% of insoluble residue present in their samples, percentages that are significantly higher than those calculated in this work. These arguments seem to confirm that the zeolite found in the new study area is much purer than that lying in the main deposit.

3.6. Chemical Analysis of Pozzolanicity

Figure 5 shows that all the samples occupied a deep position in the area below the isotherm of CaO solubility at 40 °C. This fact emphasises that all the samples analyzed were pozzolanic [34]. According to this, it appears that samples SZ–01, SZ–02, and SZ–04 were the most pozzolanic. If this fact is compared with the data shown in Table 5, the highest SiO_2 contents corresponded precisely to these samples.

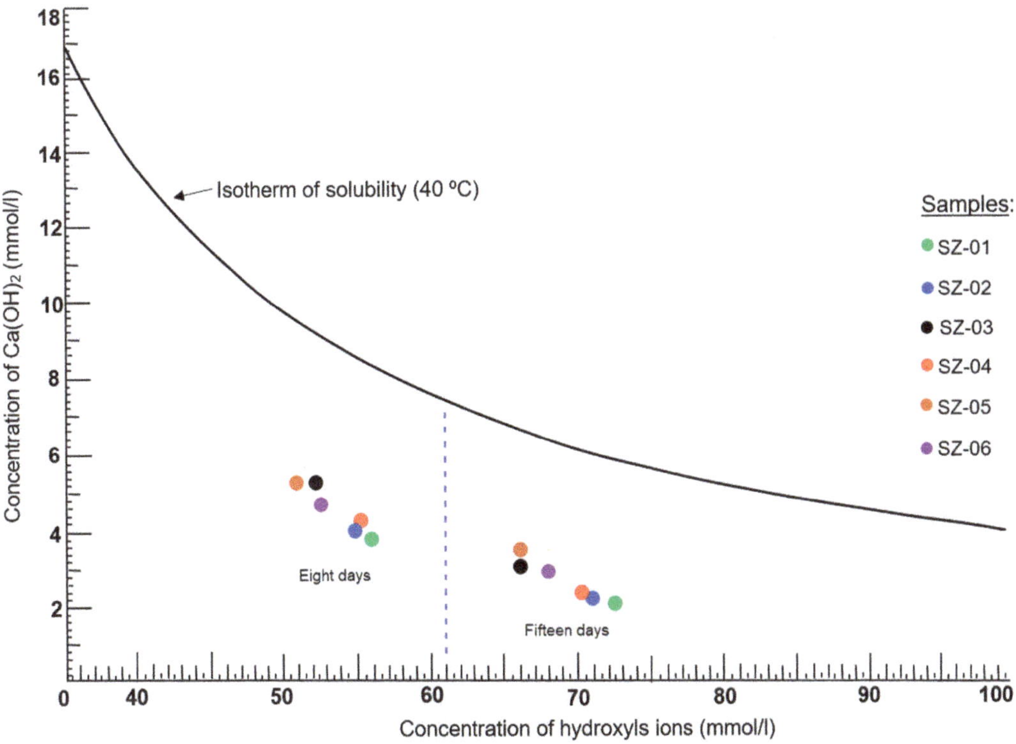

Figure 5. Evolution of the pozzolanic behavior of each sample analyzed at 8 and 15 days.

It could then be established that the SiO_2 content is a decisive factor in the pozzolanic behavior of a material [42] (Table 5 and Figure 6). Authors such as Caputo et al. [43], Saraya and Thabet [44], and Mertens et al. [45] also state that more siliceous zeolites are more pozzolanic and contribute to the gain in mechanical strength. However, other factors must be considered, such as the amount of reactive CaO present, Al_2O_3, and the relationship between the compounds SiO_2, CaO, and MgO [46]. It should be noted that the pozzolanic reactivity of the samples experienced notable increases from 8 to 15 days, according to the graph in Figure 5.

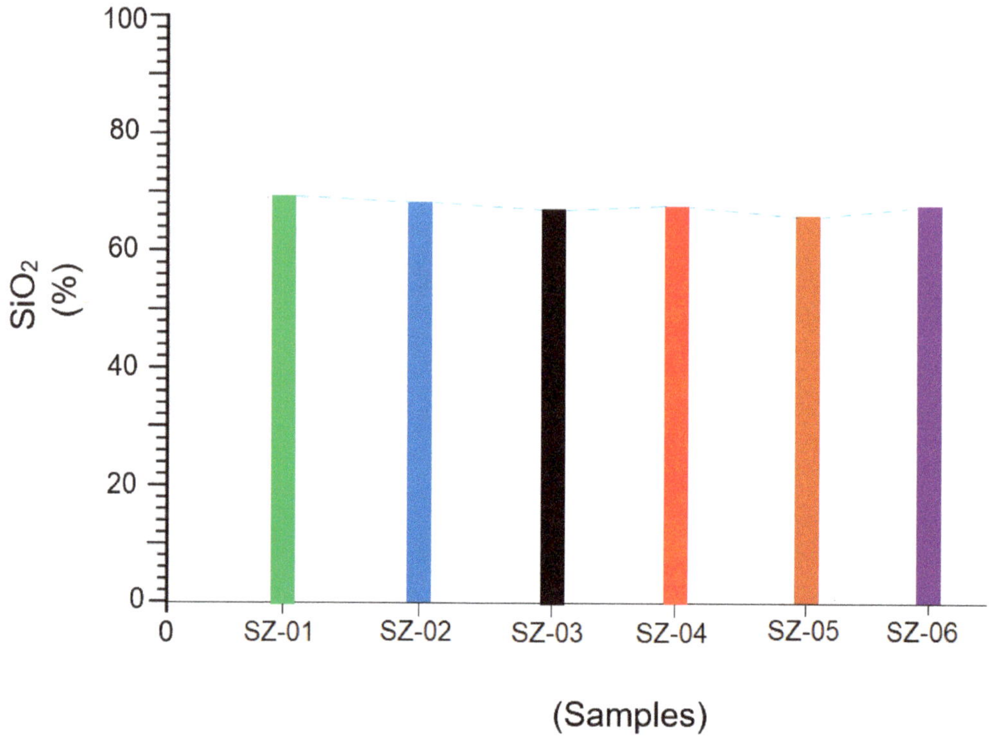

Figure 6. Behavior of SiO_2 in the analyzed samples.

Figure 6 shows the variation in the SiO_2 content for each sample analyzed. Note that those samples with higher silica contents are the same ones that in Figure 5 have higher pozzolanic reactivity, both at 8 and 15 days.

3.7. Chemical Analysis of Mortar Specimens

In this research, mortar specimens hardened at 7, 28, and 90 days were studied to monitor and compare the behavior of chemical compounds in both types of specimens over time. These specimens were made from a mixture of 75% Portland cement and 25% natural zeolite-rich tuff. In this case, the SZ-01 sample was used as a representative. In addition, reference mortars with exclusive Portland cement content were manufactured.

Table 6 shows the behavior of the different chemical compounds of anhydrous Portland cement and their mixtures in mortars set at 7, 28, and 90 days. The first detail to highlight is the variation in the SiO_2 content (47.31%) of the Portland cement specimen at 7 days of setting in relation to the initial composition (17.47%). This increase was due to the presence of normalized sand [47], which contributed silica to the paste; however, a gradual decrease was observed at 28 and 90 days, respectively. A similar but less notable case occurred with Al_2O_3. The CaO experienced a noticeable decrease after 7 days of setting, practically half of the original content. The other compounds, with the exception of MgO, which experienced a small increase, tended to decrease between 7 and 90 days of setting. As shown in Table 6, this is the normal behavior of a Portland cement mortar without pozzolanic additions [18,36].

Table 6. Chemical composition of specimens made only with Portland cement at different curing ages.

Sample	% Oxides Weight										
	S_iO_2	Al_2O_3	CaO	MgO	K_2O	Na_2O	SO_3	Fe_2O_3	TiO_2	P_2O_5	MnO
PC [1]	17.47	5.60	64.05	0.63	1.35	0.09	4.00	3.31	0.33	0.07	0.09
PCS-7 [2]	47.31	3.54	34.80	0.49	0.81	0.10	1.40	4.23	0.15	0.11	0.14
PCS-28 [3]	45.15	3.31	37.15	0.53	0.76	0.21	1.42	4.69	0.18	0.10	0.14
PCS-90 [4]	38.27	3.11	42.12	0.58	0.55	0.23	1.49	5.07	0.19	0.09	0.13
	130.73	9.96	114.07	1.60	2.12	0.54	4.31	-	-	-	-

[1] Anhydrous Portland cement sample; [2–4] specimens of Portland cement mortar and standardized sand set at 7, 28, and 90 days.

Table 7 shows the behavior of the chemical compounds of Portland cement and natural zeolite-rich tuff when both were mixed in a 75:25% ratio. After 7 days of setting, a decrease in SiO_2 was observed in the specimens. This decrease still occurred at 28 and 90 days. Al_2O_3 also decreased in relation to its original values; its presence could prevent the C_3A of the cement from reacting completely, thus preventing the progressive formation of ettringite, which would favor the hydration of bicalcium and tricalcium silicates [48].

Table 7. Chemical composition of specimens made with mixed mixtures of cement (75%) and natural zeolite-rich tuff (25%) at different curing ages.

Sample	% Oxides Weight										
	S_iO_2	Al_2O_3	CaO	MgO	K_2O	Na_2O	SO_3	Fe_2O_3	TiO_2	P_2O_5	MnO
SZ-01 [1]	65.65	15.63	1.02	1.29	4.33	2.61	0.14	1.77	0.12	0.03	0.09
PC/SZ-01-7 [2]	48.63	5.24	31.14	0.53	1.17	0.17	1.25	3.93	0.18	0.11	0.11
PC/SZ-01-28 [3]	47.30	4.89	32.51	0.55	1.23	0.16	1.21	4.17	0.19	0.10	0.13
PC/SZ-01-90 [4]	42.71	4.59	37.17	0.62	1.21	0.14	1.10	4.71	0.23	0.09	0.15
	138.64	14.72	100.82	1.70	3.61	0.47	3.56	-	-	-	-

[1] Sample of natural zeolite-rich tuff (SZ-01) in its natural state; [2–4] specimens of mixed mortars composed of cement (75%) and natural zeolite-rich tuff (25%) and set at 7, 28, and 90 days.

If the original CaO contents are compared with those observed at 7, 28, and 90 days, a notable increase is observed (Table 7). It seems then that in the hydraulic reaction process, 48.63% of SiO_2 was consumed after 7 days, 47.30% at 28 days, and 42.71% at 90 days. The sum of these percentages far exceeds 65.65% of the natural zeolite-rich tuff sample, so it follows that the system takes SiO_2 from both zeolite and normalized sand. It is evident that this excess of SiO_2 in the paste would imply a greater hydraulic reaction over time with the associated increase in the mechanical strengths of the mortar specimens. Previous studies at the main deposit have shown that PC/zeolite composite mortar specimens achieve mechanical strengths in excess of 51 and 72 MPa at 28 and 90 days, respectively [18].

The SiO_2/CaO ratio is then inversely proportional, from which it follows that to fix more lime, it is necessary to consume more silica, giving tobermorite as a reaction product [18,48]. The Na_2O/K_2O ratio in the samples was lower than that in the reference cement.

It follows that more silica was available at the cement to natural zeolite-rich tuff interface than at the interface where there was only cement. This availability of SiO_2 could favor an increase in mechanical strengths in the long term, as discussed above. Some researchers mention that the presence of pozzolans of zeolitic origin leads to an increase in the compressive strength of mortars and concretes after 28 days of setting, even exceeding the resistance of ordinary cement at older ages [49]. The SO_3 content in specimens made with natural cement to natural zeolite-rich tuff mixtures decreased significantly over time. As can be seen in Table 7, the amount of SO_3 from 7 to 90 days was comparatively lower than that recorded in specimens made of cement alone (Table 6). In the research of some authors, this fact is confirmed in a wide variety of natural pozzolans [50–52].

3.8. Mechanical Compressive Strength Tests at 2, 7 and 28 Days

Figure 7 shows the results of the mechanical compressive strengths calculated at 7, 28, and 90 days. An increasing rise in strength was observed for all the specimens made with a mixture of Portland cement and natural zeolite-rich tuff. Accordingly, it is once again confirmed that the presence of zeolite as a pozzolan favors the hydraulic reaction in the paste [45,53–55]. Furthermore, the capacity of zeolite to substitute Portland cement (PC/SZ: 75–25%) effectively without negatively altering the rheological properties of the mortars was tested.

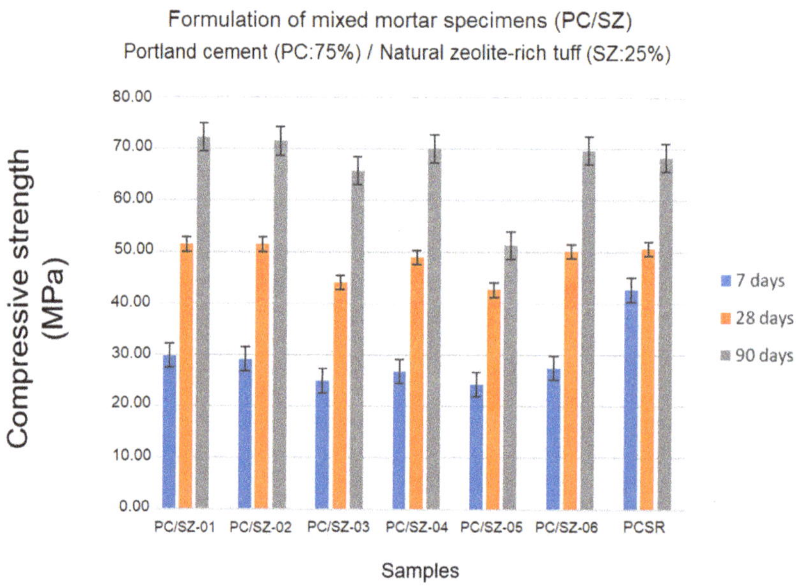

Figure 7. Behavior of the mechanical resistance in the period of 7, 28, and 90 days of curing.

As shown in Figure 7, after 7 days of curing, all the specimens had lower mechanical strengths than the standard specimen (PCSR: 42.7 MPa). However, their values were relatively close (PC/SZ-01: 29.8 MPa; PC/SZ-02: 29.1 MPa; PC/SZ-03: 24.9 MPa; PC/SZ-04: 26.8 MPa; PC/SZ-05: 24.2 MPa and PC/SZ-06: 27.4 MPa).

After 28 days of curing, the mechanical strength value of the reference specimen was reached. However, some mixed specimens significantly approached (PC/SZ-06: 50.1 MPa and PC/SZ-04: 48.9 MPa) or even surpassed it (PC/SZ-01: 51.4 MPa and PC/SZ-02: 51.3 MPa).

After 90 days of curing, the mechanical strength of the mixed specimens increased significantly (PC/SZ-01: 72.2 MPa; PC/SZ-02: 71.5 MPa; PC/SZ-04: 70.1 MPa; PC/SZ-06: 69.7 MPa), surpassing the value of the reference mortar (PCSR: 68.4 MPa). The values calculated for the remaining samples (PC/SZ-03: 65.8 MPa; PC/SZ-05: 51.3 MPa) also showed an exponential increase in compressive strength.

The analysis of the mechanical behavior of the tested specimens indicated full agreement with the pozzolanicity test results given in Section 3.6. It seems to be demonstrated that the pozzolanic reactivity of the natural zeolite-rich tuff studied has a significant influence on the process of gaining mechanical strength. This fact is confirmed by several authors in their studies on mortars and concretes made with zeolites of different species [55–58].

4. Conclusions

The following conclusions have been drawn from the study of the zeolite samples found in the study area:

1. The results presented prove that the zeolite found in the study area is composed mostly of highly crystalline mordenite and subordinately of smectite (montmorillonite), illite, halloysite, quartz, plagioclase, and muscovite.
2. The samples analyzed have high contents of SiO_2 and Al_2O_3, while the contents of alkaline compounds (Na_2O and K_2O) are significantly higher than those of alkaline–earth compounds (CaO and MgO).
3. The thermal behavior of the samples indicates that the mordenite is stable up to approximately 750 °C, after which it tends to collapse and restructure.
4. All the samples have shown a marked pozzolanic behavior both at 8 and 15 days, so their status as high-quality pozzolans is established.
5. It is concluded that more silica is available in cement to natural zeolite-rich tuff mortars than in those made exclusively with cement. This availability of SiO_2 could lead to an increase in mechanical strengths in the long term.
6. The $SiO_2/(CaO + MgO)$ ratio is high, which favors the pozzolanic reaction since the silica phase will tend to react with the alkaline phase.
7. Natural zeolite-rich tuff influences the balance of the SO_3 content in mortar samples, causing this compound to always remain in solution and prevent ettringite from forming in abnormal quantities.
8. The presence of zeolite in mixed mortar mixes (PC/ZS) favors the increase of mechanical strength from 7 to 90 days. During this period, the strength values equaled or even exceeded the reference mortar. In this study, Portland cement was replaced by 25% natural zeolite-rich tuff. However, it is possible that with formulations of PC/ZS: 70–30%, relevant results can also be obtained.
9. According to the points argued above, it is established that the natural zeolite-rich tuff studied is qualitatively suitable for the improvement of cements, mortars, and concretes.
10. Finally, the investigation of this new finding corroborates that this natural zeolite-rich tuff has similar mineral, chemical, and technical properties that are even qualitatively better than the one found in the main deposit (San José-Los Escullos). This could be advantageous when considering the expansion of geological and mining reserves from the perspective of local industry interests.

Author Contributions: Conceptualization, J.L.C., D.A.M., M.A.S., and J.L.C.-V.; methodology, J.L.C., D.A.M., M.A.S., and J.L.C.-V.; software, J.L.C., D.A.M., M.A.S., and J.L.C.-V.; validation, J.L.C., D.A.M., M.A.S., and J.L.C.-V.; formal analysis, J.L.C., D.A.M., M.A.S., and J.L.C.-V.; investigation, J.L.C., D.A.M., M.A.S., and J.L.C.-V.; resources, J.L.C., D.A.M., M.A.S., and J.L.C.-V.; data curation, J.L.C., D.A.M., M.A.S., and J.L.C.-V.; writing—original draft preparation, J.L.C., D.A.M., M.A.S., and J.L.C.-V.; writing—review and editing, J.L.C., D.A.M., M.A.S., and J.L.C.-V.; visualization, J.L.C., D.A.M., M.A.S., and J.L.C.-V.; supervision, J.L.C., D.A.M., M.A.S., and J.L.C.-V.; project administration, J.L.C., D.A.M., M.A.S., and J.L.C.-V.; funding acquisition, J.L.C., D.A.M., M.A.S., and J.L.C.-V. All authors have read and agreed to the published version of the manuscript.

Funding: This research received no external funding.

Institutional Review Board Statement: Not applicable.

Informed Consent Statement: Not applicable.

Data Availability Statement: The data are contained within the article.

Acknowledgments: The authors would like to thank the Laboratorio Oficial para Ensayos de Materiales de Construcción (LOEMCO), Spain, and the Fundación Gómez Pardo (Madrid, Spain) for the preparation of the samples, the performance of the tests and the interpretation of the results. The authors would also like to thank the laboratories of the Escuela Técnica Superior de Ingenieros de Minas y Energía of the Universidad Politécnica de Madrid (Spain), as well as the Instituto Español del Cemento y sus Aplicaciones (IECA) for their support with some analysis. Finally, the authors are very appreciative of the kindness of the Laboratorio de Tamices Moleculares of the Instituto de Catálisis y Petroleoquímica of the Centro de Investigaciones Científicas (CSIC) of Madrid (Spain).

Conflicts of Interest: The authors declare no conflict of interest.

References

1. Stamatakis, M.G.; Regueiro, M.; Calvo, J.P.; Fragoulis, D.; Stamatakis, G. A study of zeolitic tuffs associated with bentonite deposits from Almeria, Spain and Kimolos Island, Greece and their industrial potential as pozzolanas in the cement industry. *Hell. J. Geosci.* **2010**, *45*, 283–292.
2. Regueiro, M.; García-Romero, E.; Suárez, M.; Lopez-Acevedo, V.; López-García, J.A. Geología y Geoquímica del Yacimiento de Zeolita "Los Murcianos" (Cabo de Gata, Almería). In Proceedings of the XII International Congress on Energy and Mineral Resources, Oviedo, Spain, 7–11 October 2007.
3. Benito, R.; Garcia-Guinea, J.; Valle-Fuentes, F.J.; Recio, P. Mineralogy, geochemistry and uses of the mordenite–bentonite ash-tuff beds of Los Escullos, Almería, Spain. *J. Geochem. Explor.* **1998**, *62*, 229–240. [CrossRef]
4. Jiménez, R.; Cortel, A.; González, J.; Hernández, M.P.; Segura, J.M.; Soldevilla, J.A. Les zeolites d'Agua Amarga, Níjar, Almeria, Andalusia. *Mineral. Catalunya* **2022**, *14*, 24.
5. Martin Vivaldi, J.L.; López Aguayo, F. Presence de mordenite dans un gisement de bentonite de la region de Cabo de Gata, Almería. *Pascal Francis. Bol. Geol. Min.* **1975**, *86*, 187–192.
6. Mattei, M.; Riggs, N.R.; Giordano, G.; Guarnieri, L.; Cifelli, F.; Soriano, C.C.; Jicha, B.; Jasim, A.; Marchionni, S.; Franciosi, L.; et al. Geochronology, Geochemistry and Geodynamics of the Cabo de Gata volcanic zone, Southeastern Spain. *Ital. J. Geosci.* **2014**, *133*, 341–361. [CrossRef]
7. de la Fuente, S.; Cuadros, J.; Linares, J. Early Stages of Volcanic Tuff Alteration in Hydrothermal Experiments: Formation of Mixed-Layer Illite-Smectite. *Clays Clay Miner.* **2002**, *50*, 578–590. [CrossRef]
8. López-Ruiz, J.; Cebriá, J.M.; Doblas, M. *Cenozoic Volcanism I: The Iberian Peninsula*; The Geology of Spain; Geological Society: London, UK, 2002; p. 27.
9. Pelayo, M.; Marco, J.F.; Fernández, A.M.; Vergara, L.; Melón, A.M.; Pérez del Villar, L. Infrared and Mössbauer spectroscopy of Fe-rich smectites from Morrón de Mateo bentonite deposit (Spain). *Clay Miner.* **2018**, *53*, 17–28. [CrossRef]
10. Suárez, M.; Lorenzo, A.; García-Vicente, A.; Morales, J.; García-Rivas, J.; García-Romero, E. New data on the microporosity of bentonites. *Eng. Geol.* **2022**, *296*, 106439. [CrossRef]
11. De La Villa, R.V.; Cuevas, J.; Ramírez, S.; Leguey, S. Zeolite formation during the alkaline reaction of bentonite. *Eur. J. Miner.* **2001**, *13*, 635–644. [CrossRef]
12. Deon, F.; van Ruitenbeek, F.; van der Werff, H.; van der Meijde, M.; Marcatelli, C. Detection of Interlayered Illite/Smectite Clay Minerals with XRD, SEM Analyses and Reflectance Spectroscopy. *Sensors* **2022**, *22*, 3602. [CrossRef]
13. *Mapa Geológico de España a escala 1:50.000. "El Pozo de los Frailes"*; Segunda Serie-Primera Edición; Instituto Geológico y Minero de España: Madrid, Spain, 1983; pp. 1–35.
14. Calvo Pérez, B.; Costafreda Mustelier, J.L.; Estévez, E. Caracterización Preliminar de las Zeolitas del Yacimiento Los Murcianos, Al-Mería. V Iberian Congress of Geochemistry. IX Spanish Geochemistry Congress, Soria, Spain, 2005. Legal Deposit: SO-69/2005. ISBN: 84-95099-88. p. 10. Available online: https://oa.upm.es/1417/1/CALVO_PON_2005_01.pdf (accessed on 10 April 2024).
15. Calvo Pérez, B.; Estévez Fernández, E.; Costafreda Mustelier, J. Estudio de las Propiedades Puzolánicas de Materiales de Origen Volcánico Ubicados en la Zona Sureste de España. V Iberian Congress of Geochemistry. IX Spanish Geochemistry Congress, Soria, Spain, 2005. p. 10. Legal Deposit: SO-69/2005. ISBN: 84-95099-88-8. Available online: https://oa.upm.es/1418/1/CALVO_PON_2005_02.pdf (accessed on 10 April 2024).
16. García-Romero, E.; Suárez, M.; López-Acevedo, V.; Lozano, R.; Oyarzun, R.; López-García, J.A.; Regueiro, M. *Caracterización Mineralógica y Textural del Yacimiento de Zeolitas de «Los Murcianos» (Cabo de Gata, Almería)*; Resultados Preliminaries; Spanish Mineralogical Society: Madrid, Spain, 2006; MACLA 6; p. 4.
17. Suárez, M.; García-Romero, E.; Manchado, E.M. *Thermal Activation of the Mordenite from Los Murcianos Deposist (Almería)*; Geo-Temas 10; Geological Society of Spain: Madrid, Spain, 2008; p. 4. ISSN 1567-5172.
18. Costafreda, J.L. Geología, Caracterización y Aplicaciones de las Rocas Zeolíticas del Complejo Volcánico de Cabo de Gata (Almería). Ph.D. Thesis, Universidad Politécnica de Madrid, Madrid, Spain, 2008; 515p.
19. Martín, D.A.; Costafreda, J.L.; Presa, L.; Zambrano, J.; Costafreda, J.L., Jr. A New Study of the Lower Levels of the Los Frailes Caldera (Spain) for the Location and Characterisation of Pozzolans as Construction Materials. *Constr. Mater.* **2022**, *2*, 40–52. [CrossRef]
20. Tarazona, O.A. Estabilización Química de Aguas Afectadas por Metales Pesados en Áreas con Actividad Minerometalúrgica Mediante el Uso de Zeolitas. Universidad Complutense de Madrid. Facultad. De Ciencias Geológicas: Madrid, Spain, 2019; p. 46. Available online: https://purl.org/pe-repo/ocde/ford#2.07.01/ (accessed on 11 April 2024).
21. Villaseñor, J.; Rodríguez, L.; Fernández, F. Composting domestic sewage sludge with natural zeolites in a rotary drum reactor. *Bioresour. Technol.* **2011**, *102*, 1447–1454. [CrossRef]

22. Domene, M.A.; Agüera, J.M.; Uceda, M.; Buendía, D.; Racero, J.L.; Sevilla, A. *Nuevas Soluciones para la Restauración am-Biental en Zonas Áridas*; Dialnet. Cuaderno interdisciplinar de desarrollo sostenible (CUIDES): Madrid, Spain, 2011; pp. 11–89, ISSN 1889-0660.
23. Martín, D.A.; Costafreda, J.L.; Costafreda, J.L., Jr.; Presa, L. Improving the Performance of Mortars Made from Recycled Aggregates by the Addition of Zeolitised Cineritic Tuff. *Crystals* **2022**, *12*, 77. [CrossRef]
24. Costafreda, J.L.; Martín, D.A.; Presa, L.; Parra, J.L. Effects of a Natural Mordenite as Pozzolan Material in the Evolution of Mortar Settings. *Materials* **2021**, *14*, 5343. [CrossRef]
25. Jo, B.-W.; Choi, J.-S.; Yoon, K.-W.; Park, J.-H. Material characteristics of zeolite cement mortar. *Constr. Build. Mater.* **2012**, *36*, 1059–1065. [CrossRef]
26. Cobîrzan, N.; Balog, A.-A.; Aciu, C.; Iluţiu–Varvara, D.A. Sustainable Uses of Zeolitic Tuff as Building Materials. *Procedia Technol.* **2014**, *12*, 542–547. [CrossRef]
27. Iswarya, G.; Beulah, M. Use of zeolite and industrial waste materials in high strength concrete—A review. *Mater. Proceeding* **2021**, *46*, 116–123. [CrossRef]
28. Tran, Y.T.; Lee, J.; Kumar, P.; Kim, K.-H.; Lee, S.S. Natural zeolite and its application in concrete composite production. *Compos. Part B Eng.* **2019**, *165*, 354–364. [CrossRef]
29. Karakurt, C.; Kurama, H.; Topçu, I.B. Utilization of natural zeolite in aerated concrete production. *Cem. Concr. Compos.* **2010**, *32*, 1–8. [CrossRef]
30. Ghourchian, S.; Wyrzykowski, M.; Lura, P.; Shekarchi, M.; Ahmadi, B. An investigation on the use of zeolite aggregates for internal curing of concrete. *Constr. Build. Mater.* **2013**, *40*, 135–144. [CrossRef]
31. Boháč, M.; Kubátová, D.; Nečas, R.; Zezulová, A.; Masárová, A.; Novotný, R. Properties of Cement Pastes with Zeolite During Early Stage of Hydration. *Procedia Eng.* **2016**, *151*, 2–9. [CrossRef]
32. Vejmelková, E.; Koňáková, D.; Kulovaná, T.; Keppert, M.; Žumár, J.; Rovnaníková, P.; Keršner, Z.; Sedlmajer, M.; Černý, R. Engineering properties of concrete containing natural zeolite as supplementary cementitious material: Strength, toughness, durability, and hygrothermal performance. *Cem. Concr. Compos.* **2015**, *55*, 259–267. [CrossRef]
33. *UNE-EN 196-2:2014*; Métodos de Ensayo de Cementos—Parte 2: Análisis Químico de Cementos. AENOR: Madrid, Spain, 2014.
34. *UNE-EN 196-5:2006*; Métodos de Ensayo de Cementos—Parte 5: Ensayo de Puzolanicidad para Cementos Puzolánicos. AENOR: Madrid, Spain, 2006.
35. *UNE-EN 197-1:2011*; Composición, Especificaciones y Criterios de Conformidad de los Cementos Comunes. AENOR: Madrid, Spain, 2011.
36. *UNE-EN 196-1:2016*; Métodos de Ensayo de Cementos—Parte 1: Determinación de Resistencias Mecánicas. AENOR: Madrid, Spain, 2016.
37. Giannetto, G.; Montes, A.; Rodríguez-Fuentes, G. *Zeolitas. Características, Propiedades y Aplicaciones Industriales*; Edit. In-novación Tecnológica; Facultad de Ingeniería, Universidad Central de Venezuela: Caracas, Venezuela, 2000; p. 351, ISBN 980-00-1648-1.
38. Auerbach, S.M.; Carrado, K.A.; Dutta, P.K. *Handbook of Zeolite Science and Technology*; Library of Congress Catalogu-ing-in-Publication Data; CRC Press: New York, NY, USA, 2003; pp. 1–19, ISBN 0-8247-4020-3.
39. Hajash, A. Hydrothermal processes along mid-ocean ridges: An experimental investigation. *Contrib. Miner. Pet.* **1975**, *53*, 205–226. [CrossRef]
40. Fernández Soler, J.M. El Volcanismo Calco-Alcalino de Cabo de Gata (Almería). Estudio Volcanológico y Petrológico. Ph.D. Thesis, Universidad de Granada, Granada, Spain, 1992; 243p.
41. Presa, L.; Costafreda, J.L.; Martín, D.A.; Díaz, I. Natural Mordenite from Spain as Pozzolana. *Molecules* **2020**, *25*, 1220. [CrossRef]
42. Hamada, H.M.; Abdulhaleem, K.N.; Majdi, A.; Al Jawahery, M.S.; Thomas, B.S.; Yousif, S.T. The durability of concrete produced from pozzolan materials as a partially cement replacement: A comprehensive review. *Mater. Today Proc.* 2023, *in press*. [CrossRef]
43. Caputo, D.; Liguori, B.; Colella, C. Some advances in understanding the pozzolanic activity of zeolites: The effect of zeolite structure. *Cem. Concr. Compos.* **2008**, *30*, 455–462. [CrossRef]
44. Saraya, M.E.; Thabet, M.S. CHARACTERIZATION AND EVALUATION OF NATURAL ZEOLITE AS A POZZOLANIC MATERIAL. *Al-Azhar Bull. Sci.* **2018**, *29*, 17–34. [CrossRef]
45. Mertens, G.; Snellings, R.; Van Balen, K.; Bicer-Simsir, B.; Verlooy, P.; Elsen, J. Pozzolanic reactions of common natural zeolites with lime and parameters affecting their reactivity. *Cem. Concr. Res.* **2009**, *39*, 233–240. [CrossRef]
46. Costafreda, J.L.; Martín, D.A.; Presa, L.; Parra, J.L. Altered Volcanic Tuffs from Los Frailes Caldera. A Study of Their Pozzolanic Properties. *Molecules* **2021**, *26*, 5348. [CrossRef]
47. Stefanidou, M.; Papayianni, I. The role of aggregates on the structure and properties of lime mortars. *Cem. Concr. Compos.* **2005**, *27*, 914–919. [CrossRef]
48. Rosell, M.; Galloso, R.; Calvo, B. Zeolite as an active mineral additive in hight yield concrete. *Boletín Geol. Min.* **2006**, *117*, 783–792. ISSN: 0366-0176.
49. Kocak, Y.; Tascı, E.; Kaya, U. The effect of using natural zeolite on the properties and hydration characteristics of blended cements. *Constr. Build. Mater.* **2013**, *47*, 720–727. [CrossRef]
50. Małolepszy, J.; Grabowska, E. Sulphate Attack Resistance of Cement with Zeolite Additive. *Procedia Eng.* **2015**, *108*, 170–176. [CrossRef]

51. Irassar, E.; González, M.; Rahhal, V. Sulphate resistance of type V cements with limestone filler and natural pozzolana. *Cem. Concr. Compos.* **2000**, *22*, 361–368. [CrossRef]
52. Merida, A.; Kharchi, F. Pozzolan Concrete Durability on Sulphate Attack. *Procedia Eng.* **2015**, *114*, 832–837. [CrossRef]
53. Liguori, B.; Aprea, P.; de Gennaro, B.; Iucolano, F.; Colella, A.; Caputo, D. Pozzolanic Activity of Zeolites: The Role of Si/Al Ratio. *Materials* **2019**, *12*, 4231. [CrossRef]
54. Uzal, B.; Turanlı, L.; Yücel, H.; Göncüoğlu, M.; Çulfaz, A. Pozzolanic activity of clinoptilolite: A comparative study with silica fume, fly ash and a non-zeolitic natural pozzolan. *Cem. Concr. Res.* **2010**, *40*, 398–404. [CrossRef]
55. Najimi, M.; Sobhani, J.; Ahmadi, B.; Shekarchi, M. An experimental study on durability properties of concrete containing zeolite as a highly reactive natural pozzolan. *Constr. Build. Mater.* **2012**, *35*, 1023–1033. [CrossRef]
56. Markiv, T.; Sobol, K.; Franus, M.; Franus, W. Mechanical and durability properties of concretes incorporating natural zeolite. *Arch. Civ. Mech. Eng.* **2016**, *16*, 554–562. [CrossRef]
57. Raggiotti, B.B.; Positieri, M.J.; Oshiro, Á. Natural zeolite, a pozzolan for structural concrete. *Procedia Struct. Integr.* **2018**, *11*, 36–43. [CrossRef]
58. Nagrockiene, D.; Girskas, G. Research into the properties of concrete modified with natural zeolite addition. *Constr. Build. Mater.* **2016**, *113*, 964–969. [CrossRef]

materials

Article

Performance Research of Cement Concrete Pavements with a Lower Carbon Footprint

Tomasz Rudnicki [1,*] and Przemysław Stałowski [2]

[1] Faculty of Civil Engineering and Geodesy, Military University of Technology, 2 Kaliskiego St, 00-908 Warsaw, Poland
[2] Faculty of Civil and Environmental Engineering and Architecture, Bydgoszcz University of Science and Technology, 85-796 Bydgoszcz, Poland; przemyslaw.stalowski@pbs.edu.pl
* Correspondence: tomasz.rudnicki@wat.edu.pl

Abstract: The growing interest in the use of building materials with a reduced carbon footprint was the aim of this research assessing the impact of four different types of low-emission cements on the properties of cement concretes used for the construction of local roads. This research work attempted to verify the strength characteristics and assess the durability of such solutions, which used the commonly used CEM I 42.5 R pure clinker cement and three multi-component cements: CEM II/A-V 42.5 R, CEM III/A 42.5 N-LH/HSR/NA, and CEM V/A S-V 42.5 N-LH/HSR/NA. Cement was used in a constant amount of 360 kg/m^3, sand of 0/2 mm, and granite aggregate fractions of 2/8 and 8/16 mm. This research was carried out in two areas: the first concerned strength tests and the second focused on the area of assessing the durability of concrete in terms of frost resistance F150, resistance to de-icing agents, water penetration under pressure, and an analysis of the air entrainment structure in concrete according to the PN EN 480-11 standard. Analyzing the obtained test results, it can be concluded that the highest compressive strength of more than 70 MPa was obtained for CEM III concrete, 68 MPa for CEM V concrete, and the lowest for CEM I cement after 90 days. After the durability tests, it was found that the smallest decrease in compressive strength after 150 freezing and thawing cycles was obtained for CEM III (−0.9%) and CEM V (−1.4%) concretes. The high durability of concrete is confirmed by water penetration tests under pressure, because for newly designed recipes using CEM II, CEM III, and CEM V, water penetration from 17 mm to 18 mm was achieved, which proves the very high tightness of the concrete. The assessment of the durability of low-emission cements was confirmed by tests of resistance to de-icing agents and the aeration structure performed under a microscope in accordance with the requirements of the PN-EN 480-11 standard. The obtained analysis results indicate the correct structure and minimal spacing of air bubbles in the concrete, which confirms and guarantees the durability of concrete intended for road construction. Concretes designed using CEM V cement are characterized by a carbon footprint reduction of 36%, and for the mixture based on CEM III, we even observed a decrease of 39% compared to traditional concrete. Concrete using CEM II, CEM III, and CEM V cements can be successfully used for the construction of local roads. Therefore, it is necessary to consider changing the requirements of the technical specifications recommended for roads in Poland.

Keywords: road concrete; multi-component cements; reduced carbon footprint of concrete

1. Introduction

With the development of the road network in Poland, growing requirements of investors and road users are expected related to the use of materials that affect the possibility of reducing CO_2 emissions [1–4]. The need to take action to reduce the carbon footprint by reducing the emissions of individual components of concrete as a building material is one of the challenges facing the construction industry in the context of climate protection, as reflected in the very ambitious CO_2 emission reduction goals under the Green Deal Program announced by the European Commission in December 2019 [5,6].

The entire building materials industry in Europe will have to rise to this challenge by 2050. The European Green Deal assumes that over the next 10 years, Europe will be able to reduce CO_2 emissions by 55% of the amount of CO_2 emitted in 1990 and achieve full climate neutrality by 2050. The search for low-emission materials and production technologies must cover the entire supply chain because, in the case of modern, energy-efficient cement production technologies, the possibilities for reduction at this stage are significantly limited [7–10]. Additionally, reducing CO_2 emissions from cement production is a particular challenge because only approximately 40% of direct emissions come from the combustion of fuels, while the remaining 60% of CO_2 results from the decomposition of raw materials subjected to thermal treatment in a cement kiln [11].

Additionally, as in many other industrial processes, CO_2 emissions resulting from electricity consumption must also be taken into account. Only the implementation of innovative technological and product solutions and a longer-term approach can bring significant effects in this area [12–14]. The most effective technology for reducing CO_2 emissions from cement plants is the method of capturing CO_2 and then storing it in geological structures or using it as a raw material in various technologies (Carbon Capture Usage/Storage—CCUS). Pilot research on this technology has been ongoing for several years, but currently in Poland, Holcim, a leader in environmental protection, has started work on this technology and plans to implement it on an industrial scale in 2027.

An important tool in climate policy is the assessment of the carbon footprint of products, which requires an analysis of the entire product life cycle [15,16]. In the case of cement concrete, a life cycle analysis should be carried out for each stage, especially for the stage of demolition of the concrete pavement through recycling and its reinstallation, which constitutes the closure of the cycle and is in line with the circular economy strategy [16–18]. It is worth emphasizing the positive impact of CCP (Cement Concrete Pavements) on reducing the ambient temperature compared to asphalt pavements. Studies were conducted on the development of important climatic characteristics (maximum and average annual temperature and the frost index) of Central Europe (CE) used in the design of CCPs. The research clearly confirmed a significant increase in CE temperatures and the resulting changes in pavement stresses [19].

This idea assumes the optimization of the use of materials in pavement construction and the reuse of recycled materials after the end of pavement operation [20,21]. The reliability of the assessment of the carbon footprint of pavement concrete should also take into account an additional process that takes place during the use of concrete structures—carbonation, i.e., the process of absorbing CO_2 by concrete.

Analyzing the achievements in terms of reducing the carbon footprint, it should be noted that a significant reduction effect in CO_2 emissions can be achieved by limiting the amount of Portland clinker in cement [1,14,17]. Traditional Portland cement without additives contains >90% Portland clinker and is mainly made of limestone. Producing 1 ton of clinker requires, on average, 3.7 GJ of heat and emits approximately 850 kg of CO_2 [18]. Lower emissions in the case of reductions in the clinker/cement ratio (currently, in Poland, this indicator is 75%) results, on the one hand, from lower process emissions from the calcination of raw materials and, at the same time, from the lower heat demand for the production of clinker minerals [16,18]. On the other hand, the possibility of using mineral additives such as blast furnace slag, fly ash [8,22–28], and limestone, which partially replace clinker in cement, significantly reduces the carbon footprint, which was demonstrated in [11,29,30].

Another element enabling the reduction in the carbon footprint is the possibility of using alternative aggregates, especially local ones [9,30], which, due to shorter transport distances, can reduce CO_2 emissions by up to 11%. The modification of the composition of cement concrete should also take into account the impact of the use of mineral additives that are industrial waste (stone dust from aggregate dedusting or additives such as fly ash and microsilica) [23–28]. Obtaining high-strength parameters of cement concrete can be achieved by using chemical admixtures that strongly reduce the amount of water and increase workability over time [31–34], and the use of air-entraining admixtures significantly increases the durability of concrete over time.

An important element of maintaining the fluidity of road user traffic is the possibility of performing quick renovation works under traffic, which is described in detail in a previous article [35]. When performing a full life cycle analysis, the assessment of the durability of the designed solutions should be taken into account, especially the possibility of reusing the materials used in terms of recycling concrete [18]. The possibility of reusing cement concrete after recycling has been repeatedly confirmed in studies and even directly in the implementation of a concrete highway in Poland [20,21]. In this work, the authors attempted to use low-emission cements in cement concrete road surfaces in order to achieve the maximum reduction in the carbon footprint, which was confirmed by testing the frost resistance of F150 and analyzing the structure of concrete air entrainment.

2. Materials and Methods

In order to analyze the possibility of reducing the carbon footprint in concrete used to build road surfaces, the following types of cement were used in accordance with the PN-EN 197-1 standard [36]: CEM I 42.5 R, CEM II/A-V 42.5 R, CEM III/A 42.5 N-LH/HSR/NA, and CEM V/A S-V 42.5 N-LH/HSR/NA. In the requirements for the construction of road infrastructure in Poland, cements from the CEM I group are permitted for use, while the use of cements with mineral additives requires the investor to present not only an analysis of the reduction in the carbon footprint but also, above all, durability over time. Therefore, this publication presents comprehensive tests of strength characteristics after 28, 56, and 90 days, and more importantly, tests of resistance to frost and water penetration under pressure. In order to obtain the most comparable test results, granite aggregate in fractions of 2/8 and 8/16 with a density of 2.64 g/cm^3 and water absorption of 0.80% and natural non-alkaline reactive sand were used in all designed recipes, with a density of fraction of 0/2 of 2.65 g/cm^3. Due to the method of transport, installation, and compaction, a low w/c ratio of 0.4 was required. Two types of chemical admixtures were used in the form of a new-generation fluidizing admixture marked as SP in accordance with the PN EN 934-1 [37] and PN-EN 934-2 [38] standards (admixture with a high degree of water reduction with a density of 1.01 g/cm^3) and a traditionally used air-entraining admixture.

2.1. Analysis of the Properties and Microstructure of Cements

The physical and mechanical properties of the cements used are presented in Table 1.

Table 1. Properties of cements.

Property	Unit	CEM I 42.5 R	CEM II A-V 42.5 R	CEM III A 42.5 N-LH/HSR/NA	CEM V A S-V 42.5 N-LH/HSR/NA
	chemical composition of cement				
CaO	%	62.82	55.44	50.54	46.3
SiO_2	%	18.95	23.58	29.18	28.7
Al_2O_3	%	4.99	7.83	6.17	9.7
SO_3	%	3.14	2.83	2.48	2.74
Fe_2O_3	%	2.81	3.25	1.55	3.0
MgO	%	1.37	1.56	4.04	2.06
K_2O	%	0.88	0.71	0.69	1.08
$eqNa_2O$	%	0.79	0.73	0.78	1.06
Na_2O	%	0.21	0.26	0.33	0.35
Cl^-	%	0.08	0.076	0.07	0.041
	physical and mechanical properties of cement				
Compressive strength 2 days	MPa	27.8	26.1	14.2	20.0
Compressive strength 28 days	MPa	56.7	53.4	51.5	58.1
Specific surface	cm^2/g	3374	3716	4466	4800
Beginning of cement setting	minutes	208	179	238	275
End of cement setting	minutes	284	260	317	350
Consistency in volume	mm	0.8	0.88	0.8	0.81
Water demand of cement	%	27.8	27.5	29.8	32.2

The mineral additives used in individual cements comply with the EN 197-1 standard [36] and are as follows: CEM I 42.5 R 95% clinker, 5% secondary minerals, CEM II A-V 80–94% clinker, 6–20% silica fly ash, 5% secondary minerals, CEM III A 35–64% clinker, 36–65% blast furnace slag, 5% secondary minerals, CEM V A 40–64% clinker, 18–30% blast furnace slag, 18–30% silica fly ash, and 5% secondary minerals.

Analysis of the cements used was performed using a JSM-6610 scanning microscope (Joel, Tokyo, Japan), in which the samples were applied to carbon tape and dusted with graphite. Analysis of the chemical composition of the tested materials was performed in an EDS analyzer, which is part of the device. Cement images are shown in Figure 1.

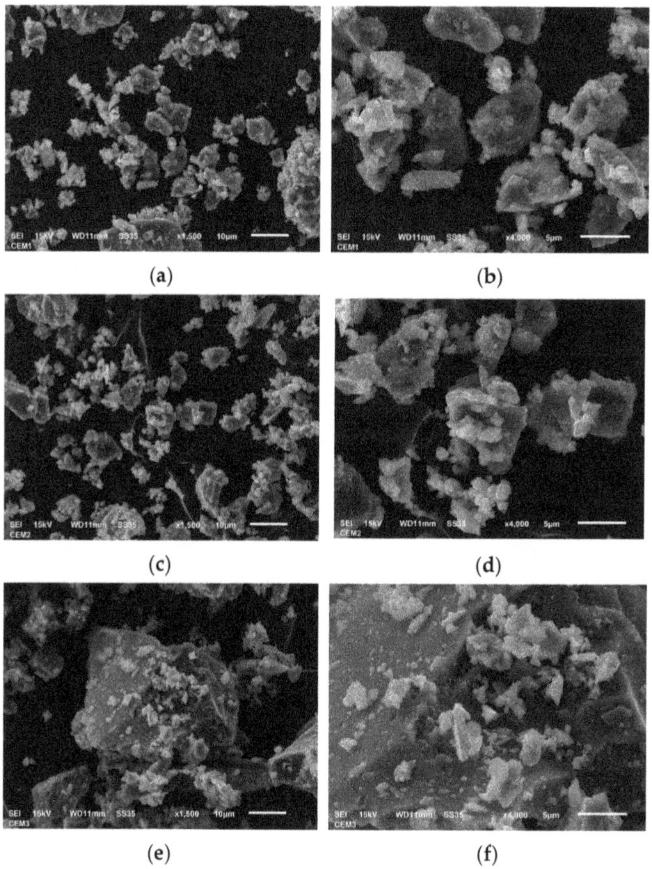

Figure 1. Cement images taken with Jeol JSM-6610 (Joel, Tokyo, Japan): (**a**) CEM I × 1500, (**b**) CEM I × 4000, (**c**) CEM II/B-V × 1500, (**d**) CEM II/A-V × 4000, (**e**) CEM III/A × 1500, and (**f**) CEM III/A × 4000.

Cement images were taken with a Jeol JSM-6610 microscope (Jeol, Tokyo, Japan) for CEM I, CEM II, and CEM III cements at magnifications of ×1500 and ×4000. Images a and b for cement CEM I show a compact clinker structure with small amounts of gypsum, which is a setting regulator. In images c and d for CEM II, ash particles can be seen, and in images e and f for CEM III, there is a high content of blast furnace slag. Additionally, chemical analysis was performed for each of the four cements (Table 2):

Table 2. The chemical analysis of the composition of CEM I, CEM II, CEM III, and CEM V.

Element	CEM I 42.5 R	CEM II A-V 42.5 R	CEM III A 42.5 N-LH/HSR/NA	CEM V A S-V 42.5 N-LH/HSR/NA	Standard Label
	wt%	wt%	wt%	wt%	
O	37.92	39.61	41.8	38.54	SiO_2
Mg	1.06	0.56	2.8	0.97	MgO
Al	3.59	1.37	3.4	3.14	Al_2O_3
Si	9.15	9.09	14.8	9.87	SiO_2
K	0.52	1.41	1.4	1.56	KBr
Ca	44.68	46.26	35.2	44.26	Wollastonite
Fe	3.07	1.70	0.6	1.66	Fe
Total:	100.00	100.00	100.00	100.00	

The results of the analysis of the cement used with a 45% share of blast furnace slag (S) are marked as CEM III/A 42.5 HSR-NA (Figure 2). The obtained results are as follows: CaO—35.2%, Si—14.8%, Al—3.4%, Mg—2.8%, S—1.3%, and Fe—0.6% [36].

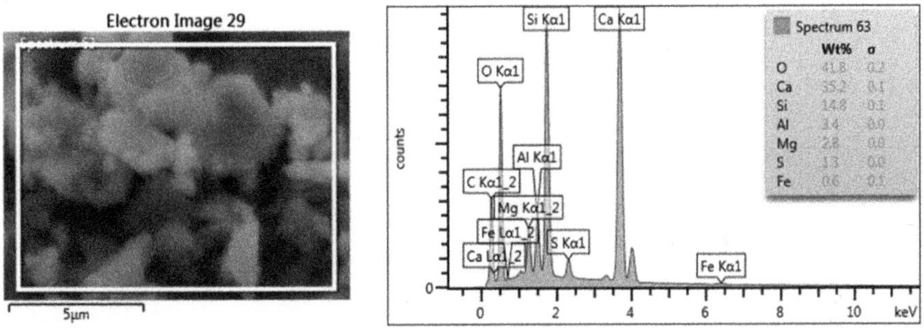

Figure 2. An example of a chemical analysis image for CEM III cement with 45% slag content created using the EDS analysis method.

2.2. *Analysis of Grain Size and Reactivity of Aggregates*

According to the requirements of technical specifications, the aggregate should be characterized by high technical parameters and meet the condition of maximum saturation [39]. The screening analysis of the granite aggregate of 2/8 and 8/16 mm and sand of 0/2 mm is given in Table 3 and Figure 3.

Table 3. Graining of materials.

Sieve [mm]	Volume of Aggregate on Individual Control Sieves [%]		
	Granite 8/16	Granite 2/8	Sand 0/2
16	1.7	0.0	0.0
8	96.3	1.2	0.0
4	2.0	59.7	0.0
2	0.0	39.1	2.5
1	0.0	0.0	15.9
0.5	0.0	0.0	25.5
0.25	0.0	0.0	43.4
0.125	0.0	0.0	11.9
0.0	0.0	0.0	0.8

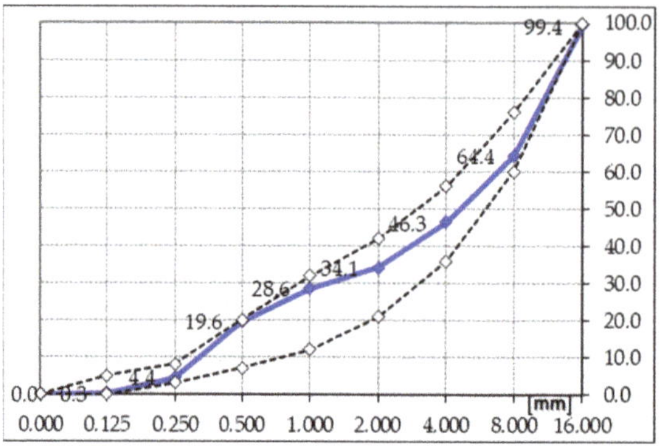

Figure 3. Graining curve of the mineral mixture.

The designed mineral mixture composition and the required specification curves for good grain size are shown in Figure 3.

Due to the requirement of using mineral materials characterized by the lack of a potential alkali–silica reaction (ASR), the potential reactivity of the aggregate was determined using two methods: PB1 (14-day test) and PB2 (365-day test). This determination involved measuring the elongation of the prepared mortar and concrete samples in accordance with the procedure. The final test result, which is presented in Tables 4 and 5, was in reactivity category R0 for both fine and coarse aggregate in 1 M NaOH solution at 80 °C. The detailed description of the test procedure and analysis of the results in this regard were discussed in more detail in [40]. Tables 4 and 5 describe the terms resulting from the procedure and are described as L_7, L_{28}, and L_{91}, which represent the linear changes in sample length during testing after 7, 28, and 91 days of testing.

Table 4. Determination of the ASR for 0/2mm aggregate.

Samples	Change in Length on Individual Days [mm]					
	L_7	L_{14}	L_{28}	L_{91}	L_{182}	L_{365}
1	−0.002	0.003	0.001	0.008	0.014	0.016
2	−0.004	0.003	0.001	0.009	0.012	0.015
3	−0.003	0.003	0.002	0.007	0.012	0.014
Average value	−0.003	0.003	0.001	0.008	0.013	0.015

Table 5. Determination of the ASR for 2/8 and 8/16 mm aggregate.

Samples	Change in Length on Individual Days [mm]					
	L_7	L_{14}	L_{28}	L_{91}	L_{182}	L_{365}
1	0.000	−0.001	−0.002	0.006	0.018	0.027
2	−0.001	0.000	−0.003	0.006	0.016	0.019
3	−0.001	−0.001	−0.001	0.006	0.016	0.022
Average value	−0.001	−0.001	−0.002	0.006	0.017	0.023

The aggregate test results obtained, performed in accordance with procedures PB1 and PB2, allowed the aggregates to be classified as R0, i.e., non-reactive.

After obtaining positive aggregate test results, the formulation compositions shown in Table 6 were designed.

Table 6. Designed composition of concrete mixtures [kg/m^3].

Materials	Concrete Mix Compositions [kg/m^3]			
	C30/37_I	C30/37_II	C30/37_III	C30/37_V
CEM I 42.5 R	360	-	-	-
CEM II/A-V 42.5 R	-	360	-	-
CEM III A 42.5 HSR/NA	-	-	360	-
CEM V/A S-V 42.5 N-LH/HSR/NA	-	-	-	360
Water	144	144	144	144
Fine aggregate sand 0/2 mm	682	682	682	682
Coarse granite aggregate 2/8 mm	565	565	565	565
Coarse granite aggregate 8/16 mm	702	702	702	702
SP PC	2.5	2.5	2.5	2.5
LPA	0.7	0.7	0.7	0.7
Density	2456	2450	2452	2449

In the described work, over 250 samples were prepared and tested in accordance with the requirements of the PN-EN 12390-2 standard [41]. The scope of individual tests and reference to the requirements for concrete intended for road construction are included in Section 2.3.

2.3. Methods

The first phase of the project concerned the testing of hardened concrete and consisted of determining the compressive, bending, and splitting strength after 28, 56, and 90 days [42–44]. The density of concrete was determined in accordance with the PN-EN 12390-7 standard [45]. The second phase consisted of determining the durability of concrete with a reduced carbon footprint by testing frost resistance after 150 cycles in accordance with the PN-B-06250:1988 standard [46]. The F150 durability test was conducted on 48 samples with dimensions of 100 × 100 × 100 mm. A total of 144 samples were prepared and tested to determine the strength properties. The starting date of the tests was 28 days for CEM I, 56 days for CEM II, and 90 days for CEM III and CEM V, i.e., at the equivalent time [39,47]. A positive result of this test as specified in the standard must not exceed a 5% loss in mass and a 20% decrease in compressive strength compared to the reference samples. Furthermore, the samples after freeze/thaw cycles must not show cracks. Furthermore, in order to check the durability, resistance to de-icing salts [48], and penetration of water under pressure for each series, the pore distribution in the hardened concrete was determined in accordance with PN-EN 480-11 [49]. The requirements for cement concrete [39,47] used for road construction are presented below in Table 7.

Table 7. Requirements for the concrete pavement for traffic categories KR1 ÷ KR4.

Properties of Pavement Concrete	Requirements	Test Method
Compressive strength	C30/37	PN-EN 12390-3 [42]
Frost resistance F150: -Loss of sample mass -Loss of compressive strength	 <5% <20%	PN-B-06250 [46]
Characteristics of air voids in concrete: -content of micropores with a diameter of less than 0.3 mm (A_{300}), % -index of pore distribution in concrete, L mm	 ≥1.5% ≤0.200 mm	PN-EN 480-11 [49]
Density, tolerance in reference to the formula	±3.0%	PN-EN 12390-7 [45]

3. Results

3.1. Results of Assessing the Properties of the Concrete Mixture after 5 and 45 min

The properties of the concrete mixture were determined 5 and 45 min after mixing the ingredients (Table 8). These time intervals were used because experience shows that 45 min is the maximum time for incorporating the mixture without significantly changing its properties and ingredients under normal conditions.

Table 8. Properties of the concrete mix.

Materials	Concrete Mix Compositions [kg/m^3]			
	C30/37_I	C30/37_II	C30/37_III	C30/37_V
Consistency after 5 min, mm	120	110	140	140
Consistency after 45 min, mm	80	95	130	125
Air content after 5 min, %	5.5	5.6	5.7	6.2
Air content after 45 min, %	4.9	5.2	5.6	6.0
Density, g/cm^3	2.456	2.452	2.454	2.448

3.2. Determination of the Strength Characteristics of Concrete for Road Surfaces

In the first phase of the tests, strength characteristics were determined, such as compressive strength after 28, 56, and 90 days in accordance with PN-EN 12390-3 [42]. Additionally, flexural strength and tensile splitting strength tests of the test specimens were performed in accordance with the PN-EN 12390-5 [43] and PN-EN 12390-6 [44] standards. All samples were prepared and stored in accordance with the EN 12390-2 standard [41]. The average test results of the 144 samples are shown in Table 9.

Table 9. Strength properties of road concrete.

C30/37 Road Concrete	Study Start Date [MPa]		
	28 Day	56 Day	90 Day
	Compressive strength		
C30/37_I with CEM I 42.5 R	54.5 ± 2.1	55.6 ± 1.8	58.4 ± 3.1
C30/37_II with CEM II/A-V 42.5 R	48.4 ± 1.8	56.1 ± 1.9	59.3 ± 1.3
C30/37_III with CEM III/A 42.5 N-LH/HSR/NA	43.2 ± 3.2	60.7 ± 2.1	70.9 ± 2.2
C30/37_V with CEM V/A S-V 42.5 N-LH/HSR/NA	42.3 ± 1.1	54.6 ± 1.1	68.6 ± 0.9
	Flexural strength		
C30/37_I with CEM I 42.5 R	5.2 ± 0.9	5.4 ± 0.7	5.3 ± 0.4
C30/37_II with CEM II/A-V 42.5 R	4.8 ± 0.5	5.2 ± 1.1	5.5 ± 0.8
C30/37_III with CEM III/A 42.5 N-LH/HSR/NA	4.6 ± 1.2	5.5 ± 0.9	6.4 ± 0.9
C30/37_V with CEM V/A S-V 42.5 N-LH/HSR/NA	4.5 ± 0.4	5.3 ± 0.6	6.6 ± 1.1
	Tensile Splitting Strength of the Test Specimens		
C30/37_I with CEM I 42.5 R	4.0 ± 0.5	4.3 ± 0.2	4.2 ± 0.8
C30/37_II with CEM II/A-V 42.5 R	4.1 ± 0.2	4.4 ± 0.8	4.7 ± 1.0
C30/37_III with CEM III/A 42.5 N-LH/HSR/NA	3.7 ± 0.7	4.3 ± 0.5	5.0 ± 0.5
C30/37_V with CEM V/A S-V 42.5 N-LH/HSR/NA	3.6 ± 0.3	4.1 ± 0.4	5.2 ± 0.3

The highest compressive strength results were obtained after 90 days for the C30/37_III formulation of 70.9 MPa, i.e., 18% more than for the C30/37_I base formulation, for which the result was 58.4 MPa. When assessing the flexural strength for 36 samples, it should be emphasized that for all recipes, the required result of >4.5 MPa required in the technical specification was achieved. The highest results were obtained for the C30/37_III and C30/37_V formulations, which were 6.4 MPa and 6.6 MPa. The last determination was the tensile splitting strength, and here, the highest results were also for the C30/37_III and C30/37_V recipes, which were 5.0 MPa and 5.2 MPa, respectively. It should be emphasized that the obtained results of strength tests for concretes with a reduced carbon footprint met the high requirements for concrete surfaces of roads with low and very high traffic intensity, KR4-KR7.

3.3. *Testing the Durability of Road Concrete Using Multi-Component Cements*

An assessment of the possibility of using concrete with a reduced carbon footprint made of low-emission cements is possible after performing comprehensive durability tests. As part of the durability assessment, 150 freezing and thawing cycles were performed to determine frost resistance, resistance to de-icing salts was checked, the depth of water penetration under pressure was determined, and the air entrainment structure in the concrete was determined as a final verification of durability.

3.3.1. Frost Resistance after 150 Cycles

The determination of frost resistance in cyclic freezing and thawing at temperatures −20 °C and +20 °C was performed in accordance with the PN-B-06250:1988 standard [46]. As part of the study, 12 cubic samples with dimensions of 100 × 100 × 100 mm were prepared for each recipe. After 150 test cycles, the % loss of compressive strength and weight loss were determined. Samples after 150 freezing and thawing cycles must not show cracks. The average results from the four tests are shown in Table 10.

Table 10. The influence of low-clinker cements on the durability of road concrete.

Materials	Concrete Mix Compositions [kg/m^3]				Requirement
	C30/37_I	C30/37_II	C30/37_III	C30/37_V	
	Frost Resistance Test F150 average result				
Loss of compressive strength ΔR, %	−4.5	−2.2	−0.9	−1.4	<20%
Weight loss of samples ΔG, %	−0.2	−0.12	−0.1	−0.1	<5%
	Degree of Defect m_{56}/m_{28} average result				
Resistance to frost in the presence of de-icing salt	1.22	0.95	0.77	0.82	<2
	Water penetration under pressure				
Depth of water penetration under pressure, mm	24	17	18	17	<45
	Characteristics of air voids in hardened concrete				
Total air content, *A*, %	4.9	5.2	5.6	6.0	4.5–6.0
Spacing factor, *L*, mm	0.18	0.17	0.14	0.13	<0.20
Micro air-void content, A_{300}, %	2.10	1.85	2.45	2.50	>1.50

3.3.2. Resistance to Deicing Salts According to EN 12390 PKN-CEN/TS 12390-9

The test was performed in accordance with the EN 12390 PKN-CEN/TS 12390-9 standard [48] and involved the cyclic freezing and thawing of concrete samples in the presence of de-icing salts (3% NaCl). After 28 and 56 days, the mass loss of the samples was determined, and the determined loss must be less than 2. A positive test result is when, for each sample, the mass loss was less than 1.5 kg/m^2, and the average mass loss was determined after 28 and 56 days.

3.3.3. Air Void Analysis

The determination of the air-entrainment structure in concrete was performed in accordance with the procedure described in the PN-EN 480-11:2008 standard [49]. Two samples were cut from the solid samples for analysis, with dimensions of 150 × 100 × 20 mm. After grinding the samples, they were polished, and contrast was added to obtain better accuracy. Automatic analysis of the number and size of air voids was performed using the microscopic method using the Nikon SMZ1270 Navitar computer automatic image analysis system (Tokyo, Japan). The image and average values from four measurements for each sample are shown in Figure 4 and Table 10.

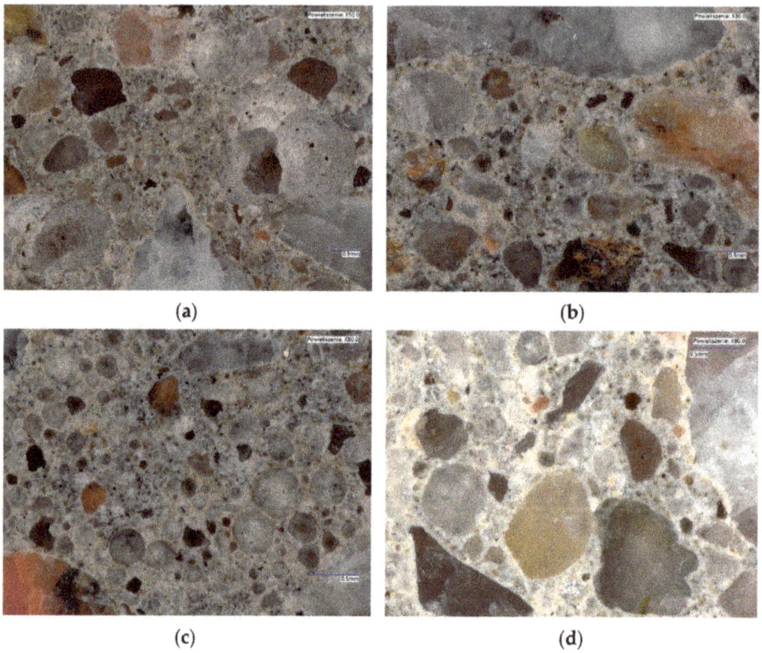

Figure 4. Images of the structure of prepared concrete samples: (**a**) C30/37_I, (**b**) C30/37_II, (**c**) C30/37_III, and (**d**) C30/37_V.

A very effective and commonly used method for assessing the durability of cement concrete over time is to assess the structure of the void distribution L (the distance between the voids), which should be less than 0.2 mm. The second assessment parameter is the content of micropores with a diameter of less than 0.3 mm, described as A_{300}, which must be greater than 1.5%. Additionally, the total amount of air in the concrete was determined (A). The results of average measurements for each of the four samples are presented in accordance with the standard [49] in Table 10. The total air content A is comparable in value to the results for the concrete mixture before forming (from 4.9% to 6.2%) because the air content in hardened concrete is 4.9 to 6.0%. The most favorable results were obtained for the C30/37_V recipe based on CEM V cement.

The results of frost resistance determination after 150 cycles indicate that the smallest decrease in compressive strength was obtained for the C30/37_III recipe on CEM III cement. This decrease was 0.9% and the decrease in sample weight was only 0.1%. In terms of determining resistance to deicing agents (salt), the most favorable results were obtained for the C30/37_V and C30/37_III formulas and were in line with the requirements of the standard. As part of the durability tests, water penetration tests were also carried out under pressure, and for all designed compositions, the results ranged from 17 to 24 mm, where the requirement was < 45 mm. The obtained results of the analysis of the air structure in hardened concrete are very similar to the results obtained using the pressure method for the concrete mix, which may indicate the appropriate amount of air-entraining admixture and ranges from 4.9% to 6.2%. The total air content in hardened concrete is in the range of 4.9–6.0%. By analyzing the results obtained to determine the characteristics of the spacing of air voids, it can be concluded that they meet the requirements for concrete used in XF conditions.

3.4. Carbon Footprint Reduction

The aim of the study was to determine the possibility of reducing the carbon footprint in concrete by using multi-component cements. These cements differed in the amount of clinker and the type and amount of mineral additives in their composition. For calculations using the LCA (Life Cycle Assessment) method according to the ISO 14067 standard [50], the actual transport distances of cement and aggregate to the construction site were assumed, which were 100 and 120 km. Having calculated CO_2 emissions for individual components of the concrete mix and knowing the distance from the construction site, the determined carbon footprint is presented in Table 11.

Table 11. Carbon footprint reduction results of paving concrete with multi-component cements [9,11].

Parameter	Unit	C30/37_I	C30/37_II	C30/37_III	C30/37_V
Carbon footprint cement	CO_2/kg	0.703	0.559	0.377	0.405
Cement transportation	CO_2/km/t	0.199	0.199	0.199	0.199
Carbon footprint granite aggregate	CO_2/kg	0.007	0.007	0.007	0.007
Aggregate transport	CO_2/km/t	0.166	0.166	0.166	0.166
The carbon footprint of concrete in cubic meters	CO_2/m^3	**298**	**247**	**181**	**191**
Carbon footprint reduction	[%]	0.0	−17.4	−39.4	−35.9

Analyzing the results obtained using the software recommended by the Global Cement and Concrete Association (GCCA), it can be seen that the use of CEM II cement allows a reduction in the carbon footprint of 17.4%, and the use of CEM V and CEM III cement in concrete intended for the construction of roads, engineering structures, and airports allowed a reduction of up to 35.9 and 39.4%. This is an important result and suggests greater use of cements with mineral additives for infrastructure construction should be considered.

4. Discussion

The CEM III and CEM V multi-component cements used in road concrete were characterized by the highest compressive strength, above 70 MPa, obtained for CEM III concrete and 68 MPa for CEM V concrete. However, CEM I cement, commonly used in infrastructure, achieved the lowest compressive strength of 58 MPa. These results confirmed the usefulness of multi-component cements in durability tests, as the smallest decrease in compressive strength after 150 freeze–thaw cycles was noted for CEM III (−0.9%) and CEM V (−1.4%) concretes. This can be confirmed by the results of water penetration under pressure, because for the recipes using CEM II, CEM III, and CEM V, 17, 18, and 17 mm of water penetration were obtained, which proves the high tightness of the concrete. The final verification took place during the analysis of the aeration structure performed under a microscope in accordance with the requirements of the PN-EN 480-11 standard [49]. The analysis results confirmed the proper structure and minimal spacing of air bubbles in the concrete, which guarantees durability. Concrete using CEM II, CEM III, and CEM V can be used to build local roads and even expressways with heavy vehicle traffic. In addition, the concrete mix based on CEM V is characterized by a carbon footprint reduced by 36%, and the mix based on CEM III by a reduction of up to 39% for the A1–A3 phase. Carbon footprint calculations were performed for the processes of extraction and transport of raw materials and the production of concrete at the plant. Considering that this A1–A3 phase could be responsible for a reduction in the carbon footprint of up to 40%, such concrete can be classified as environmentally friendly concrete, consistent with the idea of sustainable development of modern building materials. Surface concrete designed with the addition of low-clinker cements such as CEM II, CEM III, and CEM V meets the very high requirements for a modern concrete surface intended for the construction of local roads and parking lots. The results of tests on the durability of road concrete clearly indicate that concrete with a reduced carbon footprint is a sustainable solution and an alternative to the commonly used and required Portland cement CEM I 42.5 in Poland. The cements used are blast furnace slag, silica fly ash, and limestone, which are waste materials, thanks to which we can reduce our carbon footprint and reduce CO_2 emissions by over 40%. Thanks to this research, concrete with a reduced carbon footprint can be successfully used for the construction of roads and communication infrastructure, in line with the sustainable construction strategy.

5. Conclusions

The obtained test results for road concrete with a reduced carbon footprint thanks to the use of low-clinker cements allow the following conclusions to be drawn:

1. Road concrete designed with CEM V is characterized by a reduced carbon footprint of 36%, and for a mixture based on CEM III, as much as 39%, which is why such concrete can be classified as environmentally friendly concrete and materials consistent with the sustainable development of modern construction.
2. The highest compressive strength exceeding 70 MPa was obtained for the C30/37_III concrete formulation, while 68 MPa was observed for the C30/37_V concrete, and the lowest value was observed for the C30/37_I comparative formulation after 28, 56, and 90 days.
3. The most favorable results of frost resistance in cyclic freezing and thawing were obtained for the recipe using CEM III (−0.9%) and CEM V (−1.4%) cement. This is confirmed by the results of water penetration under pressure, because for the recipes using CEM II, CEM III, and CEM V, water penetration of 17, 18, and 17 mm was obtained, which proves the high tightness of the concrete.
4. Concrete using CEM II, CEM III, and CEM V can be used to build local roads and even expressways with heavy vehicle traffic.
5. The microscopic analysis of the pore structure in the hardened concrete allows us to conclude that the total air content is similar to the results obtained in the concrete mix using the pressure method, ranging from 4.9% to 6.2%. The total air content in hardened concrete ranges from 4.9 to 6.0%. The highest value of the air permeability coefficient (L) and the lowest micropore content (A_{300}) was demonstrated by concrete marked as C30/37_V based on CEM V cement.

Author Contributions: Conceptualization, T.R.; software, P.S.; formal analysis, P.S.; investigation, T.R.; resources, T.R.; data curation, P.S.; writing—original draft, T.R.; writing—review and editing, T.R.; project administration, T.R. and P.S. All authors have read and agreed to the published version of the manuscript.

Funding: This research received no external funding.

Institutional Review Board Statement: Not applicable.

Informed Consent Statement: Not applicable.

Data Availability Statement: The original contributions presented in the study are included in the article, further inquiries can be directed to the corresponding author.

Conflicts of Interest: The authors declare no conflicts of interest.

References

1. Levitt, R.E. CEM research for the next 50 years: Maximizing economic, environmental, and social value of the built environment. *J. Constr. Eng. Manag.* **2007**, *133*, 619–628. [CrossRef]
2. Ruegg, R.T.; Petersen, S.R.; Marshall, H.E. *Recommended Practice for Measuring Life-Cycle Cost for Buildings and Building Systems*; US Department of Commerce, NBS: Washington, DC, USA, 1980.
3. Latawiec, R.; Woyciechowski, P.; Kowalski, K.J. Sustainable Concrete Performance-CO_2-Emission. *Environments* **2018**, *5*, 27. [CrossRef]
4. Czarnecki, L.; Woyciechowski, P. Model of concrete carbonation as limited process—Experimental investigations of fluidal ash concrete. *Brittle Matrix Compos.* **2009**, *9*, 183.
5. EU Commission. Internal Market, Industry, Entrepreneurship and SMEs. Construction Sector. Available online: https://single-market-economy.ec.europa.eu/sectors/construction_en (accessed on 14 February 2024).
6. EU Industrial Strategy. A More Resilient, Green and Digital Construction Ecosystem. Available online: https://single-market-economy.ec.europa.eu/sectors/construction/construction-transition-pathway_enprefang (accessed on 14 February 2024).
7. Ajdukiewicz, A. Aspekty trwałości i wpływu na środowisko w projektowaniu konstrukcji betonowych. *Przegląd Bud.* **2011**, *82*, 20–29.
8. Glinicki, M.A. *Inżynieria Betonowych Nawierzchni Drogowych*; Wydawnictwo Naukowe PWN: Warszawa, Poland, 2019.
9. Rudnicki, T. The Impact of the Aggregate Used on the Possibility of Reducing the Carbon Footprint in Pavement Concrete. *Sustainability* **2022**, *14*, 16478. [CrossRef]
10. Institute of Building Technique, Department of Heat Physics, Sanitary Installations and Environment Laboratory of Environmental Protection. *Performing a Life Cycle Analysis (LCA) to Determine the Carbon Footprint for Medium Cements from the CEM I, CEM II and CEM III Groups Produced in Poland, in Accordance with PN-EN 15804: 2012 Work Number: 01929/12/ZOONF*; Institute of Building Technique: Warsaw, Poland, 2012.
11. Rudnicki, T. The Influence of the Type of Cement on the Properties of Surface Cement Concrete. *Materials* **2022**, *15*, 4998. [CrossRef] [PubMed]
12. Kim, T.; Chae, C. Evaluation analysis of the CO_2 emission and absorption life cycle for precast concrete in Korea. *Sustainability* **2016**, *8*, 663. [CrossRef]
13. Kashef-Haghighi, S.; Shao, Y.; Ghoshal, S. Mathematical modeling of CO_2 uptake by concrete during accelerated carbonation curing. *Cem. Concrete Res.* **2015**, *67*, 1–10. [CrossRef]
14. Turner, L.K.; Collins, F.G. Carbon dioxide equivalent (CO_2) emissions: A comparison between geopolymer and OPC cement concrete. *Constr. Build. Mater.* **2013**, *43*, 125–130. [CrossRef]
15. Babu, K.G.; Kumar, V.S.R. Efficiency of GGBS in concrete. *Cem. Concr. Res.* **2000**, *30*, 1031–1036. [CrossRef]
16. Sjunnesson, J. *Life Cycle Assessment of Concrete*; Lund University: Lund, Sweden, 2005.
17. Mohammadi, J.; South, W. Life cycle assessment (LCA) of benchmark concrete products in Australia. *Int. J. Life Cycle Assess.* **2017**, *22*, 1588–1608. [CrossRef]
18. Załegowski, K.; Jackiewicz-Rek, W.; Garbacz, A.; Courard, L. Carbon footprint of concrete (in Polish). *Mater. Bud.* **2013**, *12*, 34–36.
19. Decky, M.; Papanova, Z.; Juhas, M.; Kudelcikova, M. Evaluation of the Effect of Average Annual Temperatures in Slovakia between 1971 and 2020 on Stresses in Rigid Pavements. *Land* **2022**, *11*, 764. [CrossRef]
20. Rudnicki, T.; Jurczak, R. Recycling of a Concrete Pavement after over 80 Years in Service. *Materials* **2020**, *13*, 2262. [CrossRef] [PubMed]
21. Korentz, J.; Jurczak, R.; Szmatuła, F.; Rudnicki, T. Właściwości nawierzchni betonowej autostrady A18 po 82 latach eksploatacji. *Bud. Technol. Archit.* **2021**, *4*, 68–71.
22. Glinicki, M.A. Właściwości betonu nawierzchniowego z kruszywem odkrytym—Wpływ rodzaju cementu i pielęgnacji. *Drogownictwo* **2019**, *4*, 99–104.
23. Rudnicki, T.; Jurczak, R. The impact of the addition of diabase dusts on the properties of cement pavement concrete. *Arch. Civ. Eng.* **2022**, *68*, 395–411. [CrossRef]
24. Szcześniak, A.; Zychowicz, J.; Stolarski, A. Influence of Fly Ash Additive on the Properties of Concrete with Slag Cement. *Materials* **2020**, *13*, 3265. [CrossRef]
25. Nadeem, A.; Memon, S.A.; Lo, T.Y. The performance of Fly ash and Metakaolin concrete at elevated temperatures. *Constr. Build. Mater.* **2014**, *62*, 67–76. [CrossRef]
26. Dzięcioł, J.; Radziemska, M. Blast Furnace Slag, Post-Industrial Waste or Valuable Building Materials with Remediation Potential. *Minerals* **2022**, *12*, 478. [CrossRef]
27. Qi, J.; Jiang, L.; Zhu, M.; Mu, C.; Li, R. Experimental Study on the Effect of Limestone Powder Content on the Dynamic and Static Mechanical Properties of Seawater Coral Aggregate Concrete (SCAC). *Materials* **2023**, *16*, 3381. [CrossRef] [PubMed]
28. Małek, M.; Łasica, W.; Jackowski, M.; Kadela, M. Effect of Waste Glass Addition as a Replacement for Fine Aggregate on Properties of Mortar. *Materials* **2020**, *13*, 3189. [CrossRef] [PubMed]
29. Kaszuba, S. *Kształtowanie Składu Trwałego Betonu z Udziałem Cementów Wieloskładnikowych (CEM II, CEM III) do Zastosowania w Budownictwie Drogowo-Mostowym*; Praca doktorska, Politechnika Śląska: Gliwice, Poland, 2019.
30. Szcześniak, A.; Siwiński, J.; Stolarski, A. Effect of Aggregate Type on Properties of Ultra-High-Strength Concrete. *Materials* **2022**, *15*, 5072. [CrossRef] [PubMed]

31. Gołaszewski, J. *Domieszki Do Betonu: Efekt Działania, Ocena i Badania Efektywności, Stosowanie*; Wydawnictwo Politechniki Śląskiej: Gliwice, Poland, 2016.
32. Pasławski, J.; Rudnicki, T. Agile/Flexible and Lean Management in Ready-Mix concrete delivery. *Arch. Civ. Eng.* **2021**, *67*, 689–709. [CrossRef]
33. Giergiczny, Z.; Baran, T.; Dziuk, D.; Ostrowski, M. The increase of concrete frost resistance by using cement with air-entraining agent. *Cem. Wapno Beton* **2016**, *21*, 96–105.
34. Giergiczny, Z.; Glinicki, M.A.; Sokołowski, M.; Zieliński, J. Air void system and frost salt scaling of concrete containing slag blended cement. *Constr. Build. Mater.* **2009**, *23*, 2451–2456. [CrossRef]
35. Rudnicki, T.; Stałowski, P. Fast-Setting Concrete for Repairing Cement Concrete Pavement. *Materials* **2023**, *16*, 5909. [CrossRef]
36. *PN-EN 197-1:2012*; Cement—Część 1: Skład, Wymagania i Kryteria Zgodności Dotyczące Cementów Powszechnego Użytku. European Committee for Standardization: Brussels, Belgium, 2012.
37. *EN 934-1:2008*; Admixtures for Concrete, Mortar and Grout—Part 1: Common Requirements. European Committee for Standardization: Brussels, Belgium, 2008.
38. *EN 934-2:2009+A1:2012*; Admixtures for Concrete, Mortar and Grout—Part 2: Concrete Admixtures—Definitions, Requirements, Conformity, Marking and Labelling. European Committee for Standardization: Brussels, Belgium, 2012.
39. GDDKiA. *Warunki Wykonania i Odbioru Robot Budowlanych D-05.03.04 Nawierzchnia z Betonu Cementowego*; GDDKiA: Warszawa, Poland, 2017.
40. Stałowski, P.; Pawlik, M. Technical requirements for the prevention of aggregate reactivity and the availability of materials. In Proceedings of the14th International Symposium on Concrete Roads, Kraków, Poland., 25–28 June 2023.
41. *EN 12390-2:2019*; Testing Hardened Concrete—Part 2: Making and Curing Specimens for Strength Tests. European Committee for Standardization: Brussels, Belgium, 2019.
42. *EN 12390-3:2019*; Testing Hardened Concrete—Part 3: Compressive Strength of Test Specimens. European Committee for Standardization: Brussels, Belgium, 2019.
43. *EN 12390-5:2019*; Testing Hardened Concrete—Part 5: Flexural Strength When Splitting Test Specimens. European Committee for Standardization: Brussels, Belgium, 2019.
44. *EN 12390-6:2019*; Testing Hardened Concrete—Part 6: Tensile Splitting Strength of Test Specimens. European Committee for Standardization: Brussels, Belgium, 2019.
45. *EN 12350-6:2019*; Testing Fresh Concrete. Density; Comite Europeen de Normalisation: Brussels, Belgium, 2019.
46. *PN-B-06250:1988*; Polish Standard Normal Concrete. PKN: Warsaw, Poland, 1988. (In Polish)
47. Catalogue of Typical Structures of Rigid Pavements. GDDKiA, Warszawa, GDDKiA. 2014. Available online: https://www.gddkia.gov.pl/frontend/web/userfiles/articles/d/dokumenty-techniczne_8162/Dokumenty%20techniczne/KTKNS.pdf (accessed on 30 September 2020).
48. *EN 12390 PKN-CEN/TS 12390-9*; Testing Hardened Concrete—Part 9: Freeze-Thaw Resistance. Comite Europeen de Normalisation: Brussels, Belgium, 2016.
49. *EN 480-11:2008*; Admixtures for Concrete, Mortar and Grout-Test Methods—Part 11: Determination of Air Void Characteristics in Hardened Concrete. Comite Europeen de Normalisation: Brussels, Belgium, 2008.
50. *ISO 14067:2016*; Greenhouse Gases—Carbon Footprint of Products—Requirements and Guidelines for Quantification. International Organization for Standardization: Brussels, Belgium, 2016.

materials

MDPI

Article

Advancements in Heavy Metal Stabilization: A Comparative Study on Zinc Immobilization in Glass-Portland Cement Binders

Abdelhadi Bouchikhi [1,*], Amine el Mahdi Safhi [2], Walid Maherzi [3], Yannick Mamindy-Pajany [3], Wolfgang Kunther [4], Mahfoud Benzerzour [3] and Nor-Edine Abriak [3]

1 Laboratoire des Infrastructures Intelligentes et des Technologies de l'Environnement Connectés (LabI2TEC), Institut Supérieur du Bâtiment et des Travaux Publics, F-13009 Marseille, France
2 Department of Building, Civil, and Environmental Engineering, Concordia University, Montreal, QC H3G 1M8, Canada; amineelmahdi.safhi@concordia.ca
3 Institut Mines-Télécom Lille Douai, Université de Lille, ULR 4515—LGCgE, F-59000 Douai, France; walid.maherzi@imt-nord-europe.fr (W.M.); yannick.mamindy@imt-lille-douai.fr (Y.M.-P.); mahfoud.benzerzour@imt-nord-europe.fr (M.B.); nor-edine.abriak@imt-nord-europe.fr (N.-E.A.)
4 Materials and Durability, Department of Environmental and Resource Engineering, Technical University of Denmark, 2800 Kongens Lyngby, Denmark; wolku@dtu.dk
* Correspondence: abdelhadi.bouchikhi@cciamp.com

Abstract: Recent literature has exhibited a growing interest in the utilization of ground glass powder (GP) as a supplementary cementitious material (SCM). Yet, the application of SCMs in stabilizing heavy metallic and metalloid elements remains underexplored. This research zeroes in on zinc stabilization using a binder amalgam of GP and ordinary Portland cement (OPC). This study juxtaposes the stability of zinc in a recomposed binder consisting of 30% GP and 70% OPC (denoted as 30GP-M) against a reference binder of 100% CEM I 52.5 N (labeled reference mortar, RM) across curing intervals of 1, 28, and 90 days. Remarkably, the findings indicate a heightened kinetic immobilization of Zn at 90 days in the presence of GP—surging up to 40% in contrast to RM. Advanced microstructural analyses delineate the stabilization locales for Zn, including on the periphery of hydrated C_3S particles (Zn–C_3S), within GP-reactive sites (Si*–O–Zn), and amid C–S–H gel structures, i.e., (C/Zn)–S–H. A matrix with 30% GP bolsters the hydration process of C_3S vis-à-vis the RM matrix. Probing deeper, the microstructural characterization underscores GP's prowess in Zn immobilization, particularly at the interaction zone with the paste. In the Zn milieu, it was discerning a transmutation—some products born from the GP–Portlandite reaction morph into GP–calcium–zincate.

Keywords: zinc stabilization; ground glass; supplementary cementitious materials; sorption isotherms

1. Introduction

Intensive research previously conducted on recycled ground glass powder (GP) as supplementary cementitious materials (SCMs) [1–4] has led to the normalization of using these materials as additives. ASTM C1866–2020 Standard has established a specification for ground-glass pozzolan used in concrete [5]. Indeed, their use has proven to be beneficial in stabilizing the structure of concrete and decreasing its carbonation and chloride diffusion [6,7]. Ground-glass pozzolan has been used both in conventional cementitious matrixes and as a source of silicate in alkali-activated binders [8–11]. Nevertheless, the effect of GP on the stabilization of pollutants, such as metallic and metalloid trace elements (MMTEs), and on organic pollutants has not been reported to date. The reactivity and constituted phases of hydrated binders are among the keys to understanding the modes and locations of pollutants fixed on a solid support. Most aluminosilicate compounds are characterized by regular micropores in the form of connected channels, where pollutants preferentially accumulate. This structural state also provides a strong ionic exchange capacity between the alkalis present in zeolites (K^+, Na^+, and Ca^{2+}) and the metallic pollutants present in water (Ni^{2+}, Zn^{2+}, Cu^{2+}, As^{5+}, U^{6+}, Cr^{3+}, Pb^{2+}, and Cs^+) [12–15]. In this type of fixing medium, cation exchange leads to extremely stable and sometimes irreversible bonds. However, these cationic exchanges are influenced by certain parameters in the matrix, particularly the pH, nature of the binder (Portland, geopolymer, or polymer), cure time, and degree of hydration [16–18].

Other compounds, such as sepiolite, geopolymeric compounds, and polymers, can also stabilize metal pollutants [19–22]. Cement hydration phases are also known for their ability to stabilize various types of pollutants. Moudilou's work on the stabilization of MMTE using Portland CEM I 52.5 N enabled the phases involved in the binding of transition metals to be determined [23]. It appears that elements such as Ni^{2+}, Zn^{2+}, Cu^{2+}, and Pb^{2+} are extremely stable in C–S–H gel (calcium silicate hydrate), while Cr^{3+} and U^{6+} are extremely stable in ettringites. Other elements are rather soluble in the interstitial water and hydroxide phases. Cu^{2+} and Pb^{2+} are particularly well adsorbed on C–S–H. Likewise, they are involved in the crystallographic structure of C–S–H, while Cd^{2+}, Ni^{2+}, and Zn^{2+} remain attached to the hydroxide phases and in the inter-granular porosities [24]. There are three mechanisms for immobilizing MMTE, including sorption, chemical incorporation, or encapsulation [19,25,26].

These attachment modes of the elements to the hydrated cementitious phases involve modifications to the crystal structure of the formed phases, heat of hydration, setting times, and durability of the matrix, including the release of the heavy metals. The presence of water plays an important role in the transport, ionization, and hydration of the substrate surface and pollutants. This allows for other types of interaction with the solid, such as adsorption, precipitation, substitutions, and inclusion. Pollutants can interact with the cement phases in several ways, including adsorption as an outer sphere complex, formation of an internal sphere complex, diffusion in the crystal lattice and isomorphic substitution, rapid lateral diffusion and formation of a surface polymer, adsorption on a mineral growth front, and surface polymer formation and incorporation into the host matrix after crystal growth. The adsorbed ion can optionally be released back into the solution by, e.g., following surface redox reactions or dynamic equilibrium. The solvency of water also influences this balance of the presence of Zn in the different phases [27].

In France, the concentrations of MMTE in wastes (e.g., sediments and industrial scraps) indicate that zinc is present at rates exceeding 5–10 times the concentrations of other MMTEs and heavy metals. To address this issue, several studies have been conducted to examine the influence of Zn in cementitious binders [28]. This study aims to analyze the effect of GP as an SCM on the binding behavior of Zn^{2+} in different hydration phases. To do this, we evaluated a matrix with a 70:30 ratio of cement to GP and a reference matrix (100% cement). Furthermore, mechanical, absorption, isothermal, and microstructural studies were conducted as a function of time. The objective of using GP is to study its influence on both the pozzolanic activity (by monitoring the compressive strength compared to a 100% cement matrix) and the stabilization of active Zn (reactivity within the matrix, i.e., pollutant immobilization).

2. Materials and Methods

2.1. Materials

The particle size distribution (PSD) analyses were conducted using the Mastersize 2000 laser (Malvern Panalytical, Malvern, UK) in dry mode. This instrument facilitates the determination of PSD spanning from 0.04 μm to 2000 μm. The test adheres to the ISO 13320 standard [29]. The cement used in this study was CEM I 52.5 N, containing 95% clinkers and a 5% calcium sulfate additive. The GP was crushed finely on a laboratory scale (d_{90} < 15 μm) before being washed with demineralized water. To ensure high solubility of the contaminant in water, an ultra-pure zinc sulfate hexahydrate ($Zn(N_2O_6)_2,6H_2O$) was used. To ensure a homogeneous presence of Zn in the paste, the pollutant was dissolved beforehand in the mixing water. Different percentages by weight of the binder ($\frac{1}{4}$, $\frac{1}{2}$, 1, 2, and 3%) of $Zn(SO_4)_2$ were added. Normalized sand, in accordance with the standard CEN 196-1 ISO [30], was used to produce 40 × 40 × 160 mm^3 mortars.

2.2. Methods

The compressive strength was tested at different ages of curing (1, 7, 28, and 90 days) on three cubes of 40 mm^3, prepared according to EN 196-1 standard [30] and using an INSTRON press (Norwood, MA, USA). The stabilization of Zn was monitored by measuring the leached Zn in demineralized water at 20 ± 2 °C using a ROTAX instrument (VELP Scientifica Srl, Usmate Velate, Italy) programmed to run for 1 turn per minute (a leaching test on a 1:10 ratio of solid to liquid) using inductively coupled plasma mass spectrometry (ICP-MS) analysis employing an Agilent Technologies SPS4 Autosampler. Thermogravimetric analysis (TG) was performed with a QMS 403D-NETZSCH machine (NLIR ApS, Farum, Denmark) using argon as the atmosphere at a flow rate of 20 mL/min. X-ray diffraction analysis (XRD) was performed using a D8 Focus diffractometer from Bruker (Billerica, MA, USA) that was equipped with a cobalt anode ($\lambda K\alpha 1$ = 1.74 Å) and Lynx Eye detector. Scanning Electron Microscopy (SEM) analysis was conducted using a Hitachi S-4300SE/N (Hitachi, Chiyoda-ku, Tokyo). The heat of hydration evolution was measured on pastes. Figure 1 illustrates the experimental design adopted in this study.

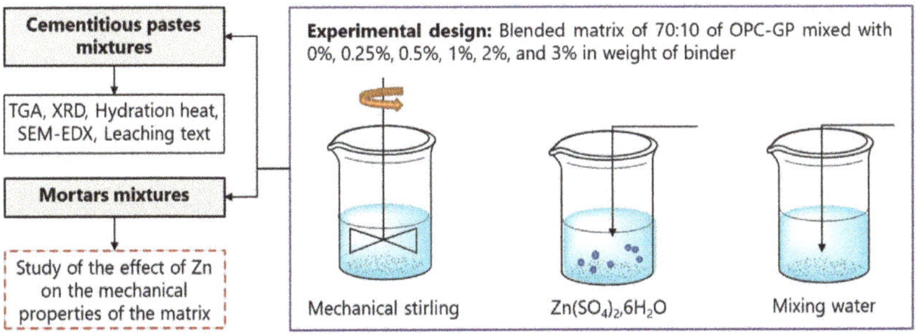

Figure 1. Experimental design that was adopted in this study.

2.3. Data Analysis

The Zn distribution between the liquid phase and the binder sorbent (adsorption on surfaces) was calculated using Equation (1). The first-order Lagergren equation, the second-order kinetics model, and the Bingham equations are presented in linear form in Equations (2) (second order) and (3) (pseudo-second order). Here, qt is the sorbed concentration in μg/mg at time t (h), qm is the pseudo-equilibrium sorbed concentration (μg/mg), and k_1 (h^{-1}), k_2 ($g{\cdot}g^{-1}\ h^{-1}$), and k_b ($g{\cdot}g^{-1}\ h^{-1}$) are the sorption rate constants. In Equation (1), K_d is the solid–liquid distribution coefficient (L/kg), q_s is the sorbed concentration (mg/kg), and q_e is the equilibrium concentration (mg/L).

$$LogK_d = log(q_s{\cdot}q_e) \tag{1}$$

$$log(qt) = log(k_d) + \frac{1}{n}log(t) \tag{2}$$

$$\frac{t}{qt} = \frac{1}{k_2 qm^2} + \frac{1}{qm}t \tag{3}$$

3. Results and Discussion

3.1. Physicochemical Properties

The physicochemical properties of GP and CEM are presented in Table 1. The GP is mostly composed of SiO_2 (71 wt.%), 11 wt.% Na_2O, and 13 wt.% CaO, with low contents of

MgO, Fe_2O_3, SO_3, and K_2O. The cement is principally composed of CaO (61 wt.%), SiO_2 (18 wt.%), Al_2O_3 (4.4 wt.%), SO_3 (3.7 wt.%), and Fe_2O_3 (3.0 wt.%). The cement phases included 59% of C_3S, 13% of C_2S, 8.8% of C_3A, and 10% of C_4AF using Bogue formulas. The PSD of those powders is presented in Figure 2.

Table 1. Physicochemical properties of the used powders.

Parameters		GP	CEM	Standard/Method
Physical Properties	Absolute density (g/cm^3)	2.54	3.15	NF EN 1097-7 [31]
	SSA BET (m^2/kg)	793	835	NF EN ISO18757 [32]
	Fire loss % (450 °C/3 h)	0.02	–	NF EN ISO 17892-12 [33]
	Fire loss % (550 °C/1 h)	0.03	–	NF EN 15169 [34]
	Fire loss % (1000 °C/1 h)	–	0.30	NF EN 1097-7 [31]
Chemical Composition wt.%	SiO_2	70.8	17.4	measured by XRF
	Al_2O_3	1.70	4.44	
	MgO	1.20	0.85	
	Fe_2O_3	0.40	2.97	
	CaO	11.5	60.7	
	Na_2O	13.0	0.31	
	K_2O	0.70	1.16	
	SO_3	–	3.75	
	Others	0.01	0.90	

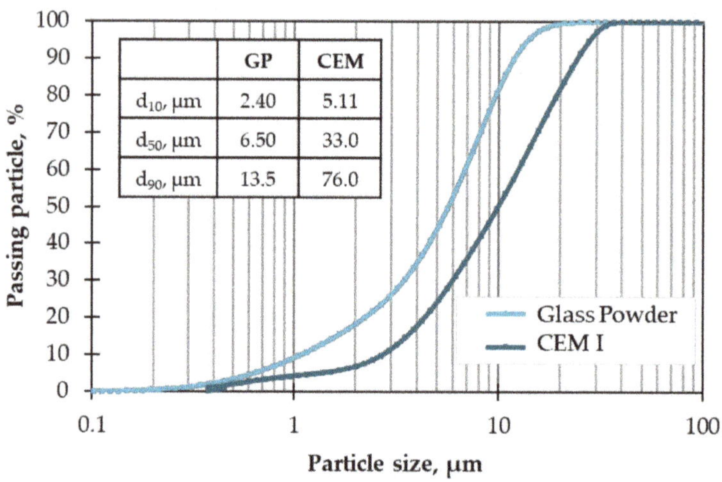

Figure 2. PSD of GP and CEM.

3.2. *Compressive Strength Evolution*

The compressive strengths obtained between 1 and 90 days indicate changes as a function of time and dosage of Zn (Figure 3). After 1 day of curing, the samples containing 2% and 3% of Zn showed large decreases in strength as low as −23 MPa compared to the reference matrix (0RM) (90% compared to 0% of Zn). In these samples, the slow hydration leads to reduced zinc stabilization in the mortars. These results are correlated with an extremely low heat of hydration and are reflected in the setting delay by more than 40 h compared to the 3 h for 0RM. A progressive decrease was observed in UCS of approximately −7 MPa between the 0.25% and 1% dosages. For comparison, 1% of Zn in a hydraulic cementitious mix affects the compressive strength of the hydraulic matrix equivalently to a 25% substitution of fly ash or glass aggregates [1,35]. After 90 days of curing, while all the samples showed improvements in compressive and flexural strengths, the 30% GP-mortars were slightly better compared with the reference matrix series.

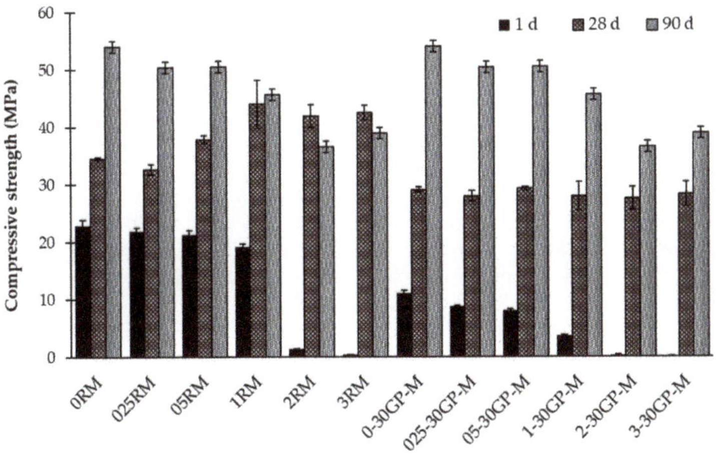

Figure 3. Compressive strength as a function of Zn-to-cement weight.

3.3. Fixation Isotherms

The leaching test results at 28 and 90 days (Figure 4) represent the importance of curing time to the fixation of Zn, which is much more important at later ages. At 20 days of curing at 20 °C and RH > 90%, Zn is more stabilized in the reference mortar (RM) than in the composed-binder mortar 30GP-M. However, at 90 days of curing, Zn was more stabilized in the composed-binder matrix. This improvement in the fixation of the blended matrix is linked to the reactivity of the GP in the cement paste after 90 days. At 28 days of curing, the RM produced more C–S–H gel compared to the 30GP-M, resulting in more Zn incorporated or adsorbed in the C–S–H gel silicate sheets in the RM. The presence of Zn in a bivalent cationic state, Zn(II), allows an immediate reaction and integrates the C–S–H gel structures. This strong reactivity is mainly linked to the solubility of $Zn(SO_4)_2$ (96.5 g/100 g of water at 20 °C) and to a very basic pH in ordinary Portland cement. Through long curing times (explained by the slow kinetics of GP), GP reacts as a pozzolan and interacts with the Portlandite, as well as with the different forms of Zn, such as $Zn(OH)_2$ and $ZnCa(OH)_2$.

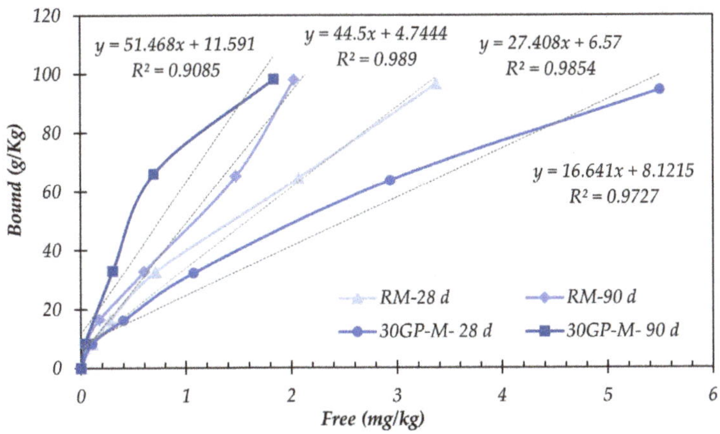

Figure 4. Fixed Zn in the function leached Zn in the samples.

The stabilization of Zn is linked to the duration of the curing time. To establish this principle, a fixation study was conducted. The graphical presentation of *log(Qt)* as a function of *log(t)* was calculated according to the secondary derivative of the Bangham equation as follows:

$$logQt = log(kb) + \frac{1}{n} * log(t) \tag{4}$$

$$logQt = log(kd) + \frac{1}{n} * log(t) \tag{5}$$

The linear relationship between (q_t) and log(t), which allows for the determination of the constant solid–liquid distribution coefficient (k_d), indicates the capacity of the binder to sorb Zn. Table 2 and Figure 5 present the linear equations, k_b coefficients, and R^2 for Equation (4) above. The low concentration of Zn in the medium led to an increase in the binding capacity of the binders hydrated with 0.25 and 0.5% Zn. High concentrations of Zn (2–3%), which delay the setting and hydration of the binders, hinder its stabilization.

Table 2. Details of the Bangham fit (solid lines) are reported in Figure 5.

	Zn/Binder w%	Y = Log(Qt)	Log(kd)	kd	R^2
Reference Mortar	0.25%	y = −0.46x − 2.82	−2.82	1.50×10^{-3}	0.98
	0.50%	y = −0.46x − 2.44	−2.44	3.64×10^{-3}	0.98
	1%	y= −0.36x − 2.24	−2.24	5.75×10^{-3}	0.88
	2%	y = −0.40x − 1.57	−1.57	2.70×10^{-2}	0.97
	3%	y = −0.49x − 1.29	−1.29	5.13×10^{-2}	0.99
Blended Mortar	0.25%	y = −0.64x − 2.34	−2.34	4.53×10^{-3}	0.98
	0.50%	y = −0.50x − 2.07	−2.07	8.47×10^{-3}	0.78
	1%	y = −0.51x − 1.81	−1.81	1.55×10^{-2}	0.98
	2%	y = −0.60x − 1.22	−1.22	6.00×10^{-2}	0.99
	3%	y = −0.60x − 1.00	−1.00	9.88×10^{-2}	0.92

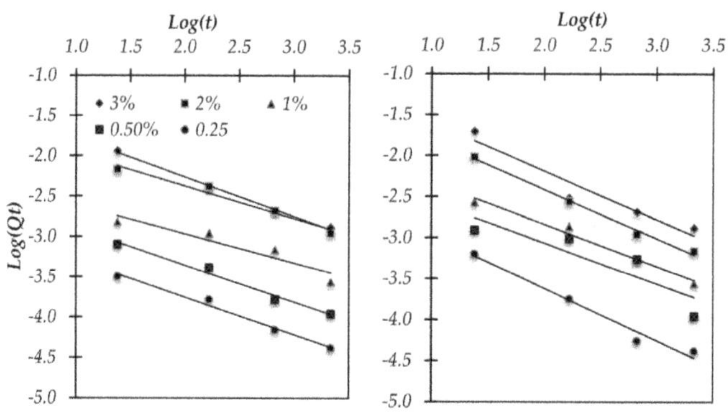

Figure 5. Bangham fits (solid lines) for adsorption of Zn as a function of time for each concentration of Zn. (**left**) in the MR hydrated binder and (**right**) in the 30GP-M hydrated binder.

The reactivity of GP in the cementitious medium is linked to the interface between the GPs and contaminated paste by Zn^{2+}. The nucleation site density at the surface determines the glass's capacity for stabilization. For this, the k_d of the glass is linked to its specific surface area rather than the quantity or mass of substitution.

The precipitation of an amorphous layer of $Zn(OH)_2$ and the hydroxide crystalline calcium zinc around the cement particles suggest very little pozzolanic activity over time. This explains the loss of mechanical properties observed with the increase in Zn dosage (Figure 6). The compression strength obtained at 28 days decreased proportionally with the addition of Zn concentration, e.g., 1RM, 2RM, and 3RM. A significant decrease in compressive strength down to −15 MPa was observed compared to the 0RM reference matrix, which indicates negative effects on the dissolution of the cement grains and, consequently, on the hydration coefficient [36,37]. A decrease in the cement amount in the mortars can also influence the sample's mechanical properties; however, this factor remains negligible compared to the influence of zinc on the hydration of cement.

Figure 6. SEM observations: (**right**) anhydrous cement, (**left**) Zn–glass reaction at 90 days.

3.4. Heat Hydration Evolution

The heat of hydration of the pastes with different Zn substations is presented in Figure 7. The calorimetric variation provides information about the reaction kinetics according to the evolution of Zn concentration. First, when observing the control sample (with 0% Zn) through the lens of conventional hydration stages, the heat calorimetry measured indicates a correlation with the early stages of clinker hydration, specifically C_3S and C_3A hydration characterized by exothermal reactions up to 2 h (stage 3) and 12 h (stage 4). Second, specifically for the Zn^{2+} concentration that varied between 0.25% and 1%, the heat of hydration decreased, and the samples passed through stages 3 and 4 more and more rapidly compared to the control sample (0% of Zn). At this concentration interval, the effect of Zn is clearly observed in the degree of hydration of C_3S and C_3A and as a delay in the start of the reaction. The initial hydration of C_3S required 5 h for 025RM, 10 h for 05RM, and 18 h for 1RM. The third interval is partially characterized by a complex perturbation in the hydration of paste, which produces an immediate exothermal reaction that decreases after the hydration of C_3S and C_3A is produced. This interval is also characterized by the hydration of C_3A and C_3S at 24 and 40 h for 2RM and 3RM, respectively. The compressive strength at 24 h shows a correlation with the hydration results (Figure 3). The reduced compressive strength and the degree of absence of heat at the hydration concentration of Zn are clearly noted.

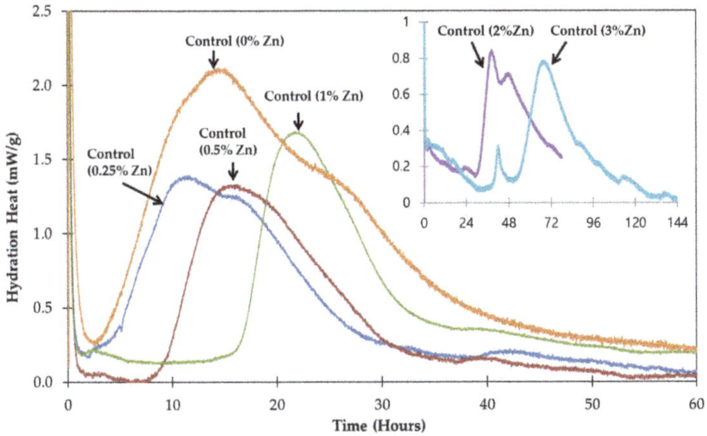

Figure 7. Heat of hydration of different cement pastes doped by 0% to 3% of Zn.

3.5. Mineralogical Properties

XRD analyses were performed at 28 and 90 days, as represented in Figure 8. The results show the occasional appearance of certain precipitation phases of $Zn(OH)_2$, $CaZn_2(OH)_6{\cdot}2H_2O$ and $Ca(OH)_2$, hydrozincate $CaZn_2(OH)_6{\cdot}2H_2O$, and hydrozincite $Zn_5(OH)_6(CO_3)_2$. Zn^{2+} takes several forms depending on its chemical ties with its environment. The presence of C_3A also promotes the formation of hydrozincate-dehydrated ($CaZn_2(OH)_6{\cdot}2H_2O$) [37,38].

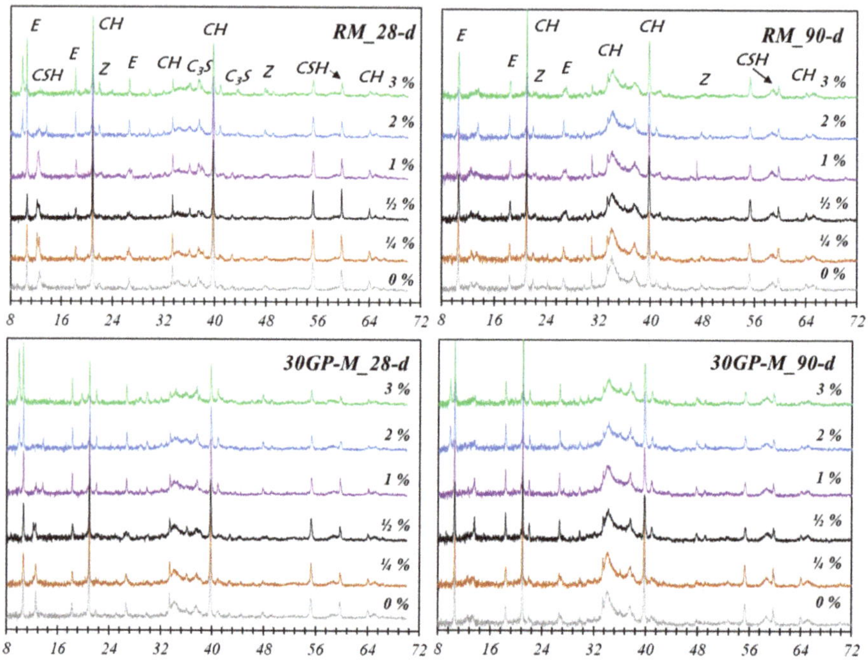

Figure 8. XRD analyses (2θ°) in different Zn concentrations for the RM and 30GP-M, where E = Ettringite and Z = $CaZn_2(OH)_6{\cdot}2H_2O$.

The hydraulic binders are characterized by a basic medium rich in Ca^{2+} and the presence of zinc-form compounds of γ-$Zn(OH)_2$ and Zn-Ca type. Curing Zn-Si bonds, such as C–S–H gel (in the interlayer), form a simple sorption as no exchange has been observed with Ca^{2+} [39]. This was confirmed by Moudilou [23], explained by trapping in the C–S–H gel interlayers with a fixing mechanism on the outer silicon of the C–S–H gel. The presence of amorphous silicon in the glass powder GP leads to a stable chemical bond with zinc (SiOZn), which explains the appearance of crystallization peaks in the samples with high Zn content. Moulin et al. [28] show that Zn stabilizes on tetrahedral silicon SiOZn. Amorphous silica in solid solutions promotes Zn-Si(GV)-type bonds (in the amorphous silicate in GP).

The amphoteric characteristic of Zn allows it to form morphological variations depending on the pH and the concentration of Zn in the solution. For a very high pH corresponding to that of cement (~13), Zn takes the form $Zn(OH)_2$; for a pH less than 13.65, Zn takes the form of $Zn(OH)_4^{2-}$ [23]. The ionic form of Zn (Zn^{2+}) exhibits strong reactivity, translated as amorphous phases formed at the aggregate surface. These phases are characterized by a waterproof surface that delays the hydration binder [36,37,40,41].

- 13.65 < pH

$$Zn(OH)_4^{2-} + 4H_3O^+ + 2e^- \rightarrow Zn + 8H_2O$$

$$E = E_0 \quad Zn(OH)_4^{2-}/Zn + 0.03\log\left[Zn\,(OH)_4^{2-}\right]\left[H_3\,O^+\right]_4$$

$$E = E_0 \quad Zn(OH)_4^{2-}/Zn - 0.12pH + 0.03\log\left[Zn\,(OH)_4^{2-}\right]$$

$$E = E_0 \quad Zn(OH)_4^{2-}/Zn - 0.12pH + 0.03\log 0.02$$

- 6.55 < pH < 13.65

$$Zn(OH)_2 + 2H_3O^+ + 2e^- \rightarrow Zn + 4H_2O$$

$$E = E_0 \quad \frac{Zn(OH)_2}{Zn} + 0.03\log\left[H_3O^+\right]^2$$

$$E = E_0 \quad \frac{Zn(OH)_2}{Zn} - 0.06\ pH$$

In basic solutions, $Zn(OH)_2$ takes several crystalline forms. For a pH less than 12, $Zn(OH)_2$ has an insoluble amphoteric character. For a pH of 12.8, some researchers suggest that Zn exhibits a contradictory effect because both Zn hydroxides are only formed in the presence of OH^-. In other words, after the hydration of the main phases of C_3S and C_2S cement in acidic solutions, the presence of Zn slows the hydration of these cementitious phases [42]. However, the rapid precipitation of $Zn(OH)_2$ on the anhydrous grains, which occurs within a few hours, creates setting delays that can continue for several days. The hydration of the anhydrous phases, therefore, depends on the permeability of this film and on the pH, or the initial formation of Ca^{2+}. Incidentally, OH^- is dislocated during the transformation of $Zn(OH)_2$ into $CaZn_2(OH)_2{\cdot}2H_2O$ and reinforces the encapsulation of the C_3S grains, preventing the transport of water, which is necessary for their continuous hydration [43]. The first has been the calcium-crystallized former [42].

$$2Zn(OH)_{3-} + Ca^{2+} + 2H_2O \rightarrow Ca\ Zn_2\ (OH)_6,\ 2H_2O$$

Intermediate appearances of $Zn(OH)_2$

$$Zn^{2+} + 2HO- \leftrightarrow Zn(OH)_2 \leftrightarrow ZnO(s) + H_2O \tag{6}$$

$$Ks = [Zn^{2+}][OH^-]^2 \tag{7}$$

3.6. TG Analyses on Hardened Pastes

TG analyses for samples cured between 28 and 90 days show that the presence of Zn^{2+} and $N_2O_6{}^{2-}$ in different concentrations influences the degree of hydration of the cement phases (Figure 9). The results show a change in the loss of mass over temperature intervals in relation to the concentration of Zn. For the two curing times, the matrices develop Zn–AFm compounds (characterized by mass losses between 250 °C and 300 °C) as a function of the Zn concentration. The formation of other hydrates decreases with the Zn concentration increase, such as the formation of C–S–H gel hydration gels (mass losses at 150 °C) and Portlandite CH (mass losses between 450 °C and 600 °C). This observation remains compatible with the blocking nature of the dissolution of C_3S. The increased amount of Zn led to small decreases in decomposition temperature at the hydrate compounds. The evolution of samples containing 0% to 3% Zn involved a decrease in temperature at 50 °C and 25 °C for the degradation of compounds presented at 50 °C and 180 °C, respectively.

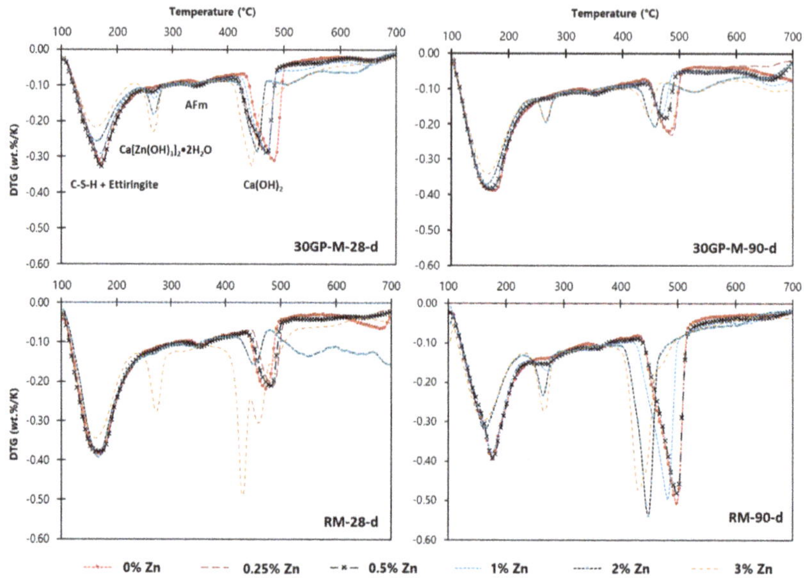

Figure 9. DTG cures of cement pates (100–700 °C).

3.7. FTIR Analyses

The presence of Zn in the hydrate products of cement leads to the appearance of bands specific to zincate compounds. In this part, the objective is to identify the stable compounds (samples cured for 90 days). Hence, four samples were analyzed, namely the hydrated pastes without Zn (100% CEM I 52.5 N and 70% CEM I 52.5 N + 30GP) and with 3% of Zn. The results are presented in Figure 10. FTIR analyses of Zn-based compounds, such as calcium zincate $Ca[Zn(OH)_3]_2{\cdot}2H_2O$, show identical results or IR curves for the hydrated cement pastes (95% clinker), such as those shown in Figure 10, with wavenumbers between 3000 and 3600 cm^{-1}, which are related to the hydration cement products [44]. The spectra characterized by peaks at 650 and 1650 cm^{-1} are shared by the hydration products and non-hydrated cement phases [28,41,45]. The 620 and 1670 cm^{-1} wavenumber boundaries are specific to the combination of hydrated cement phases and Zn.

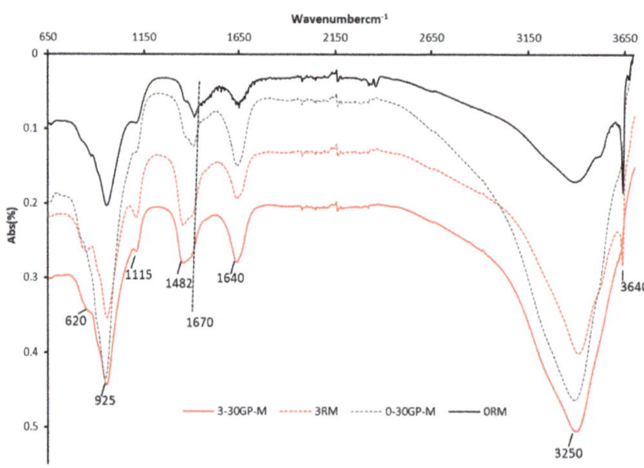

Figure 10. IR patterns (spectra) of cement paste without and with 3% of Zn.

3.8. SEM–EDS Analyses

SEM analysis was conducted to observe the GP–paste interface for sample 30GP-M at 3% Zn. As shown in Figure 9, several types of reactions were observed. The first type was in the GP particles, where nucleation sites were observed on the glass surface. The horizontal observation shows two types of morphological germination: spherical and tubular-shaped. These shapes are not related to the initial shape of the glass particles, which results in the germination of the GP. The vertical observation shows inter-diffusion between the GP and the chemical composition of the exterior phases. Beyond PG, more than the classic form of hydration of cement, the filleted structures. The presence of Ca^{2+} and OH- in high concentrations in the cement leads to the formation of a stable phase of calcium hydroxyzincate as a precipitate, which allows the Zn to pass from the amorphous structure ($Zn(OH)_2$) to the calcium hydroxyzincate-type crystalline structure, according to the reaction below:

$$2\ Zn(OH)^{2-}\ + Ca^{2+} + 2\ H_2O{\cdot}CaZn(OH)_4,\ 2\ H_2O \quad (8)$$

Calcium zincate can take several forms, depending on the treatment conditions and the concentration of elements, with shapes varying from haliatory leaves to well-structured leaves [46].

The SEM–EDS analyses were conducted on the 30GP-M-90d paste (Figures 11 and 12). These samples show rich Ca and Zn germination around GPs. The gray spots indicate anhydrate cement particles that are surrounded by Zn. Around the GP particles, the compatibility of germination between Zn and Ca around GPs. The localization of Zn in the hydrated paste is almost homogeneous, which was explained by the addition of mixing water when the hydration had started. Finally, sorption isotherms of Zn^{2+} were conducted on unhydrated (C_3S), hydrated gel (C–S–H gel), and on the GP surface (Si*–O–Zn). Two types of Zn germination are observed on the surfaces of the GP, noted as ϕ_1 and ϕ_2. Si*-($ZN(OH)_2$) was observed to be bulky compared to the compounds of the Zn–O–Si type [47].

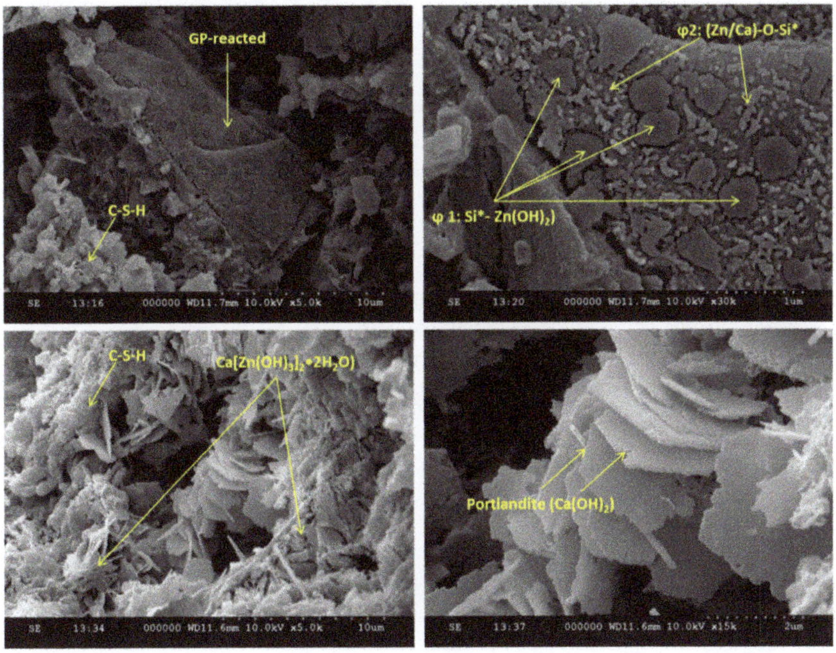

Figure 11. SEM images of sample 3MR3PG were cured for 90 days.

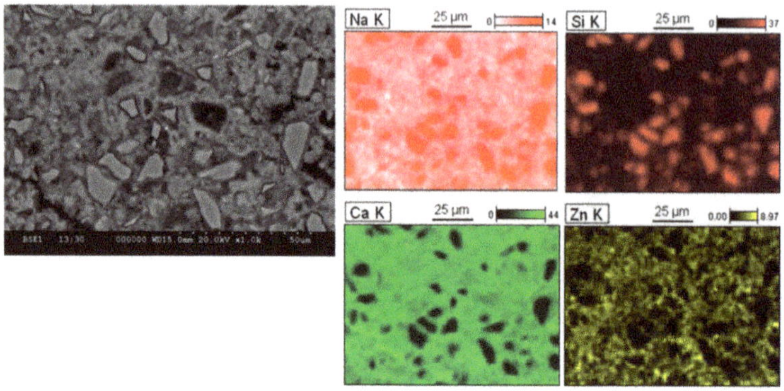

Figure 12. SEM–EDS analysis of polished surface sample 3MR3PG cured for 90 days.

4. Discussion

An approximate model of the binding of Zn^{2+} by the C–S–H gel hydration phase resulting from the hydration of the cement as well as by the glass aggregates is given in Figure 13. Microstructure studies and approaches described in the literature [28,48] make it possible to propose methods of binding Zn in the structure of C–S–H gel. Note that the glass aggregates after 90 days of cure are also involved in the fixing of Zn. The presence of GP gives a new mode of fixation of Zn^{2+}, and the GP surface can immobilize this metal cation and help the hydration of cement to start progressively. The presence of Zn^{2+} in a cementitious matrix leads to the formation of an impermeable film around the cement and glass particles, which blocks their hydration and subsequently hinders the process of forming a hydration gel (C–S–H) in large quantities. This impermeabilization primarily occurs at the interface, where Zn integrates into the C–S–H gel layers. In the presence of Ca^{2+}, glass undergoes pozzolanic reactivity, while the presence of Zn^{2+} introduces competition on the reaction surface with glass. This results in a strong dynamism in the reaction mechanism and the amount of Zn that is fixed. This mechanism is not well documented in the literature. Based on analytical results and the reaction mechanism of GP on the contact surface, Figure 13 presents an approximate mechanism of GP reactivity as well as cement grain interactions.

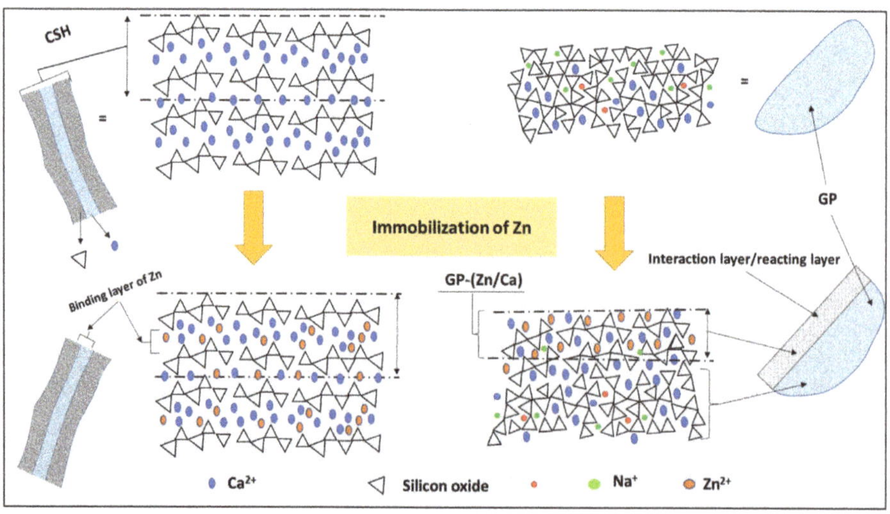

Figure 13. Approximate model of Zn^{2+} fixation on C–S–H gel and glass aggregates.

5. Conclusions

This study shows the advantage of using glass in combination with hydraulic binders for Zn stabilization. The results of this study are detailed below.

- The presence of Zn in hydroxide form influences the setting time of Portland cement. Delays in the setting time can occur up to 24 h (for Zn concentrations between 2% and 3% relative to the cement). The presence of glass only influences the setting properties after ~90 days of curing. For shorter hardening times, the reactivity of the glass is lower than that of the cement particles. The high concentration of Zn (from 2%) significantly retarded the hydration of C_3S (>40 h).
- The glass aggregates have the capacity to immobilize Zn after 90 days of curing. The glass forms irreversible phases on the layers in contact with the pastes. Stabilization only occurs through inter-diffusion or dissolution in a solid base of GP. TG analysis shows the influence of GP on the formation of hydrated phases, which is crucial for Zn remediation. As such, the presence of glass promotes hydration reactions. SEM–EDS analyses confirmed surface germination reactions, and the chemical analyses confirmed the presence of phases rich in both Zn and Ca.
- The microstructural studies and approaches described in the literature [28,48] make it possible to propose methods of binding Zn in the structure of C–S–H gel. Note that glass aggregates after 90 days of curing are also involved in the fixing of Zn.
- The presence of GP provides a new pathway for the fixation of Zn^{2+}. The GP surface can immobilize the metal cation and help the hydration of cement to start gradually.

Author Contributions: Conceptualization, A.B.; methodology, A.B.; validation, W.M., Y.M.-P., W.K., M.B., and N.-E.A.; formal analysis, A.B.; investigation, A.B. and A.e.M.S.; resources, Y.M.-P. and N.-E.A.; data curation, A.B.; writing—original draft preparation, A.B.; writing—review and editing, A.e.M.S., Y.M.-P., and W.K.; visualization, A.B. and A.e.M.S.; supervision, W.M., Y.M.-P., and M.B.; project administration, W.M. and N.-E.A.; funding acquisition, N.-E.A. All authors have read and agreed to the published version of the manuscript.

Funding: This research received no external funding.

Institutional Review Board Statement: Not applicable.

Informed Consent Statement: Not applicable.

Data Availability Statement: The original contributions presented in the study are included in the article, further inquiries can be directed to the corresponding author.

Conflicts of Interest: The authors declare no conflicts of interest.

References

1. Bouchikhi, A.; Benzerzour, M.; Abriak, N.E.; Maherzi, W.; Mamindy-Pajany, Y. Study of the Impact of Waste Glasses Types on Pozzolanic Activity of Cementitious Matrix. *Constr. Build. Mater.* **2019**, *197*, 626–640. [CrossRef]
2. Idir, R.; Cyr, M.; Tagnit-Hamou, A. Use of Fine Glass as ASR Inhibitor in Glass Aggregate Mortars. *Constr. Build. Mater.* **2010**, *24*, 1309–1312. [CrossRef]
3. Lee, H.; Hanif, A.; Usman, M.; Sim, J.; Oh, H. Performance Evaluation of Concrete Incorporating Glass Powder and Glass Sludge Wastes as Supplementary Cementing Material. *J. Clean. Prod.* **2018**, *170*, 683–693. [CrossRef]
4. Ling, T.C.; Poon, C.S.; Wong, H.W. Management and Recycling of Waste Glass in Concrete Products: Current Situations in Hong Kong. *Resour. Conserv. Recycl.* **2013**, *70*, 25–31. [CrossRef]
5. *ASTM C1866/C1866M-20*; Standard Specification for Ground-Glass Pozzolan for Use in Concrete. ASTM International: Montgomery, PA, USA, 2020. [CrossRef]
6. Zidol, A. *Durabilité En Milieux Agressifs des Bétons Incorporant la Poudre de Verre*; Scientific Research: Sherbrooke, QC, Canada, 2014.
7. Dali, J.S.; Tande, S.N. Performance of Concrete Containing Mineral Admixtures Subjected to High Temperature. In Proceedings of the 37th Conference on Our World in Concrete and Structures, Singapore, 29–31 August 2012.
8. Bouchikhi, A.; Mamindy-Pajany, Y.; Maherzi, W.; Albert-Mercier, C.; El-Moueden, H.; Benzerzour, M.; Peys, A.; Abriak, N. Use of Residual Waste Glass in an Alkali-Activated Binder—Structural Characterization, Environmental Leaching Behavior and Comparison of Reactivity. *J. Build. Eng.* **2020**, *34*, 101903. [CrossRef]

9. Bouchikhi, A.; Maherzi, W.; Benzerzour, M.; Mamindy-pajany, Y.; Peys, A. Manufacturing of Low-Carbon Binders Using Waste Glass and Dredged Sediments: Formulation and Performance Assessment at Laboratory Scale. *Sustainability* **2021**, *13*, 4960. [CrossRef]
10. Torres-Carrasco, M.; Puertas, F. Waste Glass in the Geopolymer Preparation. Mechanical and Microstructural Characterisation. *J. Clean. Prod.* **2015**, *90*, 397–408. [CrossRef]
11. Luhar, S.; Cheng, T.W.; Nicolaides, D.; Luhar, I.; Panias, D.; Sakkas, K. Valorisation of Glass Wastes for the Development of Geopolymer Composites—Durability, Thermal and Microstructural Properties: A Review. *Constr. Build. Mater.* **2019**, *222*, 673–687. [CrossRef]
12. El-Kamash, A.M. Evaluation of Zeolite A for the Sorptive Removal of Cs+ and Sr2+ Ions from Aqueous Solutions Using Batch and Fixed Bed Column Operations. *J. Hazard. Mater.* **2008**, *151*, 432–445. [CrossRef]
13. Osmanlioglu, A.E. Treatment of Radioactive Liquid Waste by Sorption on Natural Zeolite in Turkey. *J. Hazard. Mater.* **2006**, *137*, 332–335. [CrossRef] [PubMed]
14. Davis, M.E. Zeolites and Molecular Sieves: Not Just Ordinary Catalysts. *Ind. Eng. Chem. Res.* **1991**, *30*, 1675–1683. [CrossRef]
15. Katzer, J.R. Atlas of Zeolite Structure Types. W. M. Meier, and D. H. Olson, 2nd Rev. Ed., Butterworth, 1987. *AIChE J.* **1989**, *35*, 875. [CrossRef]
16. Nikolići, I.; Dsignurović, D.; Tadić, M.; Blečić, D.; Radmilović, V. Immobilization of Zinc from Metallurgical Waste and Water Solutions Using Geopolymerization Technology. *E3S Web Conf.* **2013**, *1*, 41026. [CrossRef]
17. Ziegler, F.; Johnson, C.A. The Solubility of Calcium Zincate ($CaZn_2(OH)_6 \cdot 2H_2O$). *Cem. Concr. Res.* **2001**, *31*, 1327–1332. [CrossRef]
18. Deschamps, T.; Benzaazoua, M.; Bussière, B.; Belem, T.; Mbonimpa, M. Mécanismes de Rétention Des Métaux Lourds En Phase Solide: Cas de La Stabilisation Des Sols Contaminés et Des Déchets Industriels. *VertigO* **2006**, *7*. [CrossRef]
19. Shi, C.; Fernández-Jiménez, A. Stabilization/Solidification of Hazardous and Radioactive Wastes with Alkali-Activated Cements. *J. Hazard. Mater.* **2006**, *137*, 1656–1663. [CrossRef] [PubMed]
20. Arliguie, G.; Grandet, J. Influence de la Composition D'un Ciment Portland Sur Son Hydratation en Presence de Zinc. *Cem. Concr. Res.* **1990**, *20*, 517–524. [CrossRef]
21. Hills, C.D.; Pollard, S.J.T. The Influence of Interference Effects on the Mechanical, Microstructural and Fixation Characteristics of Cement-Solidified Hazardous Waste Forms. *J. Hazard. Mater.* **1997**, *52*, 171–191. [CrossRef]
22. El-Eswed, B.I. Solidification Versus Adsorption for Immobilization of Pollutants in Geopolymeric Materials: A Review. In *Solidification*; IntechOpen: London, UK, 2016; Volume i, p. 13. [CrossRef]
23. Moudilou, E. Cinetiques et Mecanismes de Relargage des Metaux Lourds Presents en Traces dans les Matrices Cimentaires. Ph.D. Thesis, Université d'Orléans, Orléans, France, 2003.
24. Gineys, N.; Aouad, G.; Damidot, D. Managing Trace Elements in Portland Cement—Part I: Interactions between Cement Paste and Heavy Metals Added during Mixing as Soluble Salts. *Cem. Concr. Compos.* **2010**, *32*, 563–570. [CrossRef]
25. Chen, Q.Y.; Tyrer, M.; Hills, C.D.; Yang, X.M.; Carey, P. Immobilisation of Heavy Metal in Cement-Based Solidification/Stabilisation: A Review. *Waste Manag.* **2009**, *29*, 390–403. [CrossRef] [PubMed]
26. Abdulhussein Saeed, K.; Anuar Kassim, K.; Eisazadeh, A. Interferences of Cement Based-Solidification/Stabilization and Heavy Metals: A Review. *Electron. J. Geotech. Eng.* **2012**, *17*, 2555–2565.
27. Panfili, F. Etude de l'évolution de La Spéciation Du Zinc Dans La Phase Solide d'un Sédiment de Curage Contamine. Ph.D. Thesis, Université de Provence, Aix-Marseille, France, 2016.
28. Moulin, I.; Stone, W.E.E.; Sanz, J.; Bottero, J.Y.; Mosnier, F.; Haehnel, C. Lead and Zinc Retention during Hydration of Tri-Calcium Silicate: A Study by Sorption Isotherms and 29Si Nuclear Magnetic Resonance Spectroscopy. *Langmuir* **1999**, *15*, 2829–2835. [CrossRef]
29. *ISO 13320*; Particle Size Analysis—Laser Diffraction Methods. International Organization for Standardization: Geneva, Switzerland, 2020.
30. *NF EN 196-1*; Methods of Testing Cement—Part 1: Determination of Strength. AFNOR: Ile-de-France, France, 2016.
31. *NF EN 1097-7*; Tests for Mechanical and Physical Properties of Aggregates—Part 7: Determination of the Particle Density of Filler—Pyknometer Method. AFNOR: Ile-de-France, France, 2022.
32. *NF EN ISO 18757*; Fine Ceramics (Advanced Ceramics, Advanced Technical Ceramics)—Determination of Specific Surface Area of Ceramic Powders by Gas Adsorption Using the BET Method. AFNOR: Ile-de-France, France, 2006.
33. *NF EN ISO 17892-12*; Geotechnical Investigation and Testing—Laboratory Testing of Soil—Part 12: Determination of Liquid and Plastic Limits. AFNOR: Ile-de-France, France, 2018.
34. *NF ISO 15169*; Petroleum and Liquid Petroleum Products—Determination of Volume, Density and Mass of the Hydrocarbon Content of Vertical Cylindrical Tanks by Hybrid Tank Measurement Systems. AFNOR: Ile-de-France, France, 2004.
35. Snellings, R.; Cizer, Ö.; Horckmans, L.; Durdzi, T.; Dierckx, P.; Nielsen, P.; Van Balen, K.; Vandewalle, L. Applied Clay Science Properties and Pozzolanic Reactivity of Fl Ash Calcined Dredging Sediments. *Appl. Clay Sci.* **2016**, *129*, 35–39. [CrossRef]
36. Fernández Olmo, I.; Chacon, E.; Irabien, A. Influence of Lead, Zinc, Iron (III) and Chromium (III) Oxides on the Setting Time and Strength Development of Portland Cement. *Cem. Concr. Res.* **2001**, *31*, 1213–1219. [CrossRef]
37. Deschamps, T. Étude Du Comportement Physique Et Hydrogéochimique D'Un Dépôt De Résidus Miniers En Pâte Dans Des Conditions De Surface. Ph.D. Thesis, Université du Québec en Abitibi-Témiscamingue, Rouyn-Noranda, QC, Canada, 2009.

38. Brault, S. Modélisation Du Comportement à La Lixiviation à Long Terme de Déchets Stabilisés à l'aide de Liants Hydrauliques. Ph.D. Thesis, Université Pierre et Marie Curie, Paris, France, 2001.
39. Ziegler, F.; Scheidegger, A.M.; Johnson, C.A.; Dáhn, R.; Wieland, E. Sorption Mechanisms of Zinc to Calcium Silicate Hydrate: X-Ray Absorption Fine Structure (XAFS) Investigation. *Environ. Sci. Technol.* **2001**, *35*, 1550–1555. [CrossRef] [PubMed]
40. Kakali, G.; Tsivilis, S.; Tsialtas, A. Hydration of Ordinary Portland Cements Made from Raw Mix Containing Transition Element Oxides. *Cem. Concr. Res.* **1998**, *28*, 335–340. [CrossRef]
41. McWhinney, H.; Cocke, D.L.; Yu, G.S. Solidification of Hazardous Substances-a Tga and Ftir Study of Portland Cement Containing Metal Nitrates. *J. Environ. Sci. Health Part A Environ. Sci. Eng.* **1989**, *24*, 589–602. [CrossRef]
42. Arliguie, G.; Ollivier, J.P.; Grandet, J. Etude de l'effet Retardateur Du Zinc Sur l'hydratation de La Pate de Ciment Portland. *Cem. Concr. Res.* **1982**, *12*, 79–86. [CrossRef]
43. Šiler, P.; Kolárová, I.; Másilko, J.; Novotný, R.; Opravil, T. The Effect of Zinc on the Portlands Cement Hydration. *Key Eng. Mater.* **2018**, *761*, 131–134. [CrossRef]
44. Li, S.; Zhou, Y. Preparation of Tetragonal and Hexagonal Calcium Zincate. *Appl. Mech. Mater.* **2012**, *130–134*, 1454–1457. [CrossRef]
45. Farcas, F.; Touzé, P. La Spectrométrie Infrarouge à Transformée de Fourier (IRTF). *Bull. Lab. Ponts Chaussées* **2001**, *230*, 77–88.
46. Shangguan, E.; Li, L.; Wu, C.; Fu, P.; Wang, M.; Li, L.; Zhao, L.; Wang, G.; Li, Q.; Li, J. Microemulsion Synthesis of 3D Flower-like Calcium Zincate Anode Materials with Superior High-Rate and Cycling Property for Advanced Zinc-Based Batteries. *J. Alloys Compd.* **2021**, *853*, 156965. [CrossRef]
47. Reddy, V.A.; Solanki, C.H.; Kumar, S.; Reddy, K.R.; Du, Y.J. New Ternary Blend Limestone Calcined Clay Cement for Solidification/Stabilization of Zinc Contaminated Soil. *Chemosphere* **2019**, *235*, 308–315. [CrossRef] [PubMed]
48. Yang, Y.; Zhao, T.; Jiao, H.; Wang, Y.; Li, H. Potential Effect of Porosity Evolution of Cemented Paste Backfill on Selective Solidification of Heavy Metal Ions. *Int. J. Environ. Res. Public Health* **2020**, *17*, 814. [CrossRef] [PubMed]

materials MDPI

Article

Effect of Carbon Nanotubes on Chloride Diffusion, Strength, and Microstructure of Ultra-High Performance Concrete

Mahdi Rafieizonooz, Jang-Ho Jay Kim *, Jin-Su Kim and Jae-Bin Jo

School of Civil and Environmental Engineering, Yonsei University, Yonsei-ro 50, Seodaemun-gu, Seoul 03722, Republic of Korea; mahdirafiei@yonsei.ac.kr (M.R.); kjinsu@yonsei.ac.kr (J.-S.K.); whwoqls@yonsei.ac.kr (J.-B.J.)

* Correspondence: jjhkim@yonsei.ac.kr

Abstract: This study delved into the integration of carbon nanotubes (CNTs) in Ultra-High Performance Concrete (UHPC), exploring aspects such as mechanical properties, microstructure analysis, accelerated chloride penetration, and life service prediction. A dispersed CNT solution (0.025 to 0.075 wt%) was employed, along with a superplasticizer, to ensure high flowability in the UHPC slurry. In addition, the combination of high-strength functional artificial lightweight aggregate (ALA) and micro hollow spheres (MHS) was utilized as a replacement for fine aggregate to not only reduce the weight of the concrete but also to increase its mechanical performance. Experimental findings unveiled that an increased concentration of CNT in CNT1 (0.025%) and CNT2 (0.05%) blends led to a marginal improvement in compressive strength compared to the control mix. Conversely, the CNT3 (0.075%) mixture exhibited a reduction in compressive strength with a rising CNT content as an admixture. SEM analysis depicted that the heightened concentration of CNTs as an admixture induced the formation of nanoscale bridges within the concrete matrix. Ponding test results indicated that, for all samples, the effective chloride transport coefficient remained below the standard limitation of 1.00×10^{-12} m^2/s, signifying acceptable performance in the ponding test for all samples. The life service prediction outcomes affirmed that, across various environmental scenarios, CNT1 and CNT2 mixtures consistently demonstrated superior performance compared to all other mixtures.

Keywords: carbon nanotubes (CNTs); ultra-high performance concrete (UHPC); mechanical properties; ponding test; life service prediction

1. Introduction

Concrete is the foremost construction material globally, appreciated for its widespread availability, affordability, and versatility. Nevertheless, conventional concrete faces inherent constraints when aspiring to achieve structures that are simultaneously ultra-high-strength, lightweight, and durable. Consequently, the production of ultra-high-performance (UHP) concrete stands as a pivotal concern within the construction industry [1–3]. The specific compressive strength requirements for UHP concrete are contingent upon the structural demands of a project and the targeted performance characteristics [4,5]. However, UHP concrete, characterized by compressive strengths ranging from 80 MPa to 150 MPa or higher, has garnered significant attention in recent years [6–8]. This attention is driven by its remarkable load-bearing capacity, ability to diminish structural dimensions and weight, and heightened durability [9–13]. Attaining high-strength concrete involves employing diverse strategies, including optimizing mix designs, incorporating supplementary cementitious materials, maintaining a low water-to-cementitious material ratio (w/c), and utilizing chemical admixtures. Consequently, integrating certain admixture materials, such as Carbon nanotubes (CNTs), presents a distinctive opportunity to further enhance the performance of UHP concrete Top of Form [14,15].

CNTs have emerged as a promising material across diverse fields of science and engineering, owing to their exceptional mechanical, electrical, and thermal properties. As highlighted earlier, one notable domain where CNTs have demonstrated substantial potential is in the manufacturing of UHP concrete [16–19]. Integrating CNTs into concrete holds the promise of elevating its mechanical properties, enhancing durability, and facilitating the creation of inventive structural designs. The incorporation of CNTs in UHP concrete paves the way for the construction of resilient and sustainable infrastructures, capable of withstanding challenging conditions and ensuring long-term structural integrity. [20–22].

CNTs exhibited extraordinary mechanical properties, including high tensile strength, exceptional stiffness, and remarkable aspect ratios [23]. These characteristics position them as ideal candidates for reinforcing concrete at the nanoscale level. When uniformly dispersed within the concrete matrix, CNTs can bridge microcracks, contribute additional tensile strength, and enhance the overall mechanical performance of the material. Moreover, the excellent electrical conductivity of CNTs opens avenues for applications in sensing and monitoring structural health [24–26]. The conductive nature of CNTs allows for the development of intelligent and self-sensing concrete structures capable of detecting stress, strain, and temperature variations [27,28].

However, utilizing CNTs in UHP concrete effectively entails several significant challenges. Key issues include achieving uniform dispersion of CNTs within the concrete matrix to prevent agglomeration and ensuring efficient load transfer between the CNTs and the surrounding matrix [25,29]. Additionally, there are important considerations regarding the compatibility of CNTs with cementitious materials, long-term durability, and the economic feasibility of large-scale production [26]. Moreover, the incorporation of CNTs with various materials like silica fume, micro hollow spheres, and artificial lightweight aggregates in the production of UHP concrete could result in diverse characteristics of UHP concrete that warrant further investigation. Hence, further investigation is necessary to tackle this challenge and gain a better understanding of the performance of CNTs as additive materials in UHP concrete production.

In recent years, significant research efforts have been dedicated to comprehending the behavior and potential applications of CNTs in UHP concrete. These studies have delved into the mechanical properties, durability considerations, and structural performance of CNT-reinforced concrete. The objective is to surmount the challenges associated with CNT integration and lay the groundwork for the practical application of CNT-reinforced high-strength and UHP concrete in real-world construction projects. In addition to investigating mechanical properties, prior research has thoroughly explored the durability aspects of UHP concrete, with a specific focus on chloride penetration. Numerous studies have scrutinized the chloride ingress and penetration behavior of both conventional and UHP concrete [30,31]. These investigations have incorporated a variety of materials as replacements or admixtures, such as blast-furnace slag [32], polyvinyl-alcohol fiber [33], fly ash [34], and steel fibers [35]. Various approaches exist for examining chloride migration and chloride concentration in the concrete matrix [36]. A commonly employed method for long-term assessment is NT Build 443 [37], akin to ASTM C1543 [38] and ASTM C1556 [39]. Notably, the key distinction lies in the concentration of the ponding solution, where NT Build 443 employs a higher concentration (16.5 wt% NaCl) compared to the ASTM methods (15 wt% and 3 wt%, respectively). Hence, in the present study, NT Build 443 is employed to assess the chloride concentration profile following a 90-day ponding in a NaCl solution.

Dehghan et al. [36] conducted a study to assess the suitability of X-ray microfluorescence for determining chloride diffusion coefficients in concrete chloride penetration experiments. The study employed two distinct test methods, namely NT Build 443 and NT Build 492 [40]. The results unveiled consistent trends across both test methods. In another study [41], ASTM C1556 is utilized to evaluate chloride binding and diffusion in slag-blended concrete mixtures. The findings of this investigation indicated that the incorporation of ground-granulated blast furnace slags enhanced chloride binding and reduced chloride diffusivity. Both NT Build 443 and NT Build 492 test methods were applied to ascertain the chloride diffusion coefficient in mortars incorporating blended cement [42]. The outcomes of this study demonstrated that the silver nitrate colorimetric technique provided adequate accuracy for determining the chloride penetration front. Additionally, it is established that the NT Build 492 and NT Build 443 test methods were suitable approaches for determining the chloride diffusion coefficient in mortars containing blended cement. Furthermore, a separate investigation [43] also employed both NT Build 443 and NT Build 492 methods to assess how microstructural properties impact the chloride diffusion resistance of alkali-activated materials. The findings indicated that both quantifying reaction products and establishing a correlation between chloride penetration and pore surface area suggested that physical chloride adsorption on the surfaces of C-A-S-H/N-A-S-H gels took precedence over chemical chloride binding.

Recognizing that incorporating Carbon nanotubes (CNT) as an admixture in the production of ultra-high-strength (UHP) concrete alongside other replacement materials, such as silica fume (SF), silica powder (SPW), micro hollow sphere (MHS), and artificial lightweight aggregate (ALA), may exert varied influences on the mechanical properties and durability of the concrete, this study aims to assess the impacts of different CNT amounts as admixtures on the mechanical performance and chloride diffusion resistance of UHP concrete. The primary innovations of this paper include (1) an evaluation of the compatibility of CNTs with other cementitious materials and their long-term durability implications and (2) an economic feasibility analysis for the large-scale production of CNT-reinforced UHP concrete.

2. Materials and Mix Design

2.1. Materials

Ultra-high-performance concrete is manufactured utilizing Ordinary Type I Portland cement sourced from Asia Cement Co. Ltd. in Seoul, Republic of Korea. The OPC's characteristics include a specific gravity of 3.15, a soundness value of 10 mm, and a specific surface area of 2800 cm^2/g. The chemical composition of OPC, micro silica powder (SPW), and micro hollow sphere (MHS) is outlined in Table 1. Upon examining the table, it is evident that MHS and SPW are predominantly composed of SiO_2 (>81%), with minor quantities of other oxides. In contrast, CaO is the predominant oxide in OPC. For this study, SF (Grade 940U, Elkem, Norway) is employed as a 20% substitute for OPC, possessing a composition of 95% SiO_2, a Loss On Ignition (LOI) at 950 °C of 1.9%, and a specific gravity of 2.2. Additionally, micro SPW (S-SIL 10, SAC, Ulsan, Republic of Korea), with a diameter ranging from 1 to 5 μm, served as a fine filler, the chemical composition of which is detailed in Table 1. Moreover, high-strength functional artificial lightweight aggregate (ALA), created by embedding TiO_2 epoxy into the internal pores (~0.3–300 μm) of pristine ALA and coating the external surface simultaneously (detailed coating process information is available in [44]), was utilized as a partial replacement for fine aggregate. The ALA exhibited a density of 1.4 g/cm^3 and a water absorption rate of 14.1%. Figure 1 displays the SEM images of ALA and MHS particles. The MHS employed in this investigation is a type of hollow glass micro-solid bubble, hereinafter referred to as the micro-hollow sphere, produced and marketed by 3M. This micro-solid bubble is a lightweight material utilized in various applications such as aircraft bodies, automobile bodies, and construction, owing to its low density, high strength, and outstanding insulation properties. Being a factory-manufactured product ensures consistent quality and a reliable supply, making it well-suited for construction materials. The physical properties of MHS are displayed in Table 2. Also, in the current work for achieving high compressive strength, reducing the amount of water, and having appropriate workability, a superplasticizer (SP) was utilized.

Table 1. Chemical composition of MHS, OPC, and SPW.

Chemical Component	SiO_2	Fe_2O_3	Al_2O_3	TiO_2	CaO	MgO	SO_3	MnO	ZnO_2	Na_2O	P_2O_5	K_2O
MHS	81.34	0.04	0.17	0.03	13.3	0.06	0.2	0.01	0.02	4.02	0.81	0.02
OPC	18.78	2.76	4.72	0.31	65.97	3.14	2.56	0.13	0	0	0.19	0
SPW	99.59	0.023	0.315	0.042	0.013	0.006	0	0	0	0.008	0	0.004

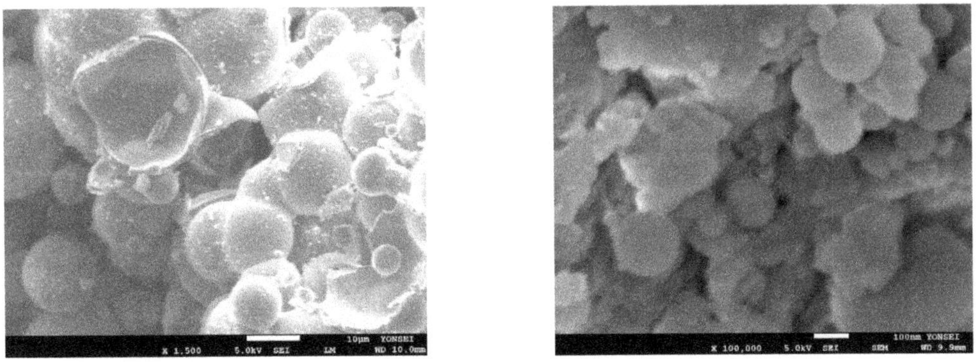

(a) (b)

Figure 1. SEM Images of (**a**) MHS and (**b**) ALA.

Table 2. Physical properties of MHS.

Physical Properties of MHS	
Shape	A hollow sphere with thin walls
Composition	Soda-lime borosilicate glass
Color	pale white powder
Pressure resistance (90% survival)	18,000 psi
Medium particle size	30 microns
Softening point	600 °C (1112 °F)
Specific Gravity	0.6
Thermal conductivity	0.05–0.20 W·m-1·K-1 @ 20 °C
Permittivity (@100 MHz)	1.2–1.9

2.2. Carbon Nanotube

Diverse nanomaterials are being employed to enhance concrete performance, with carbon nanotubes (CNT) being one such material. Studies have indicated that incorporating an optimal amount of CNT can create nucleation sites during early hydration, leading to the formation of a dense hydrate structure and increased strength. Achieving uniform dispersion of nanomaterials, including CNTs, within the cement matrix is crucial for ensuring concrete performance. A previous method involved pre-dispersing CNTs and nanomaterials uniformly in a solvent (such as water or surfactant) before application [24,45–48]. This approach, utilized in this research, involved the use of CNTs at ratios of 0%, 0.025%, 0.05%, and 0.075% relative to the weight of Ordinary Portland Cement (OPC). The present study employed a unified CNT solution. The procedure for creating this unified CNT solution, developed in a prior study, involves combining water, CNT powder, and dispersant, detail information is available in Siahkouhi et al. [48].

2.3. Mix Proportion

The procedure employed in this study involved several steps. Initially, dry materials including OPC, SF, SPW, MHS, and ALA were mixed for 5 min. Subsequently, a solution comprising water, SP, and CNT solution is added and mixed for an additional 5 min. The resulting mixture is then shaped into samples: cubic molds measuring 50 mm × 50 mm × 50 mm for compressive strength tests and cylindrical molds sized 200 mm × 100 mm for indirect tensile strength tests. These samples were left to cure at a temperature of 25 ± 3 °C for 24 h.

Curing in this study is conducted using two methods. Following the initial 24-h curing period, the samples underwent either (1) high-temperature curing at 90 °C and 95% humidity for 2 days, followed by subsequent curing in normal room air conditions at 25 ± 3 °C, or (2) high-temperature curing at 90 °C and 95% humidity for 2 days, succeeded by curing in a 25 °C sodium chloride (NaCl) solution, mimicking the conditions of a ponding test. The latter method aimed to replicate ponding test conditions and assess its impact on the mechanical properties and microstructure of the concrete mixtures.

The samples were subjected to compressive and tensile strength tests on days 3, 7, and 28. After conducting the compressive strength tests on days 3 and 28, further analyses, including X-ray Diffraction (XRD) and Scanning Electron Microscopy (SEM) image evaluations, were performed to investigate the effects of incorporating Carbon Nanotubes (CNTs) into the concrete mixtures. Table 3 displays the current study's concrete mix design. According to this table, the control mix serves as the baseline reference, containing standard proportions of water, OPC, SF, SPW, MHS, ALA, and SP in concrete. No CNTs are added to this mix, making it the control group. Similar to the control mix, the CNT 1 batch maintains the same proportions of water, OPC, SF, SPW, MHS, ALA, and SP. However, a small amount of Carbon Nanotubes (0.21 kg/m^3) is introduced into the mix. Like the previous mixes, CNT 2 maintains consistent quantities of water, OPC, SF, SPW, MHS, ALA, and SP. In this mix, a moderate amount of Carbon Nanotubes (0.43 kg/m^3) is included. The CNT 3 mix, akin to the others, has the same amounts of water, OPC, SF, SPW, MHS, ALA, and SP as the control mix. Notably, a higher concentration of Carbon Nanotubes (0.64 kg/m^3) is added in this batch. Varied quantities of CNT in the CNT 1, 2, and 3 mixtures enable the investigation of diverse CNT concentrations' impact on the concrete's properties, potentially influencing its strength and durability.

Table 3. Development of Mix Design.

Lable	Water (kg)	Binder			Aggregate		CNT (kg)	SP (kg)
		OPC (kg)	SF (kg)	SPW (kg)	MHS (kg)	ALA (kg)		
Control	212.5	739	111	208	211	111	0	7.85
CNT 1	212.5	739	111	208	211	111	0.21	7.85
CNT 2	212.5	739	111	208	211	111	0.43	7.85
CNT 3	212.5	739	111	208	211	111	0.64	7.85

3. Experimental Methods

3.1. Flow Test

The flow table test is a method used to measure the workability and consistency of concrete. It is performed to measure the ability of the concrete mix to flow and self-compact under its weight without the need for mechanical vibration. The flow test is essential to ensure that the concrete can fill the formwork completely and uniformly without segregation or excessive bleeding. The flow test provides a quantitative measurement of the flow properties of concrete. Unlike the slump test, which provides a qualitative measure of workability, the flow table test gives a numerical value that can be compared and monitored over time. The measured diameter of the concrete spread on the flow table provides a numerical value representing the flow of the concrete mix. Hence, the flow test is conducted to analyze the workability of CNT mixtures.

3.2. Compressive and Tensile Strength

The compressive strength test of concrete is one of the most common and important tests performed to ensure the quality of concrete. It measures the ability of a concrete specimen to withstand axial loads or forces applied along the axis of the specimen without any significant deformation. The compressive strength of concrete is a crucial factor in determining its structural capacity and durability. In this study, to analyze the compressive strength of CNT concrete three cube concrete specimens (50 × 50 × 50 mm) were cast and cured for each batch at three different ages of curing including 3, 7, and 28 days. Two distinct curing regimes were employed for the concrete specimens. Following demolding, all specimens were placed in an oven at a high temperature (90 °C) with 95% humidity for 48 h for both curing regimes. In the first curing method, the samples were subsequently cured at room temperature (23 °C) in the air until the testing day. Conversely, the second method involved curing the samples in a salt solution comprising 165 g of NaCl in 1 L of distilled water until the testing day. This salt solution is chosen to replicate conditions akin to the salt ponding test, allowing for an analysis of compressive strength in a challenging environmental scenario.

The indirect split tensile strength of concrete is a vital property that influences the durability, crack resistance, and overall performance of concrete structures. Therefore for analyzing the tensile strength, in this study, three cylindrical specimens (200 × 100 mm) were cast and cured for each concrete mixture at three different ages of curing including 3, 7, and 28 days. The curing method for the tensile test mirrored the initial curing process used for the compressive strength test. After demolding, the specimens were placed in an oven for 48 h and subsequently stored at room temperature in the open air until the testing day.

3.3. Microstructure and Phase Composition Analysis

Scanning Electron Microscopy (SEM), coupled with energy-dispersive X-ray spectroscopy (EDS), provides powerful tools to identify different phases and chemical elements present in the concrete. This information is crucial for understanding the composition of concrete materials and their interactions, which can affect the concrete's properties and durability. Hence, in this work, SEM with EDS is utilized to analyze the microstructure of concrete mixtures at 3 and 28 days of both curing conditions. Concrete powders were collected after conducting the compressive strength test. In this research, the Field Emission Scanning Electron Microscope (FE-SEM) 7800F-Prime machine is used for SEM and EDS.

Knowledge of the mineralogical composition provides insights into the concrete's behavior under various conditions, such as shrinkage, cracking, and long-term strength development. X-ray Diffraction (XRD) is a powerful analytical technique used to determine the chemical composition and crystallographic structure of materials, including concrete. In the context of concrete analysis, XRD is primarily employed to identify the mineralogical composition of the crystalline phases present in the concrete. In this study, the Rigaku XRD (Ultima IV) machine is used to analyze the chemical composition and crystallographic structure of CNT concrete.

3.4. Accelerated Chloride Penetration

The salt ponding test, following NT Build 443, is a widely recognized procedure for evaluating the chloride permeability of concrete [2,8,30,36,42,43,49]. This test assesses the ability of concrete to resist the penetration of chloride ions, which is crucial for the durability and longevity of concrete structures, especially in harsh environmental conditions [32]. The primary purpose of the salt ponding test is to determine the chloride permeability characteristics of concrete surfaces. It helps in identifying areas prone to chloride ingress, which can lead to various issues such as corrosion of reinforcement, deterioration of concrete, and compromised structural integrity [50]. Therefore, in this study, a salt ponding test based on NT Build 443 [37] is selected and conducted to analyze the performance of UHP concrete containing CNT against chloride penetration.

Following the standard specifications, three cylindrical samples (100 × 200 mm) were meticulously cast for each concrete mixture. Following demolding, these samples underwent a high-temperature curing process in an oven set at 90 °C, coupled with high humidity (95%) for a duration of 2 days. Subsequently, the cylindrical specimens were meticulously divided into halves by making a perpendicular cut along the axis of each cylinder. One of these halves is designated as the testing specimen, exposing its cut surface to a solution of sodium chloride (NaCl). After this initial exposure, the test samples were submerged in a saturated solution of calcium hydroxide ($Ca(OH)_2$) at a controlled temperature of approximately 23 °C. To prevent carbonation, the samples were placed in tightly sealed plastic containers, ensuring that the containers were filled to the brim with the solution. The following day, the mass in a surface-dry condition (m_{sd}) is meticulously determined by weighing the test samples, marking a crucial step in the evaluation process. A section, at least 20 mm thick, is sliced from the remaining half of the cast cylinder, extending the test samples. This section is utilized to assess the initial chloride concentration (C_i).

The immersion of the test samples in the saturated calcium $Ca(OH)_2$ solution persisted until the mass in a surface-dry condition (m_{sd}) exhibited minimal changes, not exceeding 0.1 mass % per 24 h. Subsequently, all faces of the test samples, except the one designated for exposure, were air-dried at room temperature until they reached a stable white-dry condition. A uniform epoxy coating, approximately 1 mm thick, is then meticulously applied to these surfaces. Following a curing period of 48 h, in adherence to the epoxy's specified requirements, the coated test samples were re-immersed in the $Ca(OH)_2$ solution until the m_{sd} stabilized, following the previously described criteria.

Following the standard specifications, an aqueous sodium chloride (NaCl) solution is meticulously prepared, maintaining a concentration of 165 g $\pm$ 1 g NaCl per dm^3 solution. This solution is utilized for a duration of 35 days (equivalent to 5 weeks) before being replaced by a fresh and pure NaCl solution. The concentration of NaCl in the solution is verified both before and after usage. The water bath, where the samples were placed, is carefully controlled, maintaining a temperature range between 21–25 °C, with an average target temperature of 23 °C (Figure 2). Additionally, the solution's temperature is monitored once daily to ensure consistency throughout the testing period. Subsequently, the $Ca(OH)_2$ solution in the container used for water saturation was exchanged with the designated exposure liquid (NaCl solution), and the test samples were submerged in a surface-dry condition within the saline solution. It's crucial to emphasize that the container is filled with the exposure liquid and tightly sealed. The ratio between the exposed surface area in square centimeters and the volume of the exposure liquid in cubic decimeters is maintained within a range of 20 to 80.

The sealed container is then placed inside a temperature-controlled cupboard throughout the exposure period. This exposure phase lasted for a minimum of 90 days, during which the container is shaken once every week and the solution is replaced by a fresh, pure NaCl solution per 5 weeks (35 days). Detailed records, including the date and time of exposure commencement and conclusion, were diligently maintained. While the ponding test is capable of modeling chloride diffusion into concrete, it is still a long-term test. For low-quality concretes, the minimum exposure period is 35 days. For higher quality concretes, however, this period must be extended to 90 days or longer, just as for the salt ponding test [51].

Following a 90-day immersion process, the chloride profile is obtained by milling the material parallel to the exposed surface, necessitating a minimum of eight layers to be milled. The thickness of these layers should be adjusted based on the anticipated chloride profile, ensuring that at least six locations encompass the profile between the exposed surface and the depth reached by the chlorides. The chloride content soluble in acid within the specimens is evaluated following the NT BUILD 208 [52] standard. To determine the test results, including the superficial chloride concentration (C_s) and the non-steady-state chlorides diffusion coefficient (D_{ns}), Equation (1) is adjusted for the measured chloride content using linear regression analysis based on the least squares method.

$$C_{(x,t)} = C_s - (C_s - C_i)*\mathrm{erf}\left(\frac{x}{\sqrt{4.D_{ns}.t}}\right) \quad (1)$$

In the given context: $C_{(x,\,t)}$ represents chloride concentrations (in mass %) measured at a depth x" during the exposure time t"; C_s stands for the boundary condition at the exposed surface (in mass %); C_i denotes the initial measured chlorides concentration (in mass %); x" signifies the depth beneath the exposed surface (in meters); D_{ns} represents the non-steady-state chlorides diffusion coefficient (in m^2/s); t″ is the exposure time (in seconds), and erf refers to the Gauss error function.

(a)

(b)

(c)

(d)

Figure 2. Ponding Test Procedure (NT-BUILD 443); (**a**) Samples in Ca $(OH)_2$ solution, (**b**) Before Epoxy, (c) Applying the Epoxy, (**d**) Samples in NaCl solution.

3.5. Prediction of Service Life Model

The chloride profiles obtained from concrete specimens subjected to an intense environmental condition (as in the ponding test) over 90 days offer a valuable chance to authenticate prevailing models predicting the service life. Several software packages currently exist for modeling service life in concrete structures, with Life-365 being one such model that forecasts chloride ingress and service life for reinforced concrete under chloride exposure [53]. Like certain other service-life prediction models, Life-365 operates under the assumption that concrete is entirely saturated [54]. However, in practice, reinforced concrete often undergoes wetting and drying cycles, and chloride penetration may not exclusively occur through diffusion. In this investigation, the Life-365 software is employed to formulate durability against salt damage, utilizing the results derived from the ponding test (NT Build 443). The design is customized for reinforced concrete (column dimensions: 600 mm × 600 mm), featuring cover thicknesses of 80 mm, 70 mm, and 60 mm. The assumed location is adjacent to the coastline. Furthermore, the external temperature to which the structure is subjected is presumed to align with the average monthly temperature in Korea for the year 2022.

As the migration of chloride ions in concrete follows a diffusion process, the prediction of chloride ion diffusion employs the diffusion equation derived from KDS 14 20 40 [55] and durability exceeding 30 years, as depicted in Equation (2). Equation (3) represents the most comprehensive form of the equation, incorporating adjustments for environmental conditions and chloride ion concentration on the surface.

$$t \geq t_c = \text{over 30 years}$$

$$D_p = \frac{D_R}{1-m}\left[(1-m) + m\left(\frac{t_c}{t}\right)\right]\left(\frac{t_R}{t_c}\right)^m \tag{2}$$

where: D_R denotes the chloride ion diffusion coefficient at the reference time (t_R), t_R typically represents the reference time (commonly 28 days or 0.077 years), t_c signifies the diffusion coefficient reduction limit (typically set at 30 years), and m stands for the age coefficient relative to the age constant, representing its influence.

$$C_d - C_i = (C_s - C_i)\left(1 - \text{erf}\left(\frac{x}{2\sqrt{D.t}}\right)\right) \tag{3}$$

where: C_d: chloride ion concentration (mass %) at the depth of x (m), C_i: the initial chloride concentration (mass %), C_s: the boundary condition at the exposed surface (mass %), D: diffusion coefficient of chloride ion (m^2/s), t: the exposure time (s), and erf: error function

In this investigation, considering durability life spans of 200 years, the diffusion coefficient results obtained from the ponding test are incorporated into Equation (3), utilizing the standard chloride ion amounts specified for each exposure environment in KDS 14 20 40. This process is undertaken to determine the coating thickness for each environment. Durability-related considerations adhere to the concrete specifications outlined by the Korean Society of Civil Engineers (KSCE), as detailed in Tables 4 and 5 [55]. It's essential to highlight that, from both conditions (in the sea and near the sea), the maximum chloride concentration (as emphasized in Tables 4 and 5) was chosen for each coastline to simulate the worst-case scenario. The recommended surface chloride amount and the limit chloride concentration of 1.2 (kg/m^3) are taken into account. Additionally, by applying a safety factor, as per Equation (4) in KDS, considering the critical chloride concentration of 0.93 (kg/m^3), the ensuing section presents the coating thickness for each environment.

$$\gamma_P \times C_d \leq \varnothing_K \times C_{lim} \tag{4}$$

where: γ_P is an environmental factor for salt damage, generally 1.11, ϕ_K is the durability reduction factor for salt damage, generally 0.86, C_{lim} is critical chlorine ion concentration at the start of reinforcing bar corrosion (1.2 kg/m^3), and C_d is the predicted value of chlorine ion concentration at the reinforcing bar location.

Table 4. Suggested surface chloride ion concentration in the sea [55].

Coastline	Expose Condition	Surface Chloride Concentration (kg/m^3)
West and south coast	**Tidal Zone**	**17.0**
	Splash Zone	**7.5**
	Near the coast	5.0
East coast	Tidal & Splash Coast	13.0
	Near the coast	7.0

Table 5. Suggested surface chloride ion concentration near the sea [55].

Coastline	Expose Condition	Surface Chloride Concentration (kg/m^3)
West and south coast	Near the coast	5.0
	100 m	2.0
	250 m	1.5
East coast	**Near the coast**	**7.0**
	100 m	**4.5**
	250 m	**3.0**
	500 m	2.5
	1000 m	1.5

4. Results and Discussion

4.1. Fresh Properties

The outcomes of the flow test, illustrated in Figure 3, indicated that the incorporation of CNT as an admixture in the concrete matrix had a marginally positive impact on the fresh performance and flowability of the CNT concrete mixtures. All samples exhibited a flow test result higher than 100 cm; nevertheless, mixtures containing CNT as an admixture displayed slightly improved flowability compared to the control mixture. This enhancement may be attributed to the effective dispersion of CNTs, which can enhance concrete flowability by acting as a lubricating agent [56]. This reduces internal friction among particles, facilitating smoother flow in the concrete mix. However, the influence of adding CNTs on the workability of the CNT concrete mixtures is negligible. It's important to note that substituting MHS for natural fine aggregate, given its spherical shape and micro particle size, led to an improvement in the flowability of all mixtures. Conversely, incorporating ALA as a partial replacement for fine aggregate, owing to its sharp edges and irregular shapes, caused a decrease in the flowability of concrete mixtures. However, in this study, through the use of sufficient superplasticizers, a well-designed mix proportion, and careful mixing procedures, suitable flowability is attained for all mixes.

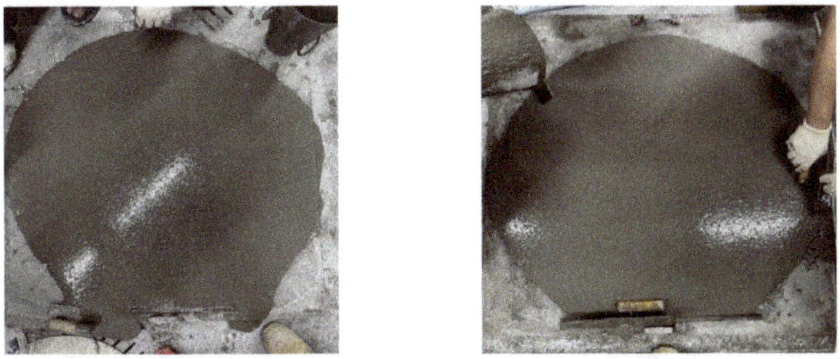

(a) (b)

Figure 3. Flow test results of (**a**) Control, and (**b**) CNT2.

4.2. Compressive Strength

The compressive strength results under standard curing conditions for three different curing times are depicted in Figure 4. At the 3-day mark of standard curing, the compressive strength values for the control, CNT1, CNT2, and CNT3 mixtures were 103 MPa, 100 MPa, 107 MPa, and 100 MPa, respectively. Notably, there were no significant changes in compressive strength after the initial 3-day period of standard curing. Upon reaching the 7-day point with the standard curing method, the compressive strength values for the same mixes were 105 MPa, 106 MPa, 108 MPa, and 100 MPa, respectively. Comparing these results to the 3-day data, it becomes evident that all samples experienced a slight increase in compressive strength. This enhancement can be attributed to the ongoing hydration process and the formation of calcium silicate hydrate gel (C-S-H gel) within the concrete matrix, contributing to strength development over time [54,57]. The data from Figure 4 indicates that the compressive strength values for the same mixtures at the 28-day mark of the standard curing period were 106 MPa for the control mix, 107 MPa for CNT1, 109 MPa for CNT2, and 102 MPa for CNT3 mixture, respectively. Upon analysis of these results, it is observed that there was a modest increase in the compressive strength of all mixtures after a 28-day curing period compared to the results at 7 days. This improvement also can be attributed to the ongoing process of hydration extending up to the 28-day curing time.

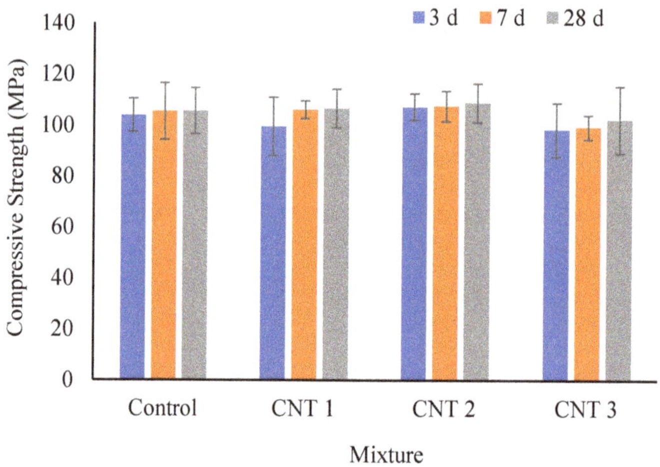

Figure 4. Compressive strength test results in normal curing conditions.

Nevertheless, an examination of the data reveals that increasing the CNT content in CNT1 and CNT2 mixtures led to a slight improvement in compressive strength compared to the control mix. This enhancement may be attributed to the formation of bridges by CNTs between particles and cement paste within the concrete matrix. Conversely, the CNT3 mixture exhibited a decrease in compressive strength with the rise in CNT content as an admixture. It needs to be considered that the tendency of CNTs to form agglomerates, driven by strong van der Waals forces, becomes apparent [58–60]. When CNTs are inadequately dispersed, these agglomerates can create localized weak points in the concrete, diminishing its overall strength. This phenomenon likely accounts for the reduction in compressive strength observed in the CNT3 mixture compared to the other mixes. It suggests that the dispersion process of CNT in the CNT3 mixture was incomplete, resulting in the formation of localized weak points in the concrete matrix and a subsequent decrease in compressive strength.

Figure 5 illustrates the compressive strength results under salty curing conditions. At the 3-day mark of exposure to the salty curing environment, the compressive strength values for the control, CNT1, CNT2, and CNT3 mixes were 92 MPa, 90 MPa, 92 MPa, and

80 MPa, respectively. Notably, these findings indicate a decrease in compressive strength for the CNT3 mixes compared to the other formulations, aligning with observations from the normal curing condition. Additionally, after 7 days of exposure to salty curing conditions, there was no significant change in the compressive strength of all mixtures. Specifically, the compressive strength values for the same mixes were 91 MPa, 90 MPa, 92 MPa, and 80 MPa, respectively. Furthermore, upon reaching the 28-day mark in the salty curing conditions, the compressive strength values for the same mixes were 94 MPa, 90 MPa, 93 MPa, and 82 MPa, respectively.

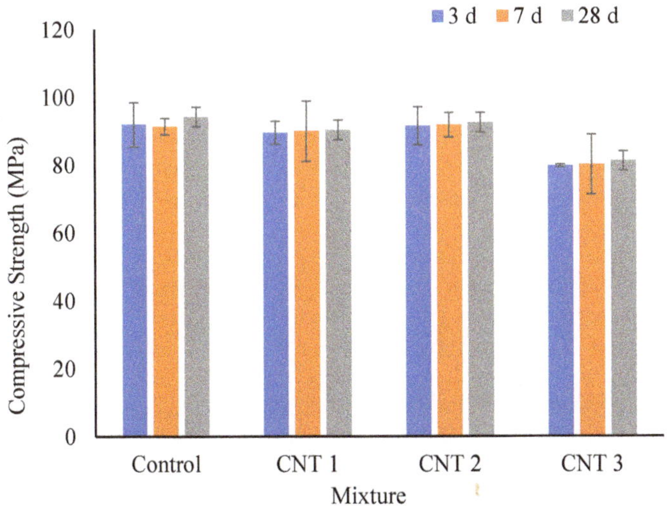

Figure 5. Compressive strength test results in salty curing condition.

From the results, it is evident that there is no substantial difference in compressive strength at various salty curing periods. However, it is crucial to emphasize that the compressive strength values at 3, 7, and 28 days in the salty curing environment were considerably lower than those observed under similar normal curing periods. This notable difference can be attributed to the presence of salts, particularly chloride ions [41]. These salts have the potential to interfere with the formation of C-S-H gel, a critical factor for the strength development of concrete. In the context of salty curing conditions, the impact on the formation of C-S-H gel is more pronounced, contributing to the observed reduction in compressive strength when compared to the results obtained under normal curing conditions. Moreover, salts have the capability to directly attack the cement paste, causing the leaching of calcium hydroxide and disrupting the paste's structure. Consequently, this can result in a weaker matrix and lower compressive strength [43]. Additionally, salty curing conditions may elevate the porosity of the concrete. The presence of salts can induce the formation of additional pores or capillaries, thereby weakening the concrete structure and diminishing its overall strength [34,61]. Lastly, the presence of salts can disrupt the hydration reactions of cement, impeding the formation of desired crystalline phases. This interference can have a negative impact on the development of strength over time.

4.3. Tensile Strength

Figure 6 presents the outcomes of the indirect spilite tensile strength test. At the 3-day mark of normal curing, the tensile strength values for the control, CNT1, CNT2, and CNT3 mixtures were 6.77 MPa, 6.65 MPa, 6.62 MPa, and 6.60 MPa, respectively. The results indicate a slight decrease in tensile strength with an increase in the amount of CNT as an admixture. This trend persists during the 7-day curing period, where the indirect tensile strength for the same mixtures was 7.38 MPa, 6.81 MPa, 6.97 MPa, and 6.65 MPa, respectively. The findings affirm the influence of elevated CNT content in diminishing the slight tensile strength of the concrete mixtures. At the 28-day mark of normal curing conditions, the indirect tensile strength for the same mixtures was 7.44 MPa, 7.38 MPa, 7.36 MPa, and 6.81 MPa, respectively. Similarly, it can be observed that the same trend persisted for tensile strength after 28 days of normal curing conditions.

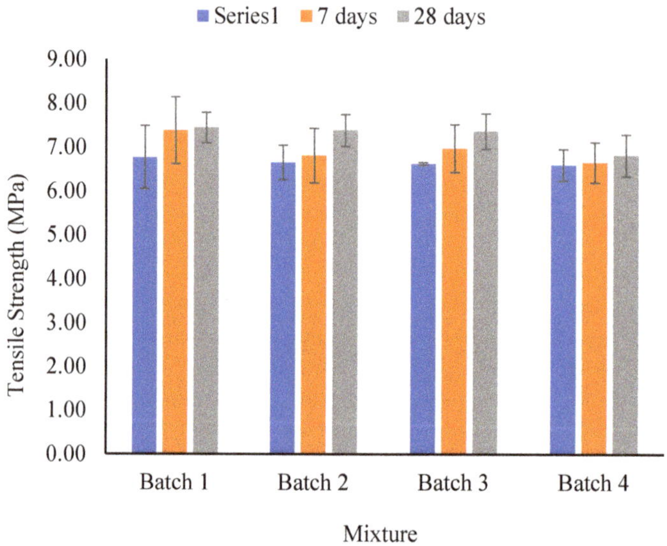

Figure 6. Indirect tensile strength test results in normal curing conditions.

The impact of incorporating CNT as an admixture into high-performance concrete on compressive and tensile strength is subject to various factors. The observation of a slight increase in compressive strength alongside a slight decrease in tensile strength can be attributed to the following considerations [16,18,33]: Firstly, achieving a uniform dispersion and alignment of CNTs within the concrete matrix is paramount. Suboptimal dispersion or agglomeration of CNTs may lead to non-uniform properties. In terms of tensile strength, the presence of agglomerates or uneven distribution can create weak points, resulting in an overall decrease in tensile performance [14]. Secondly, while CNTs can augment compressive strength by serving as additional bridges and reinforcing the concrete structure, their interaction with the cementitious matrix may vary. The establishment of effective bonding between CNTs and the matrix is crucial for load transfer. If this bonding is not well-established, it may contribute to a reduction in tensile strength [15]. Lastly, the addition of CNTs can influence the porosity of the concrete. In cases where nanotubes are not well-dispersed, they may form clusters or induce the creation of additional pores, particularly at the interfaces between the nanotubes and the cement matrix. This, in turn, can adversely affect tensile strength.

Table 6 presents the results of the current study alongside findings from previous research. It is important to note that all properties of the UHP concrete in these studies are compared to those of the control mix concrete. As shown in this table, an increase in the water-to-binder ratio led to a slight reduction in compressive strength. An increase in the water-to-binder ratio typically results in a slight reduction in compressive strength due to dilution of the cement paste and increased porosity within the concrete matrix. This diminishes the binding properties of the cement paste and creates pathways for crack propagation, ultimately weakening the overall strength of the concrete. However, the type of nanomaterial and additive admixture significantly influenced the mechanical properties of UHP concrete.

Table 6. Suggested surface chloride ion concentration near the sea.

Ref.	w/b Ratio	Compressive Strength (MPa)		Tensile Strength (MPa)	
		3 Days	28 Days	3 Days	28 Days
This study	0.25	100	105	6.5	7.0
[6]	0.18	80	130	-	-
[7]	0.225	100	139	-	-
[9]	0.27	55	-	6	-
[11]	0.18	-	180	-	8
[13]	0.18	80	140	-	-
[14]	0.22	60	120	5.8	6.3

4.4. SEM Investigation

The results obtained from Scanning Electron Microscopy (SEM) and energy dispersive X-ray spectroscopy (EDS) conducted during standard curing conditions over 3 days are illustrated in Figure 7. SEM analysis revealed that an increased content of CNTs as an admixture to the concrete resulted in the formation of bridges within the concrete matrix. These nanoscale bridges serve as reinforcement, creating a network within the cementitious matrix [47]. This network significantly improves the distribution and transfer of loads, thereby enhancing the overall strength performance of the concrete. Moreover, CNTs contribute to the reduction of microcracks and the enhancement of material toughness. Additionally, CNTs play a role in expediting the hydration process of cement particles, leading to the formation of a denser and more compact matrix. This accelerated curing process results in achieving higher compressive strength within a shorter timeframe. It can be asserted that, during the initial 3 days of curing, the presence of CNTs facilitates the formation of C-S-H gel in CNT1 and CNT2 mixtures. Conversely, in the case of the CNT3 mix, improper dispersion processes may lead to the creation of agglomerates. These agglomerates can form localized weak points in the concrete, diminishing its overall strength and durability [26]. Therefore, the effectiveness of CNTs in enhancing the concrete's properties is contingent on a thorough dispersion process, emphasizing the importance of optimized mixing techniques for optimal results.

Based on the outcomes derived from the energy dispersive X-ray spectroscopy (EDS) map sum spectrum, it is evident that O, Si, C, and Ca constitute the primary elements in the structure of all mixtures, albeit in varying proportions. These findings align seamlessly with the X-ray diffraction (XRD) results, wherein Quartz (SiO_2) and Hatrurite (Alite: Tricalcium silicate: Ca_3SiO_5) emerge as the predominant crystalline phases. Furthermore, the results indicated a marginal increase in the amount of C with the escalation of CNT content as an admixture to the concrete matrix [15,62]. In summary, the SEM results, following a 3-day curing period, reveal a subtle structural divergence in CNT mixtures attributable to the presence of CNTs. However, in contrast, the EDS results suggest no significant disparity in the composition of all mixtures.

Consistent findings were observed in both the scanning SEM images and EDS maps after a 28-day curing period (Figure 8). This suggests that, under standard curing conditions for 28 days, there was a modest increase in compressive strength attributed to the formation of bridges between the cement paste and particles (ALA and MHS). Notably, SEM images indicate a higher prevalence of bridges in samples subjected to a 28-day curing period compared to those cured for 3 days.

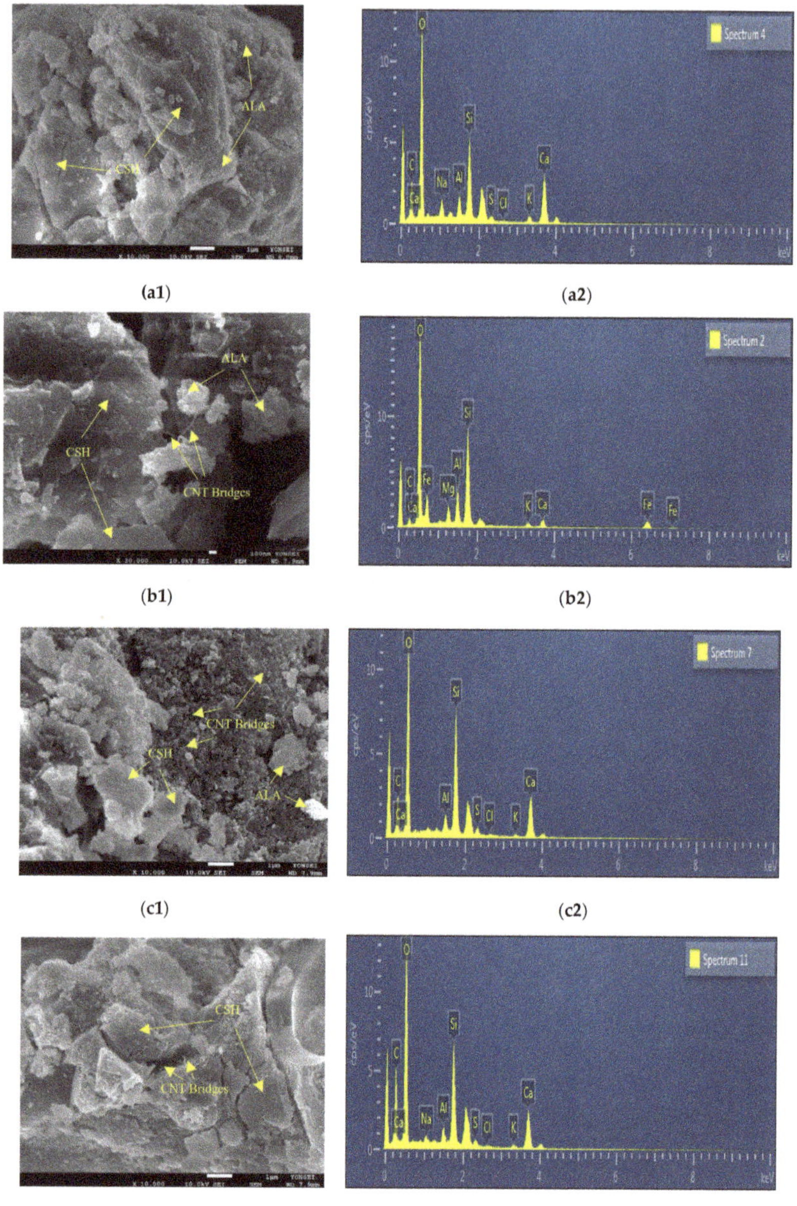

Figure 7. SEM images and EDS spectrum after 3 days of curing period for samples (**a**) control, (**b**) CNT1, (**c**) CNT2, and (**d**) CNT3.

The SEM images further indicated that with an increase in the quantity of CNT as an admixture, there was a noticeable augmentation in both the number and volume of bridges formed within the concrete matrix. Additionally, the inclusion of silica fume (SF) as a partial substitute for OPC enhances the potential for the secondary formation of C-S-H gel, particularly noticeable after extended curing periods (beyond 90 days). Consequently, the production of additional C-S-H gel becomes imperceptible beyond the 28-day mark.

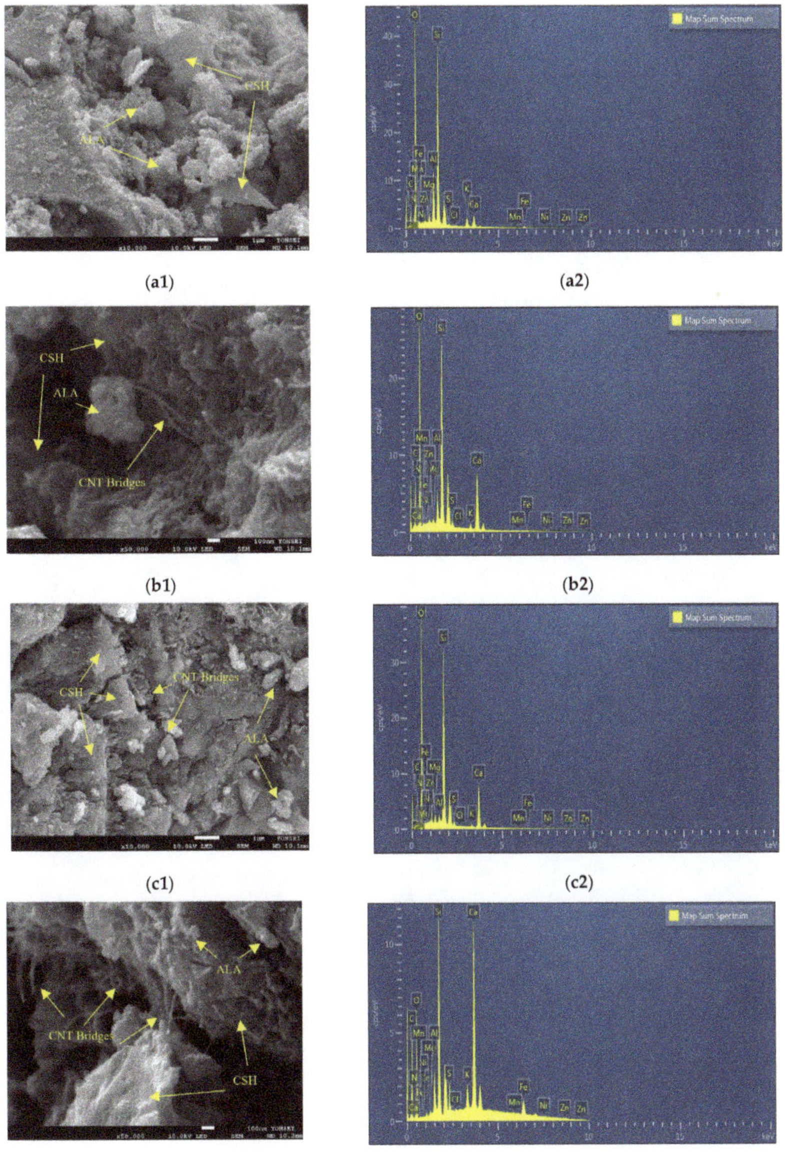

Figure 8. SEM images and EDS maps after 28 days of curing period for samples (**a**) control, (**b**) CNT1, (**c**) CNT2, and (**d**) CNT3.

The energy dispersive EDS map spectrum results for all mixtures after a 28-day curing period demonstrated a composition nearly identical to their 3-day counterparts. This uniformity is likely attributable to the absence of substantial changes in the structure and composition of all mixtures following the specified curing period. In each sample, O, Si, C, and Ca remained the primary elements, in line with the EDS map spectrum results. However, there is a slight increase in the percentage of C with the escalation of CNT content as an admixture [47,56]. These outcomes are consistent with the findings of XRD analysis, where quartz and hatrurite persist as the principal crystalline phases in the structure of all mixtures.

4.5. XRD Analysis

The X-ray Diffraction (XRD) patterns depicted in Figures 9 and 10 showcase concrete mixtures incorporating varying amounts of carbon nanotubes (CNT) at both 3 days and 28 days of the curing period. In all samples, after both 3 days and 28 days of curing, the primary crystalline phases identified include Quartz (SiO_2) and Alite (Tricalcium silicate: Ca_3SiO_5).

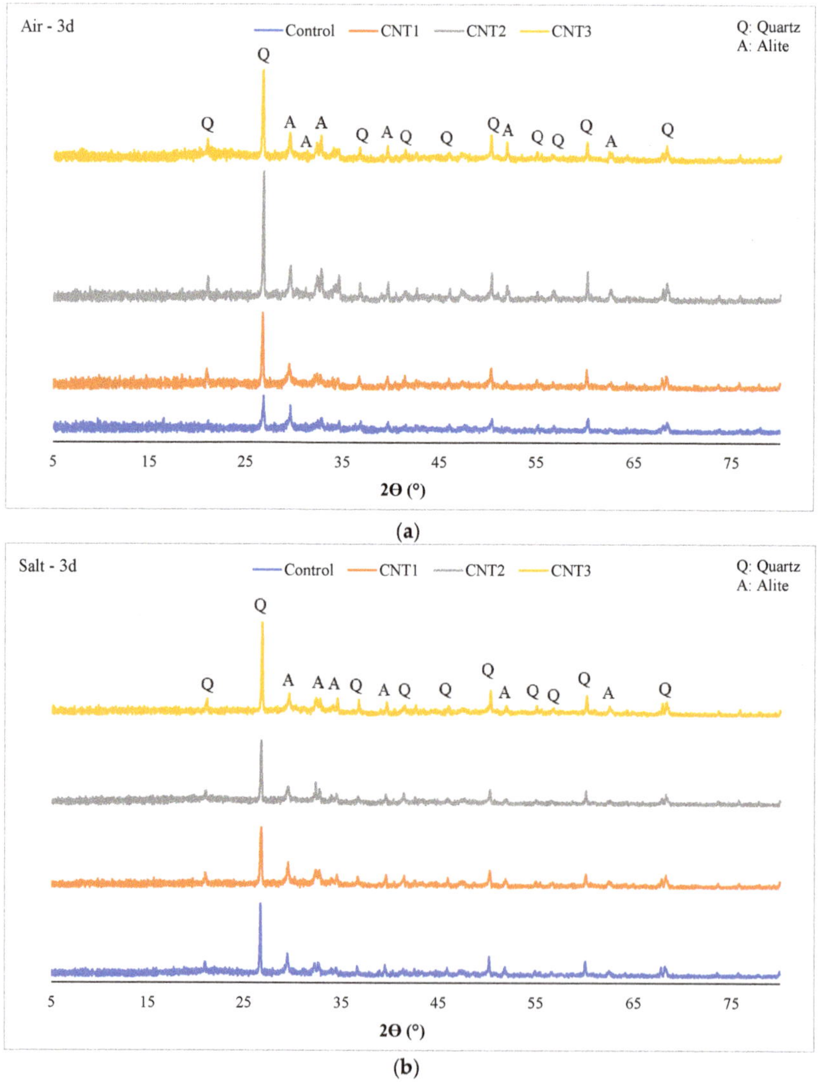

Figure 9. XRD Analysis of (**a**) 3d—Air curing, and (**b**) 3d—Salt curing condition.

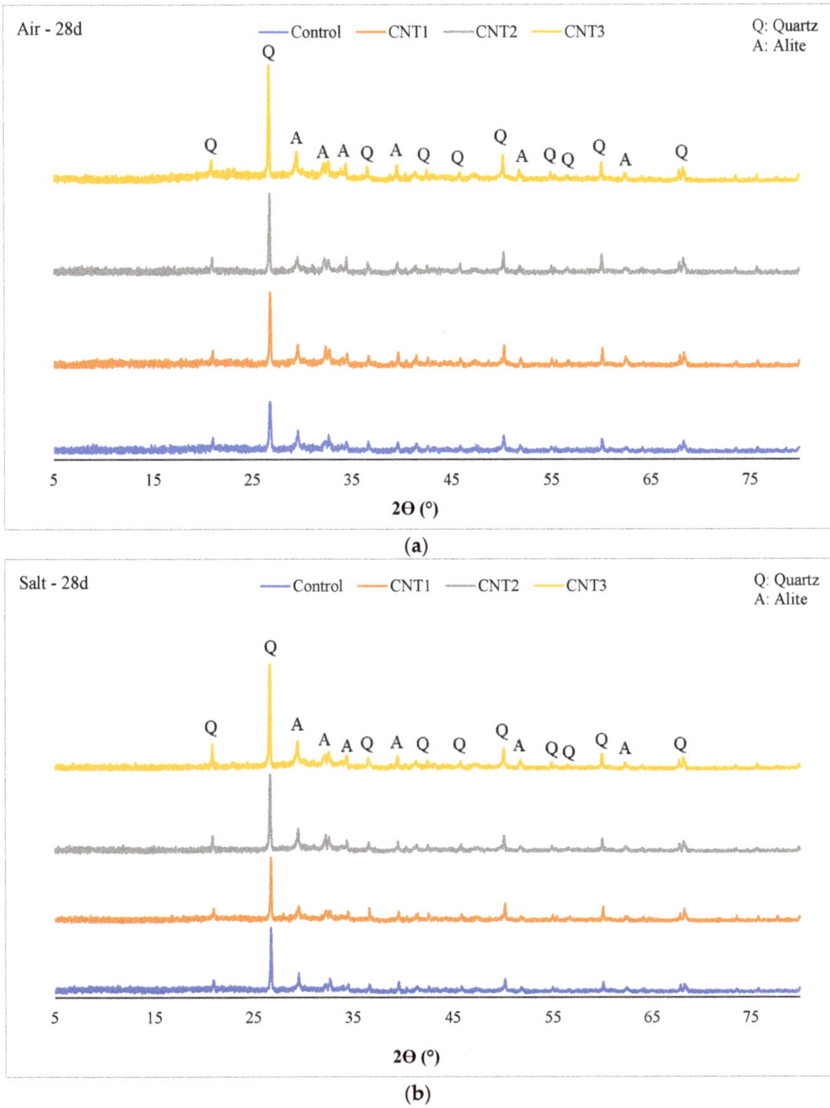

Figure 10. XRD Analysis of (**a**) 28d—Air curing, and (**b**) 28d—Salt curing condition.

Several factors may contribute to the presence of only these two phases:

1. Quartz is a common crystalline phase found in aggregates, particularly if natural aggregates or silica-based materials are utilized in the concrete mix.
2. Alite (Tricalcium silicate), a fundamental component of OPC, is expected in UHPC containing OPC.
3. Some materials in the mix may exist in an amorphous or non-crystalline form, and XRD is less sensitive to such phases.
4. The materials employed, including silica fume and silica powder, consist of extremely fine particles. Consequently, XRD may not identify them as distinct crystalline phases.
5. The mix composition may favor non-crystalline or amorphous forms, resulting in fewer distinguishable crystalline peaks.

6. XRD's sensitivity limits might preclude the detection of phases present in small quantities or exhibiting weak diffraction signals.
7. A homogenized mix design could lead to the dominance of main crystalline phases, posing challenges in identifying minor phases.

It's essential to recognize that the absence of certain phases in the XRD analysis doesn't necessarily imply their complete nonexistence in the concrete mix. Some phases may exist in lower concentrations, or their diffraction peaks may be overshadowed by dominant phases.

The inclusion of SF as a partial substitute for OPC in concrete has the potential to influence the XRD pattern, impacting the 2θ values of diffraction peaks. However, these alterations are typically subtle, particularly for primary crystalline phases like quartz. Notably, the prevalence of Alite, a significant contributor to early strength development in concrete, was higher in all mixes after 3 days of air curing compared to salt curing, as illustrated in Figure 9. In contrast, the abundance of quartz was greater in 3-day salt-cured samples than in their air-cured counterparts. This pattern persisted across all mixtures and likely serves as a key factor contributing to the augmented compressive strength observed in air-cured samples relative to their salt-cured counterparts.

The XRD analysis results at the 28-day mark under both curing conditions (as illustrated in Figure 10) displayed negligible differences compared to the 3-day curing period. Across all mixtures, the prevalence of quartz during the 28-day salt curing conditions surpassed that in the 28-day air curing conditions. In contrast, the concentrations of Alite in all mixtures were higher in the 28-day air curing conditions compared to the 28-day salt curing conditions. This disparity sheds light on the compressive strength results for these mixtures, where, after 28 days of air curing, the compressive strength notably exceeded that of the 28-day salt curing. The increased content of Alite in the air curing conditions likely contributed to the enhanced compressive strength through this curing method. Furthermore, it is essential to acknowledge that after 28 days of both curing methods, no new crystalline phases emerged in any of the mixtures. This suggests the absence of pozzolanic activity within the initial 28-day curing period. To induce pozzolanic activity, an extended curing period exceeding 90 days is required.

4.6. Ponding Test Results

The findings from the ponding test are summarized in Table 7 and visually represented in Figure 11. In this context, C_s (mass%) denotes the boundary condition at the exposed surface, C_i (mass%) signifies the initial chloride concentration measured on the concrete slice, D_a (m^2/s) represents the effective chloride transport coefficient, and t (days) denotes the exposure time. Analysis of the ponding test results reveals that, for all samples, the effective chloride transport coefficient (D_a) remains below the standard limitation of 1.00×10^{-12} m^2/s. This suggests that the performance of all samples in the ponding test falls within an acceptable range. However, a closer examination of the results reveals that the chloride transport coefficient decreases with an increase in the amount of CNT up to 0.05% (as observed in CNT1 and CNT2 mixes compared to the control mix). Conversely, for the CNT3 mixture, there is an increase relative to the control mix. This trend aligns with the compressive and tensile strength outcomes, where the strength of the CNT1 and CNT2 mixtures surpassed that of the CNT3 mixture.

Table 7. Ponding test results (NT Build 443).

Sample	C_s	C_i	D_a	t
Control	0.78	0.037	8.18×10^{-13}	90
CNT 1	0.813	0.037	7.1×10^{-13}	90
CNT 2	0.758	0.037	6.5×10^{-13}	90
CNT 3	0.862	0.037	8.3×10^{-13}	90

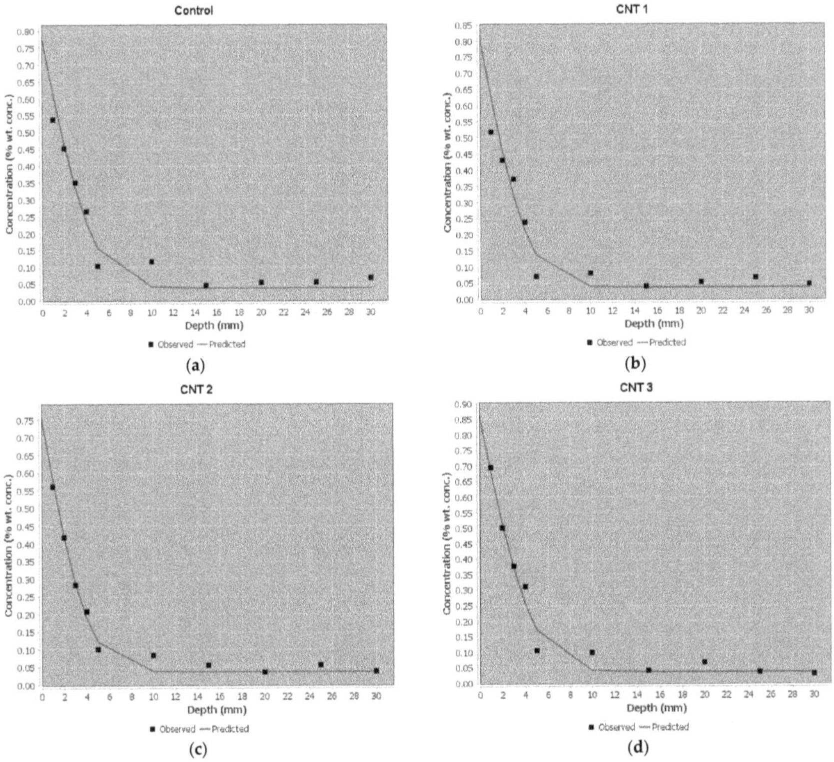

Figure 11. The regression Analysis based on the ponding test results for (**a**) Control, (**b**) CNT1, (**c**) CNT2, and (**d**) CNT3.

The primary factor influencing this trend is attributed to the incomplete dispersion process of CNT in the CNT3 mixture, resulting in the formation of localized weak points in the concrete matrix and an ensuing increase in permeability. Conversely, the well-dispersed CNT in the CNT1 and CNT2 mixes contributes to an elevated number of bridges between the cement paste and ALA and MHS particles, leading to decreased permeability and, ultimately, a reduced effective chloride transport coefficient in these two mixtures.

Considering the C_s value, derived from the contact point of the predicted curve with the Y-axis in Figure 11, which represents the chloride mass on the surface of each mixture, the results correlate with those of the chloride transport coefficient. In summary, all mixtures demonstrate acceptable performance when compared to standard limitations. Notably, the CNT2 mixture stands out, exhibiting the most favorable performance among all mixes.

4.7. Life Service Prediction

The life service prediction results under diverse environmental conditions, as determined from the ponding test outcomes, are depicted in Figure 12a–e, with a consistent concrete cover thickness of 80 mm in all scenarios. Notably, in the most challenging environmental setting, the tidal zone featuring a chloride ion concentration of 17 kg/m^3 (as shown in Figure 12a), the life service prediction model indicates ultimate resistance ages of 119, 136, 147, 118, and 19 years for the control, CNT1, CNT2, CNT3, and OPC mixtures, respectively, against chloride ions penetration. In practical terms, this implies that, in the tidal zone, all samples, excluding the OPC mix (included for comparative purposes), exhibit durability against chloride ion attack for over 100 years. Further analysis of the data underscores the superior performance of the CNT1 and CNT2 mixtures in the tidal zone. Noteworthy is the observation that, for CNT1 and CNT2 mixtures, it requires 136 and 147 years, respectively, before chloride ions can breach the concrete cover thickness of 80 mm to reach the internal reinforcement.

Derived from the life service prediction results and referencing Figure 12b, representing the second most challenging condition (splash zone with a chloride ion concentration of 7.5 kg/m^3), the control, CNT1, CNT2, CNT3, and OPC mixtures exhibit ultimate ages of 172, 197, 216, 170, and 24 years, respectively, in their resistance against chloride ion attack. In a similar vein, noteworthy is the exceptional performance of CNT1 and CNT2, displaying longevity of around 200 years in resisting chloride ions, surpassing the other mixtures. It is essential to highlight that the OPC result is drawn from prior research and is included for comparative purposes.

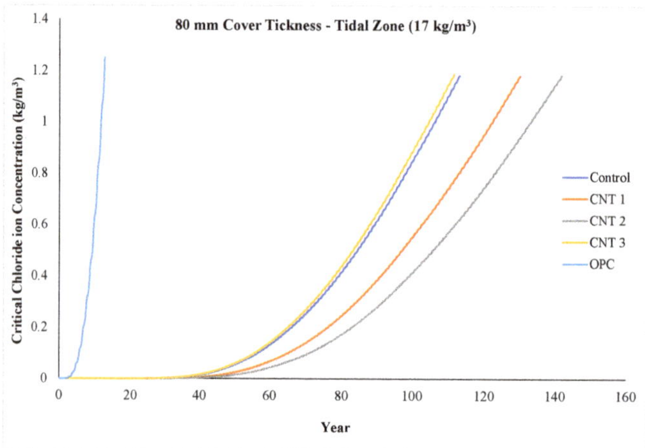

(**a**) Tidal Zone (17 kg/m³)

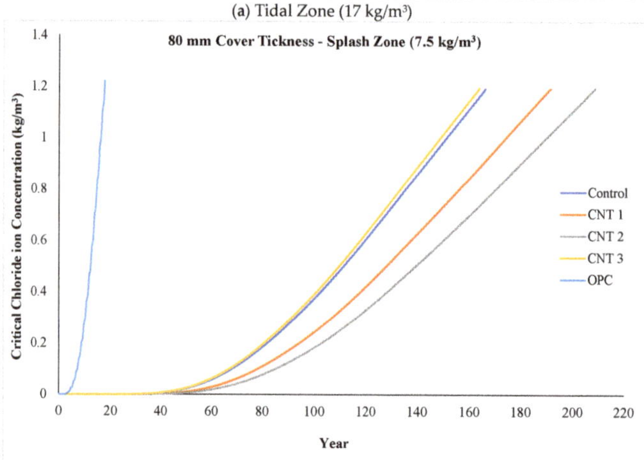

(**b**) Splash Zone (7.5 kg/m³)

Figure 12. *Cont.*

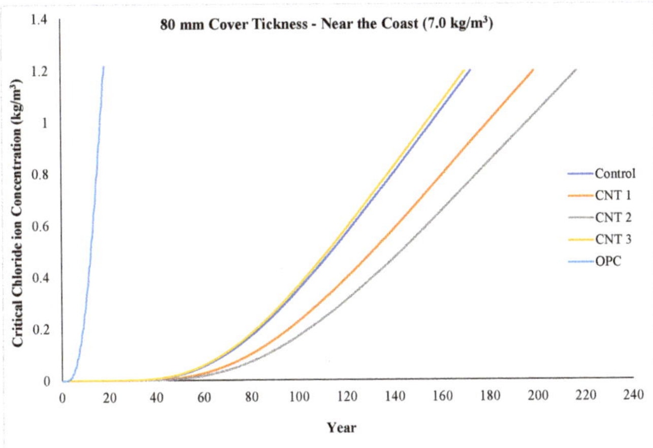

(**c**) Near the Coast (7.0 kg/m³)

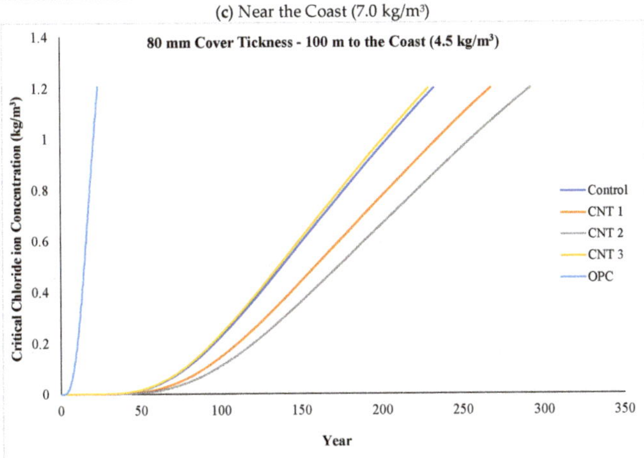

(**d**) 100 m to the Coast (4.5 kg/m³)

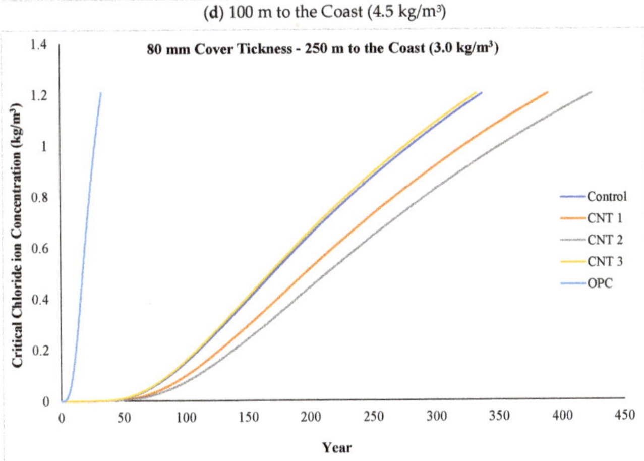

(**e**) 250 m to the Coast (3.0 kg/m³)

Figure 12. Life service prediction for all mixtures in various environmental conditions including (**a**) tidal zone, (**b**) splash zone, (**c**) near to the coast, (**d**) 100 m to the coast, and (**e**) 250 m to the coast.

Illustrated in Figure 12c are the outcomes for the near-coast condition, featuring a chloride ion concentration of 7.0 kg/m^3 for all mixtures in this investigation. The ultimate resistance age against chloride ions penetration, as indicated by the life service prediction model, for the control, CNT1, CNT2, CNT3, and OPC mixtures in this near-coast setting were 179, 205, 223, 176, and 25 years, respectively. These findings suggest that, in terms of performance, there is minimal disparity between the splash zone and the near-coast area, given that the chloride ion concentrations in these areas were 7.5 and 7.0 kg/m^3, respectively. Once again, CNT1 and CNT2 emerge as the top performers among all mixtures in this particular condition.

Figure 12d illustrates the life service prediction results in an environmental condition situated 100 m from the coast, with a chloride ion concentration of 4.5 kg/m^3. According to these outcomes, the ultimate resistance age against chloride ion penetration for the control, CNT1, CNT2, CNT3, and OPC mixtures in the 100-m coastal condition were 239, 274, 298, 235, and 30 years, respectively. Furthermore, Figure 12e represents the outcomes in the most favorable environmental condition, located 250 m from the coast, with a chloride ion concentration of 3.0 kg/m^3. In this scenario, the ultimate resistance age against chloride ion penetration for the control, CNT1, CNT2, CNT3, and OPC mixtures were 344, 396, 431, 339, and 40 years, respectively. Notably, in both scenarios, CNT1 and CNT2 exhibited superior performance compared to all other mixtures.

In summary, the life service prediction results align closely with the strength performance outcomes for all mixtures. Consistently, CNT1 and CNT2 mixtures exhibited superior performance in both compressive and tensile strength, as well as life service prediction. This can be attributed to the effective dispersion of CNT in these two mixtures, fostering the creation of additional bridges between the cement paste and ALA and MHS particles. Consequently, this contributes to the development of a more impermeable concrete matrix. As a result, these two mixtures demonstrated greater resistance against chloride ion penetration compared to the other formulations.

Ultimately, considering the required cover thickness (mm) for 200 years, detailed in Table 8, observations showed that in the tidal zone (the most challenging environmental condition), the necessary cover thickness for control, CNT1, CNT2, CNT3, and OPC mixtures were 98.71 mm, 91.90 mm, 87.99 mm, 99.49 mm, and 325.07 mm, respectively. Similarly, in the second-worst environmental condition (splash zone), the requisite cover thickness for the aforementioned mixtures was 66.72 mm, 62.11 mm, 59.47 mm, 67.25 mm, and 219.72 mm, respectively. It's noteworthy that an enhancement in the environmental condition resulted in a reduced required cover thickness for all mixtures, aligning with the positive correlation observed with service life prediction results. Conversely, in the near-coast condition, the necessary cover thickness for control, CNT1, CNT2, CNT3, and OPC mixtures were 63.98 mm, 59.56 mm, 57.03 mm, 64.48 mm, and 210.69 mm, respectively. Furthermore, at 100 m from the coast, the values for the same mixtures were 47.53 mm, 44.25 mm, 42.37 mm, 47.90 mm, and 156.51 mm, respectively. Lastly, in the optimal environmental condition (250 m from the coast), the mandated cover thickness for the same mixtures was 32.45 mm, 30.21 mm, 28.92 mm, 32.70 mm, and 106.85 mm, respectively.

Table 8. Required cover thickness (mm) over 200 years in different exposing conditions.

Expose Condition	Tidal Zone	Splash Zone	Near the Coast	100 m	250 m
Control (mm)	98.71	66.72	63.98	47.53	32.45
CNT 1 (mm)	91.90	62.11	59.56	44.25	30.21
CNT 2 (mm)	87.99	59.47	57.03	42.37	28.92
CNT 3 (mm)	99.49	67.25	64.48	47.90	32.70
OPC (mm)	325.07	219.72	210.69	156.51	106.85

5. Conclusions

This study explored the mechanical characteristics, microstructure, chloride ion permeability, and service life prediction of nano-concrete utilizing ALA and MHS as a replacement for fine aggregate, along with the addition of CNT as an admixture. The impact of diverse environmental conditions on the performance of CNT concrete mixtures was systematically examined. The evaluation of CNT mixtures involved an analysis of compressive and tensile strength, XRD, SEM, accelerated chloride penetration, and service life prediction. Based on the findings presented, the following conclusions can be derived:

1. Elevating the concentration of CNT in CNT1 and CNT2 blends resulted in a marginal enhancement in compressive strength compared to the control mix. Conversely, the CNT3 mixture demonstrated a reduction in compressive strength with an increasing content of CNT as an admixture. Furthermore, the compressive strength values at 3, 7, and 28 days in the salty curing environment were significantly inferior to those recorded under air curing periods.
2. The outcomes of tensile strength testing revealed that as the quantity of CNT increased as an admixture, the tensile strength of CNT mixtures experienced a slight decline across all curing periods. This phenomenon can be ascribed to the nuanced interaction between CNTs and the cementitious matrix. While CNTs have the potential to enhance compressive strength by acting as additional bridges and reinforcing the concrete structure, the nature of their bonding with the cementitious matrix plays a critical role in load transfer. In instances where this bonding is not optimally established, a reduction in tensile strength occurs, as observed in the present study.
3. Examination through SEM indicated that an augmented concentration of CNTs as an admixture in the concrete led to the formation of nanoscale bridges within the concrete matrix. These bridges, at the nanoscale level, act as reinforcement, establishing a network within the cementitious matrix. This network plays a crucial role in significantly enhancing the distribution and transfer of loads, thereby slightly improving the overall strength performance of the concrete during both the 3-day and 28-day normal curing periods. Furthermore, an analysis of the EDS map sum spectrum outcomes revealed that the primary elements in the structure of all mixtures include O, Si, C, and Ca, albeit in varying proportions.
4. Drawing insights from the XRD analysis outcomes, it is discerned that, across all samples, the principal crystalline phases identified after both the 3-day and 28-day curing periods consist of Quartz and Alite. This occurrence can be attributed to the potential presence of certain materials in the mix in an amorphous or non-crystalline form, a condition where XRD exhibits diminished sensitivity. Furthermore, a well-homogenized mix design may result in the prevalence of dominant crystalline phases, presenting difficulties in discerning minor phases.
5. Upon scrutinizing the ponding test results, it was evident that, for all samples, the effective chloride transport coefficient remained below the standard limitation of 1.00×10^{-12} m^2/s, indicating that the performance of all samples in the ponding test fell within an acceptable range. However, upon a more detailed examination, it was noted that the chloride transport coefficient experienced a decrease with an increase in the amount of CNT up to 0.05% (as observed in CNT1 and CNT2 mixes in comparison to the control mix). In contrast, for the CNT3 mixture, there was an increase relative to the control mix. This observed trend aligned with the compressive strength outcomes, where the strength of the CNT1 and CNT2 mixtures surpassed that of the CNT3 mixture.
6. The life service prediction outcomes indicate that in every environmental scenario, CNT1 and CNT2 mixtures demonstrated superior performance when compared to all other mixtures. These findings closely mirror the strength performance results, where CNT1 and CNT2 mixtures showcased excellence in both compressive strength and life service prediction. The effective dispersion of CNT in these two mixtures is credited for this superior performance, facilitating the formation of additional bridges

between the cement paste and ALA and MHS particles resulting in the creation of more impermeable concrete.

Author Contributions: M.R.: Conceptualization, Data curation, Writing—original draft, Writing—review, Resources Methodology. J.-H.J.K.: Supervision, Editing, Funding acquisition, Project administration. J.-S.K.: Data Collection, Validation, Investigation. J.-B.J.: Data curation, Methodology, Investigation. All authors have read and agreed to the published version of the manuscript.

Funding: This work is supported by the Korea Agency for Infrastructure Technology Advancement (KAIA) grant funded by the Ministry of Land, Infrastructure and Transport (Grant RS-2020-KA156177).

Institutional Review Board Statement: Not applicable.

Informed Consent Statement: Not applicable.

Data Availability Statement: Data will be made available on request.

Conflicts of Interest: The authors declare that they have no known competing financial interests or personal relationships that could have appeared to influence the work reported in this paper.

References

1. Saleh, S.; Mahmood, A.H.; Hamed, E.; Zhao, X.L. The mechanical, transport and chloride binding characteristics of ultra-high-performance concrete utilising seawater, sea sand and SCMs. *Constr. Build. Mater.* **2023**, *372*, 130815. [CrossRef]
2. Bogas, J.A.; Gomes, A. Non-steady-state accelerated chloride penetration resistance of structural lightweight aggregate concrete. *Cem. Concr. Compos.* **2015**, *60*, 111–122. [CrossRef]
3. Cai, R.; Tian, Z.; Ye, H. Durability characteristics and quantification of ultra-high strength alkali-activated concrete. *Cem. Concr. Compos.* **2022**, *134*, 104743. [CrossRef]
4. Rafieizonooz, M.; Mirza, J.; Salim, M.R.; Hussin, M.W.; Khankhaje, E. Investigation of coal bottom ash and fly ash in concrete as replacement for sand and cement. *Constr. Build. Mater.* **2016**, *116*, 15–24. [CrossRef]
5. Rafieizonooz, M.; Khankhaje, E.; Rezania, S. Assessment of environmental and chemical properties of coal ashes including fly ash and bottom ash, and coal ash concrete. *J. Build. Eng.* **2022**, *49*, 104040. [CrossRef]
6. Luo, Z.; Zhi, T.; Liu, X.; Yin, K.; Pan, H.; Feng, H.; Song, Y.; Su, Y. Effects of different nanomaterials on the early performance of ultra-high performance concrete (UHPC): C–S–H seeds and nano-silica. *Cem. Concr. Compos.* **2023**, *142*, 105211. [CrossRef]
7. Yu, L.; Bai, S.; Guan, X. Effect of graphene oxide on microstructure and micromechanical property of ultra-high performance concrete. *Cem. Concr. Compos.* **2023**, *138*, 104964. [CrossRef]
8. Song, B.; Liu, S.; Hu, X.; Ouyang, K.; Li, G.; Shi, C. Compressive strength, water and chloride transport properties of early CO_2-cured Portland cement-fly ash-slag ternary mortars. *Cem. Concr. Compos.* **2022**, *134*, 104786. [CrossRef]
9. Afzal, M.T.; Khushnood, R.A. Influence of carbon nano fibers (CNF) on the performance of high strength concrete exposed to elevated temperatures. *Constr. Build. Mater.* **2021**, *268*, 121108. [CrossRef]
10. Oh, T.; Chun, B.; Lee, S.K.; Kim, G.W.; Banthia, N.; Yoo, D.-Y. Effect of high-volume substituted nanosilica on the hydration and mechanical properties of Ultra-High-Performance Concrete (UHPC). *Cem. Concr. Res.* **2024**, *175*, 107379. [CrossRef]
11. Zhang, G.; Peng, G.F.; Zuo, X.Y.; Niu, X.J.; Ding, H. Adding hydrated lime for improving microstructure and mechanical properties of mortar for ultra-high performance concrete. *Cem. Concr. Res.* **2023**, *167*, 107130. [CrossRef]
12. Wang, X.; Yu, R.; Song, Q.; Shui, Z.; Liu, Z.; Wu, S.; Hou, D. Optimized design of ultra-high performance concrete (UHPC) with a high wet packing density. *Cem. Concr. Res.* **2019**, *126*, 105921. [CrossRef]
13. Jiang, J.; Qin, J.; Chu, H. Improving mechanical properties and microstructure of ultra-high-performance lightweight concrete via graphene oxide. *J. Build. Eng.* **2023**, *80*, 108038. [CrossRef]
14. Esmaeili, J.; Romouzi, V.; Kasaei, J.; Andalibi, K. An investigation of durability and the mechanical properties of ultra-high performance concrete (UHPC) modified with economical graphene oxide nano-sheets. *J. Build. Eng.* **2023**, *80*, 107908. [CrossRef]
15. Rosa, R.E.M.; Corr, D.J.; Espinosa, H.D.; Shah, S.P. Characterization of adhesion strength between carbon nanotubes and cementitious materials. *Cem. Concr. Compos.* **2023**, *138*, 104953. [CrossRef]
16. Hussain, A.; Xiang, Y.; Yu, T.; Zou, F. Nanocarbon black-based ultra-high-performance concrete (UHPC) with self-strain sensing capability. *Constr. Build. Mater.* **2022**, *359*, 129496. [CrossRef]
17. Zhang, P.; Su, J.; Guo, J.; Hu, S. Influence of carbon nanotube on properties of concrete: A review. *Constr. Build. Mater.* **2023**, *369*, 130388. [CrossRef]
18. Narasimman, K.; Jassam, T.M.; Velayutham, T.S.; Yaseer, M.M.M.; Ruzaimah, R. The synergic influence of carbon nanotube and nanosilica on the compressive strength of lightweight concrete. *J. Build. Eng.* **2020**, *32*, 101719. [CrossRef]
19. Li, L.; Wang, B.; Hubler, M.H. Carbon nanofibers (CNFs) dispersed in ultra-high performance concrete (UHPC): Mechanical property, workability and permeability investigation. *Cem. Concr. Compos.* **2022**, *131*, 104592. [CrossRef]

20. Li, C.; Li, J.; Ren, Q.; Zheng, Q.; Jiang, Z. Durability of concrete coupled with life cycle assessment: Review and perspective. *Cem. Concr. Compos.* **2023**, *139*, 105041. [CrossRef]
21. Li, K.; Han, J.; Wang, S.; Lian, H.; Xiong, J.; Wang, J.; Fan, Z.; Xu, L.; Zhu, H. Long-term performance of structural concretes in China southeast coastal environments exposed to atmosphere and chlorides. *Cem. Concr. Res.* **2023**, *164*, 107064. [CrossRef]
22. Jung, M.; Lee, Y.S.; Hong, S.G.; Moon, J. Carbon nanotubes (CNTs) in ultra-high performance concrete (UHPC): Dispersion, mechanical properties, and electromagnetic interference (EMI) shielding effectiveness (SE). *Cem. Concr. Res.* **2020**, *131*, 106017. [CrossRef]
23. Jayakumari, B.Y.; Swaminathan, E.N.; Partheeban, P. A review on characteristics studies on carbon nanotubes-based cement concrete. *Constr. Build. Mater.* **2023**, *367*, 130344. [CrossRef]
24. D'Alessandro, A.; Ubertini, F.; Materazzi, A.L.; Laflamme, S.; Porfiri, M. Electromechanical modelling of a new class of nanocomposite cement-based sensors for structural health monitoring. *Struct. Health Monit.* **2015**, *14*, 137–147. [CrossRef]
25. Aydın, A.C.; Nasl, V.J.; Kotan, T. The synergic influence of nano-silica and carbon nano tube on self-compacting concrete. *J. Build. Eng.* **2018**, *20*, 467–475. [CrossRef]
26. Yoo, D.Y.; Oh, T.; Banthia, N. Nanomaterials in ultra-high-performance concrete (UHPC)—A review. *Cem. Concr. Compos.* **2022**, *134*, 104730. [CrossRef]
27. Mudasir, P.; Naqash, J.A. Impact of carbon Nano tubes on fresh and hardned properties of conventional concrete. *Mater. Today Proc.* **2023**, *80*, 1920–1925. [CrossRef]
28. Shen, P.; Lu, L.; He, Y.; Wang, F.; Hu, S. The effect of curing regimes on the mechanical properties, nano-mechanical properties and microstructure of ultra-high performance concrete. *Cem. Concr. Res.* **2019**, *118*, 1–13. [CrossRef]
29. Jung, M.; Park, J.S.; Hong, S.G.; Moon, J. Micro- and meso-structural changes on electrically cured ultra-high performance fiber-reinforced concrete with dispersed carbon nanotubes. *Cem. Concr. Res.* **2020**, *137*, 106214. [CrossRef]
30. Shafikhani, M.; Chidiac, S.E. Quantification of concrete chloride diffusion coefficient—A critical review. *Cem. Concr. Compos.* **2019**, *99*, 225–250. [CrossRef]
31. De Weerdt, K.; Wilson, W.; Machner, A.; Georget, F. Chloride profiles—What do they tell us and how should they be used? *Cem. Concr. Res.* **2023**, *173*, 107287. [CrossRef]
32. Sanjuán, M.Á.; Rivera, R.A.; Martín, D.A.; Estévez, E. Chloride Diffusion in Concrete Made with Coal Fly Ash Ternary and Ground Granulated Blast-Furnace Slag Portland Cements. *Materials* **2022**, *15*, 8914. [CrossRef] [PubMed]
33. Dong, H.; Wang, J. Carbon nanofibers and polyvinyl-alcohol fiber hybrid-reinforced high-performance concrete: Mechanical property, chloride penetration resistance, and material characterization. *Constr. Build. Mater.* **2023**, *399*, 131891. [CrossRef]
34. Liu, Q.F.; Cai, Y.; Peng, H.; Meng, Z.; Mundra, S.; Castel, A. A numerical study on chloride transport in alkali-activated fly ash/slag concretes. *Cem. Concr. Res.* **2023**, *166*, 107094. [CrossRef]
35. Lehner, P.; Konečný, P.; Ponikiewski, T. Comparison of material properties of scc concrete with steel fibres related to ingress of chlorides. *Crystals* **2020**, *10*, 220. [CrossRef]
36. Dehghan, A.; Peterson, K.; Riehm, G.; Bromerchenkel, L.H. Application of X-ray microfluorescence for the determination of chloride diffusion coefficients in concrete chloride penetration experiments. *Constr. Build. Mater.* **2017**, *148*, 85–95. [CrossRef]
37. *NT BUILD 443*; Concrete, Hardened: Accelerated Chloride Penetration (NT BUILD 443). NORDTEST: Oslo, Norway, 1995. Available online: https://www.nordtest.info/wp/1995/11/22/concrete-hardened-accelerated-chloride-penetration-nt-build-443/ (accessed on 27 November 2023).
38. *ASTM C1543*; Standard Test Method for Determining the Penetration of Chloride Ion into Concrete by Ponding (Withdrawn 2019). American Standard Test Method: West Conshohocken, PA, USA, 2019. Available online: https://www.astm.org/c1543-10a.html (accessed on 30 November 2023).
39. *ASTM C1556*; Standard Test Method for Determining the Apparent Chloride Diffusion Coefficient of Cementitious Mixtures by Bulk Diffusion. American Standard Test Method: West Conshohocken, PA, USA, 2016. Available online: https://www.astm.org/c1556-11ar16.html (accessed on 30 November 2023).
40. *NT BUILD 492*; Nordic Cooperation, Concrete, Mortar and Cement-Based Repair Materials: Chloride Migration Coefficient from Non-Steady-State Migration Experiments (NT BUILD 492). NORDTEST: Oslo, Norway, 1999. Available online: https://www.nordtest.info/wp/1999/11/21/concrete-mortar-and-cement-based-repair-materials-chloride-migration-coefficient-from-non-steady-state-migration-experiments-nt-build-492/ (accessed on 30 November 2023).
41. Mapa, D.G.; Zhu, H.; Nosouhian, F.; Shanahan, N.; Riding, K.A.; Zayed, A. Chloride binding and diffusion of slag blended concrete mixtures. *Constr. Build. Mater.* **2023**, *388*, 131584. [CrossRef]
42. Elfmarkova, V.; Spiesz, P.; Brouwers, H.J.H. Determination of the chloride diffusion coefficient in blended cement mortars. *Cem. Concr. Res.* **2015**, *78*, 190–199. [CrossRef]
43. Runci, A.; Provis, J.; Serdar, M. Microstructure as a key parameter for understanding chloride ingress in alkali-activated mortars. *Cem. Concr. Compos.* **2022**, *134*, 104818. [CrossRef]
44. Lim, T.; Lee, J.H.; Mun, J.H.; Yang, K.H.; Ju, S.; Jeong, S.M. Enhancing functionality of epoxy–TiO_2-embedded high-strength lightweight aggregates. *Polymers* **2020**, *12*, 2384. [CrossRef]
45. Sun, S.; Han, B.; Jiang, S.; Yu, X.; Wang, Y.; Li, H.; Ou, J. Nano graphite platelets-enabled piezoresistive cementitious composites for structural health monitoring. *Constr. Build. Mater.* **2017**, *136*, 314–328. [CrossRef]
46. Ubertini, F.; Materazzi, A.L.; D'Alessandro, A.; Laflamme, S. Natural frequencies identification of a reinforced concrete beam using carbon nanotube cement-based sensors. *Eng. Struct.* **2014**, *60*, 265–275. [CrossRef]

47. García-Macías, E.; D'Alessandro, A.; Castro-Triguero, R.; Pérez-Mira, D.; Ubertini, F. Micromechanics modeling of the uniaxial strain-sensing property of carbon nanotube cement-matrix composites for SHM applications. *Compos. Struct.* **2017**, *163*, 195–215. [CrossRef]
48. Siahkouhi, M.; Razaqpur, G.; Hoult, N.A.; Baghban, M.H.; Jing, G. Utilization of carbon nanotubes (CNTs) in concrete for structural health monitoring (SHM) purposes: A review. *Constr. Build. Mater.* **2021**, *309*, 125137. [CrossRef]
49. Qu, Z.Y.; Yu, Q.L.; Brouwers, H.J.H. Relationship between the particle size and dosage of LDHs and concrete resistance against chloride ingress. *Cem. Concr. Res.* **2018**, *105*, 81–90. [CrossRef]
50. Guignone, G.C.; Vieira, G.L.; Zulcão, R.; Mion, G.; Baptista, G. Analysis of the chloride diffusion coefficients by different test methods in concrete mixtures containing metakaolin and high-slag blast-furnace cement. *Rev. Mater.* **2019**, *24*, e12512. [CrossRef]
51. Stanish, K.; Hooton, D.; Thomas, M. *Testing the Chloride Penetration Resistance of Concrete: A Literature Review*; Federal Highway Administration: Washington, DC, USA, 1997.
52. *NT BUILD 208*; Concrete, Hardened: Chloride Content by Volhard Titration (NT BUILD 208). NORDTEST: Oslo, Norway, 1996. Available online: https://www.nordtest.info/wp/1996/11/28/concrete-hardened-chloride-content-by-volhard-titration-nt-build-208/ (accessed on 27 November 2023).
53. Ehlen, M.A.; Thomas, M.D.A.; Bentz, E.C. Life-365 Service Life Prediction ModelTM Version 2.0. *Concr. Int.* **2009**, *31*, 41–46.
54. Moffatt, E.G.; Thomas, M.D.A.; Fahim, A. Performance of high-volume fly ash concrete in marine environment. *Cem. Concr. Res.* **2017**, *102*, 127–135. [CrossRef]
55. *KDS 14 20 40*; Concrete Structure Durability Design Standards. KCSC: Seoul, Republic of Korea, 2021.
56. Sun, Z.; Wan, S.G. Sachs, Compatibility of carbon nanotubes in concrete with air entrainer and superplasticizer. *Constr. Build. Mater.* **2023**, *364*, 129944. [CrossRef]
57. Liang, T.; Zhou, J.; Wu, Q. Experimental investigation on leaching behavior of ultra-high performance concrete submitted to a flow environment. *Constr. Build. Mater.* **2023**, *372*, 130843. [CrossRef]
58. Xu, Z.; Liu, Q.; Long, H.Y.; Deng, H.; Chen, Z.; Hui, D. Influence of nano-SiO_2 and steel fiber on mechanical and microstructural properties of red mud-based geopolymer concrete. *Constr. Build. Mater.* **2023**, *364*, 129990. [CrossRef]
59. Han, K.; Guo, T.; Shu, X.; Ran, Q.; Guo, Y.; Shi, J. Insight into the role of early C3A hydration in structural build-up of cement paste. *Cem. Concr. Res.* **2024**, *175*, 107354. [CrossRef]
60. Bao, J.; Zheng, R.; Zhang, P.; Cui, Y.; Xue, S.; Song, Q.; Ma, Y. Thermal resistance, water absorption and microstructure of high-strength self-compacting lightweight aggregate concrete (HSSC-LWAC) after exposure to elevated temperatures. *Constr. Build. Mater.* **2023**, *365*, 130071. [CrossRef]
61. Liu, C.; Zhang, M. Microstructure-based modelling of chloride diffusivity in non-saturated cement paste accounting for capillary and gel pores. *Cem. Concr. Res.* **2023**, *168*, 107153. [CrossRef]
62. Moon, J.; Bae, S.; Celik, K.; Yoon, S.; Kim, K.-H.; Kim, K.S.; Monteiro, P.J.M. Characterization of natural pozzolan-based geopolymeric binders. *Cem. Concr. Compos.* **2014**, *53*, 97–104. [CrossRef]

materials

MDPI

Article

Carbonation Resistance of Ternary Portland Cements Made with Silica Fume and Limestone

Miguel Ángel Sanjuán [1,*], Esperanza Menéndez [2] and Hairon Recino [2]

[1] Spanish Institute of Cement and Its Applications (IECA), C/José Abascal, 53, 28003 Madrid, Spain
[2] The Eduardo Torroja Institute for Construction Science (Spanish National Research Council, CSIC), C/Serrano Galvache, 4, 28033 Madrid, Spain; emm@ietcc.csic.es (E.M.); h.recino@ietcc.csic.es (H.R.)
* Correspondence: masanjuan@ieca.es; Tel.: +34-914429166

Abstract: Ternary blended cements, made with silica fume and limestone, provide significant benefits such as improved compressive strength, chloride penetration resistance, sulfates attack, etc. Furthermore, they could be considered low-carbon cements, and they contribute to reducing the depletion of natural resources in reference to water usage, fossil fuel consumption, and mining. Limestone (10%, 15%, and 20%) with different fineness and coarse silica fume (3%, 5%, and 7%) was used to produce ternary cements. The average size of coarse silica fume used was 238 μm. For the first time, the carbonation resistance of ternary Portland cements made with silica fume and limestone has been assessed. The carbonation resistance was assessed by natural carbonation testing. The presence of coarse silica fume and limestone in the blended cement led to pore refinement of the cement-based materials by the filling effect and the C-S-H gel formation. Accordingly, the carbonation resistance of these new ternary cements was less poor than expected for blended cements.

Keywords: sustainable materials; service life; carbonation; coarse silica fume; limestone

1. Introduction

Cement manufacturing is a major contributor to anthropogenic global warming [1], which accounts for approximately 7.4% [2–4] of worldwide carbon dioxide emissions. Consequently, life cycle analyses have demonstrated that Portland cement is responsible for 60–80% of carbon dioxide emissions from concrete manufacturing [5] since the production of one ton of Portland cement clinker releases 800–1100 kg of carbon dioxide into the atmosphere for fuel combustion (white clinker: 480–560 kg CO_2; gray clinker: 280–330 kg CO_2) and for the calcination process (white clinker: 540–520 kg CO_2; gray clinker: 530–520 kg CO_2) [6]. Accordingly, we should be able to reduce carbon dioxide release associated with Portland cement production as far as possible with current know-how. Efforts to lower the carbon dioxide emissions from Portland cement manufacturing include alternative fuels and raw materials during Portland cement clinker production and low-carbon cements with a low clinker factor. In addition, for climate change mitigation purposes, a more specific and complete inventory may be needed by considering the carbon dioxide uptake by cement-based materials [7]. Through this, it would be possible to establish a more complete life cycle inventory of concrete in terms of carbon footprint [8–12]. Currently, some low-carbon concretes with embodied carbon dioxide below 100 kg/m^3 $CO_{2\,eq.}$ are available. This means a reduction of carbon dioxide by over 70% versus standard concrete made with CEM I.

Carbon dioxide uptake is the amount of mentioned gas that has been chemically bound by the cement paste constituents and pore solution contained in the hardened concrete. It should be expressed as the mass of bound carbon dioxide per square meter of the considered structure [13].

According to Pade and Guimaraes [14], concrete structures built in 2003 will be able to bind around 28% of the carbon dioxide emissions from cement production during the next 70 years. By contrast, Yang et al. [15] reported that concrete structures could bind, for a period of a hundred years, about 18–21% of the process carbon dioxide emissions. This percentage was lowered by Fitzpatrick et al. [16] to about 16% for a period of carbonation of one hundred years (concrete structures produced in 1972). The highest percentage of bound carbon dioxide reported in the literature was 43% [17] from a global study considering worldwide data from 1930 to 2013. A detailed comparison of some studies was reported elsewhere [6].

Carbonation of cement-based materials is a natural aging physicochemical process widely studied in the literature [11,18–25], in which carbon dioxide diffuses from the atmosphere through the mortar or concrete capillary pores and reacts with carbonatable products present in the pore solution or in the solid phase, such as calcium hydroxide known as portlandite, calcium silicate hydrate, named as C-S-H gel, ettringite, and to a lesser extent, anhydrous phases of clinker, specifically tricalcium silicate, C_3S, and dicalcium silicate, C_2S [24]. The main reaction is the reaction of calcium hydroxide, $Ca(OH)_2$, with carbon dioxide, CO_2, to form calcium carbonate, $CaCO_3$ [10,18,19,26,27]. Then, the pH of the concrete pore solution decreases, and the risk of steel rebar corrosion increases [18,24–26]. Consequently, this process can negatively affect the durability of reinforced concrete on the one hand, but it can be a way to mitigate climate change impacts on the other [27–30]. In a nutshell, carbonation is a significant carbon dioxide sink that is not yet included in life cycle inventories.

Only a few studies were focused on the bound carbon dioxide quantification [18] and therefore, the estimation of carbon dioxide uptake remains a very challenging task. Andersson et al. proposed a simplified estimation method named Tier 1 [31], whereas the European standard EN 16757 defines an advanced method or Tier 2 in its Annex G, formerly known as Annex BB [13]. This European standard estimates the carbon dioxide uptake considering a direct relationship between carbonation and reactive CaO content in concrete. Several steps would have to be taken in order to estimate the carbon dioxide uptake, more specifically, (i) degree of carbonation, (ii) carbonation rate, which can be estimated from both the concrete compressive strength class and the field exposure conditions, and (iii) maximum theoretical carbon dioxide uptake in fully carbonated concrete estimated from the reactive CaO content in the binder.

Yang et al. [15] and Fitzpatrick et al. [16] found a good agreement between the results obtained by applying the experimental procedure and those calculated by applying the method defined in the European standard EN 16757. By contrast, Younsi et al. [32] found that the European standard's method underestimates the carbon dioxide uptake of concretes made with additions since the carbonation depth was underestimated. In particular, the model underestimates both the carbonation rate of up to 61% [33] and the maximum theoretical carbon dioxide uptake of up to 77% for high content of ground granulated blast-furnace slag in the concrete. This fact was highlighted by Andrade et al. [30]. Furthermore, the degree of carbonation under indoor exposure was underestimated, while under outdoor exposure was overestimated [33]. Summing up, the method defined in the European standard EN 16757 works best under outdoor exposure [33]. Although full consensus is highly desirable, it would be particularly valuable to continue the comparison in natural carbonation, experimental versus estimated results. Currently, there is a lack of availability of experimental data to perform the comparison with theoretical models [18]. This is a very difficult challenge, indeed, particularly in the case of concrete made with blended cements.

Blended cements with a high content of Portland cement constituents, other than clinker, will contribute to the drawdown of carbon dioxide emissions and, therefore, the negative environmental impact of Portland cement production. In an attempt to produce low-carbon and cost-effective blended cements, new formulations are currently being investigated [34]. Zeraoui et al. [35] studied ternary binders (Portland cement-ground granulated blast-furnace slag-flash-calcined sediment) and reported that 10% flash-calcined sediment plus 40% ground granulated blast-furnace slag can replace 50% of Portland cement. However, the water demand increases with the use of flash-calcined sediment, but it enhances the compactness and density of the mortar.

The rheological properties of ternary cements made of coal fly ash, silica fume or limestone powder and Portland cement were assessed by Srinivas et al. [36]. They found that silica fume and limestone, ranging from 5% to 10%, improved the buildability. Zhao et al. [37] reported the enhancement of the packing density of ground slag-silica fume-cement pastes and their increase of 28-days compressive strength.

Silica fume, fly ash, and limestone powder were used to produce Ultra-high-performance concrete by Li et al. [38] to model their pozzolanic reactions. Silica fume showed the strongest effect on the compressive strength, followed by the coal fly ash and limestone powder. In addition, nanosilica (2%), calcined clay (23%), and Portland cement (75%) promote the conversion of macropores into mesopores [39]. However, Papatzani and Paine [40] found that 10% of silica fume addition to ternary systems Portland cement-limestone-fly ash could be excessive since unreacted particles were observed by SEM. Ultra-high-performance fiber-reinforced concrete (UHPFC) can be made by using ternary cement with silica fume and limestone, as reported by Kang et al. [41]. According to Li et al. [42], the limestone promotes a plasticization effect; furthermore, they reported that the optimal content of limestone powder for UHPFC is 50% by volume.

Natural carbonation, also known as (re-)carbonation, permanently stores carbon dioxide. One point that should be underlined, particularly within the climate change context, is that the carbonation rate of blended cements is faster than that of CEM I cements. Therefore, this fact should be clearly considered in *The Seventh Assessment Report of the Intergovernmental Panel on Climate Change* (IPCC). Figure 1 shows the first attempt made by the *Global Carbon Budget* report [43] to estimate the (re-)carbonation from 1960 to 2023, i.e., carbon dioxide uptake by cement-based materials (line blue in Figure 1). It means that above 700 Mtons/year could be attributed to the carbonation sink. This estimation is compared to the carbon dioxide uptake estimated by the Tier 1 methodology (dashed line green in Figure 1) defined and calculated in reference [7].

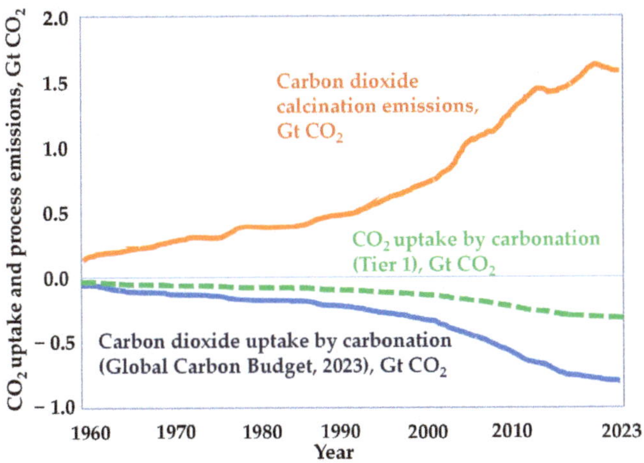

Figure 1. Global carbon dioxide uptake by cement-based materials from 1960 to 2023, reported by the *Global Carbon Budget* report, and the carbon dioxide uptake estimated by the Tier 1 methodology «source: Ref. [7]».

Accordingly, a ternary cement made with Portland cement clinker, limestone powder, and a third cement constituent, for example, coarse silica fume, can be a promising construction product due to the good experience [34,44], the adequate availability of the constituents, and the low environmental impact.

Ground limestone is already being used worldwide in Portland cement, providing some characteristics to the cement depending on its particle size distribution (PSD) and its content in the blended cement, i.e., dilution, filler, nucleation, and chemical effects [34].

Silica fume is the world's most widely used ultrafine particles for cement and concrete production because of their good durability and mechanical properties. This is explained by the optimization of the packing density [45,46] and, above all, by the pozzolanic reaction, i.e., portlandite is consumed to produce additional C-S-H gel with a lower Ca/Si ratio [44,47], and providing nucleation sites to C-S-H gel.

In view of these considerations, the carbonation resistance of clinker–limestone–silica fume ternary cements is assessed. Furthermore, this paper provides data on carbon dioxide uptake due to the natural carbonation of clinker–limestone–silica fume ternary cements. This kind of information is currently lacking in the scientific literature, particularly for concretes made with blended cements. These data are required to evaluate the actual environmental impact of concrete more accurately in civil engineering and building.

In the following, the natural carbonation results of this experimental research program are discussed to assess the effect of both ternary cement constituents' nature and curing conditions on carbonation resistance, carbon dioxide uptake of ternary cement mortars and service life estimation. For the first time, the carbonation resistance of ternary Portland cements made with silica fume and limestone has been assessed.

2. Materials and Methods

2.1. Materials

The experimental study was carried out on concretes designed with CEM I as per the European standard EN 197–1 [48] from Holcim España, Villaluenga de la Sagra, Toledo, Spain, and blended cements containing silica fume or limestone manufactured by mixing silica fume (H) from Ferroglobe PLC, Sada, Spain, and limestone (L) supplied by Holcim España, Villaluenga de la Sagra, Toledo, Spain with the Portland cement (C). The limestone was ground to reach three different fineness (given as percentage retained on the 120 μm sieve): 10% (8001 cm^2/g), 20% (25,857 cm^2/g), and 50% (25,954 cm^2/g). Grinding times were 10, 20, and 50 min, respectively. Table 1 shows the chemical analysis of the samples. Most of the elements were analyzed using the molten pearl X-ray fluorescence technique, with a wavelength scattering X-ray spectrometer, Bruker's S8 Tiger.

Table 1. Chemical composition of silica fume (H), limestone (L), and CEM I (%) [34].

Chemical Composition (%)	CEM I	H	L	Physical Properties of CEM I	
SiO_2	20.0	96.1	3.4	Specific gravity (g/cm^3)	3.11
Al_2O_3	4.5	0.2	1.6	Initial setting time (min)	160
Fe_2O_3	2.7	0.1	0.4	Final setting time (min)	240
CaO	63.0	0.4	46.3	Volume expansion (mm)	0.0
MgO	1.9	0.1	0.3	Specific surface Blaine (cm^2/g)	3811
SO_3	3.1	0.1	0.1		
K_2O	0.9	0.4	0.2	**Compressive Strength (MPa)**	
Ti_2O_5	0.2	0.0	0.1	1 days	14.32
P_2O_5	0.1	0.0	0.0	7 days	50.50
LOI	3.2	2.4	47.5	14 days	55.28
Na_2O	0.3	0.2	0.1	28 days	59.25
Cl^-	0.1	0.0	0.0		

Loss on ignition (LOI) and sulfate content determination are described in the European standard EN 196-2 [49]. The alkali content (Na^+ and K^+) in cement, limestone, and coarse silica fume was determined by inductively coupled plasma optical emission spectrometry (ICP-OES) with a Varian model 725-ES equipment.

In addition, a siliceous sand (0/4 mm) from IETcc, as per the European standard EN 196-1 [50], was used to manufacture the mortars.

The ternary mix design is given in Table 2, and the criterion of coding is as follows:

- Reference: CEM I 42.5 R (100wt% CEM I).
- H: Silica fume content (0wt%, 3wt%, 5wt%, 7wt%).

Table 2. Ternary cement mix design: silica fume (H), limestone (L) and cement (CEM I 42.5 R).

Cement Mix Code	CEM I (%)	H (%)	Total of Limestone (%)	10% Retained Limestone (%) (8001 cm^2/g)	20% Retained Limestone (%) (25,857 cm^2/g)	50% Retained Limestone (%) (25,954 cm^2/g)
Reference	100	0	0	0	0	0
H3L15-0-0	82	3	15	15	0	0
H3L20-0-0	77		20	20	0	0
H3L0-15-0	82		15	0	15	0
H3L10-10-10	67		30	10	10	10
H5L0-0-0	95	5	0	0	0	0
H5L0-15-0	80		15	0	15	0
H5L15-0-0	80		15	15	0	0
H7L0-15-0	78	7	15	0	15	0
H7L15-0-0	78		15	15	0	0
H7L20-0-0	73		20	20	0	0
H7L10-10-0	68		25	10	10	5

LX1-X2-X3: Limestone content corresponding to a sieve non-passing fraction at a mesh width of 120 μm of not more than 10% by weight (X1), 20% (X2), and 50% (X3).

Since the silica fume has a high SiO_2 content (96%), most standards limit its content in cement to less than 10%. Accordingly, the coarse silica fume content has been distributed between 0% and 10% (3%, 5%, and 7%). Regarding the limestone, it was pretended to simulate CEM II/A-L and CEM II/A-M and, therefore, the following amounts have been chosen: 15%, 20%, and 30%.

These new cements were used to manufacture prismatic mortar specimens (40 mm × 40 mm × 160 mm) with a cement-to-sand ratio of 1:3 and a water-to-cement ratio of 0.5 with distilled water and CEN standard sand [49]. Mortar mixing, molding, and curing are defined in the European standard EN 196-1 [49].

Mortar specimens were cured under lime water for 0, 3, or 28 days, rich in alkaline ions. The reason for adding lime to the curing water is to prevent the leaching out of the mortar pore solution. Limewater is a saturated aqueous solution of calcium hydroxide, which is sparsely soluble at room temperature in water (1.5 g/L at 25 °C).

2.2. Natural Carbonation Testing

The mortars were subjected to 2 years of natural carbonation (Figure 2) under sheltered conditions in the lab (40–50% RH and 25 °C), following the procedure defined in reference [24]. During the natural carbonation exposure period, the progress of carbonation was monitored by regular carbonation depth measurements of 28, 90, 180, 270, 365, and 730 days.

The mortar specimens were cut into strips of 20 mm size. Cutting with a saw in a horizontal position is suitable for determining the carbonation depth in a reliable, flexible and inexpensive manner [23]. The depth of carbonation was measured on the freshly sawn surface, which was previously cleared of dust and loose particles by spraying a mist of the phenolphthalein indicator solution (1% by weight) [51,52], which is colorless below pH 8.5 and attains a purple hue above pH 9.0.

Figure 2. Carbonation depth testing made by exposing specimens in natural atmosphere.

3. Results and Discussion

3.1. Carbonation Depth

Following the procedures given by RILEM [51] and provided in the CEN/TS 12390-10 Technical Specification [52], it is proposed to assess several aspects related to the carbonation resistance of these ternary cements, such as the carbonation depth, carbonation rate, and CO_2 uptake.

The mortars were subjected to two years of natural carbonation under sheltered conditions. During the natural carbonation exposure period, the progress of the mortar carbonation was monitored by regular carbonation depth measurements of 28, 90, 180, 270, 365, and 730 days (Figures 2–4).

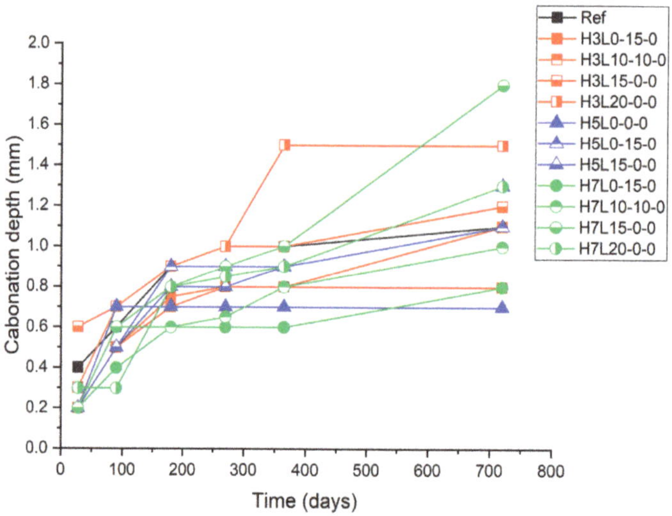

Figure 3. Carbonation depth of uncured samples.

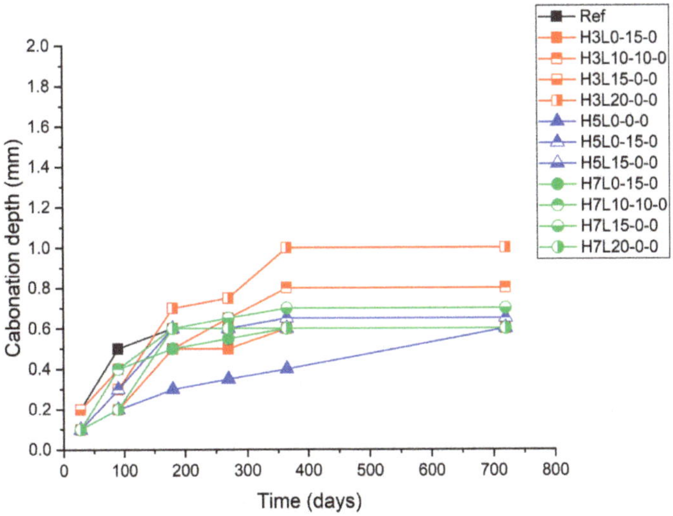

Figure 4. Carbonation depth of samples cured for 3 days.

Figure 3 shows the carbonation depth increase with the time of uncured samples, while Figures 4 and 5 show the carbonation depth increase with the time of cured samples for 3 days and 28 days, respectively. As expected, with regard to the effect of the curing time on the carbonation resistance, the longer, the better [23,24].

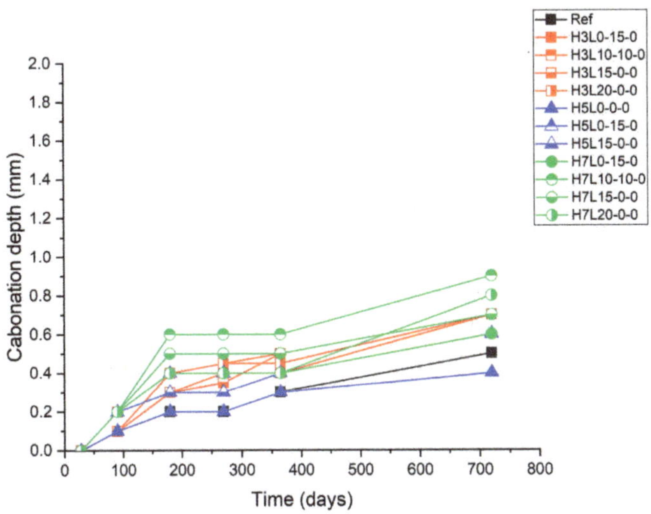

Figure 5. Carbonation depth of samples cured for 28 days.

For samples without curing, the lowest carbonation depth was found in specimens made with the highest silica fume content (5% and 7%), i.e., H5L0-0-0 and H7L0-0-0, followed by the reference cement made without any addition. This finding confirms the positive effect of the pozzolanic reaction of the silicon oxide present in the silica fume and the calcium oxide formed in the cement hydration [39]. This reaction leads to a denser structure [34]. However, the worst performance was found in the sample with the highest content of limestone (20%), H7L20-0-0 and H5L20-0-0, independently of the silica fume content (7% and 5%, respectively). This fact shows that bad curing affects limestone cements more negatively than silica fume ones. Accordingly, the use of silica fume in ternary cements made with limestone cannot improve the performance of limestone cements regarding the carbonation resistance of ternary cements.

Figure 4 shows that a three-day curing significantly enhances the carbonation resistance. In this case, cement with the lowest content of silica fume (3%) and the highest amount of limestone (20%), H3L10-0-0 and H3L20-0-0, presents the largest carbonation depth. This means that the increase of the silica fume content from 3% to 5% or 7% improves the performance (carbonation resistance).

Finally, samples cured for 28 days exhibited carbonation depths between 0.4 mm and 1 mm after 730 days of natural carbonation, as well as the samples cured for 3 days.

Test results for ternary cement carbonation have shown that curing conditions, particularly curing time, have a significant effect on the performance of samples. Accordingly, curing conditions should be optimized in relation to their performance [24]. Regarding the present results, the absence of wet-curing affects in a different way to the tested samples; indeed, at 28 days of curing for non-wet-curing, carbonated depths vary several tenths of a millimeter. Increasing the curing period to three days is sufficient for mortars with a high silica fume content; for others, curing must be longer.

Furthermore, the curing effect also depends on the ternary cement mix design. For example, increasing the curing period improves the carbonation resistance of mortars made with limestone. Accordingly, three days of curing could be enough for ternary cements made with silica fume and limestone.

Figure 6 shows the carbonation depth measured on samples exposed to 365 days of natural carbonation. The phenolphthalein indicator solution was applied to the fresh fracture surface of the mortar. When the pH is above 8.6, the phenolphthalein indicator solution turns purple. By contrast, the pH of the mortar is below 8.6, where the phenolphthalein indicator remains colorless, suggesting the carbonation of the mortar.

The pH of the CEM I mortar pore solution is normally about 13–14 since it is saturated with calcium hydroxide and also consists of sodium and potassium hydroxide. Ternary cements are less alkaline than sound mortar (CEM I); therefore, they should have lower pH values of around 10–12 [16,18]. However, it is clearly a strong color change with the phenolphthalein indicator, as shown in Figure 6. Nevertheless, the phenolphthalein indicator solution procedure is frequently said to underestimate the actual carbonation depth since the color change occurs only when the pH drops below 9.8 for phenolphthalein [53]. In addition, Schultheiß et al. [54] found that some models, such as crack influence factor (CIF) approaches, underestimate the carbonation depth with increasing exposure time.

According to Zhang et al. [55], confocal Raman microscopy (CRM)-$CaCO_3$ maps for measuring carbonation depth in cement-based materials present the same carbonation depth results to phenolphthalein solution, i.e., the average difference between both methods, about 1.2% [55]. Nevertheless, phenolphthalein is less reliable for low carbonation depths since it underestimates the true value. This fact mainly happens in short-term natural carbonation tests. Furthermore, Zhang et al. [55] suggested that phenolphthalein color change correlates well with the depletion of portlandite measured with confocal Raman microscopy (CRM) and other techniques. By contrast, Shi et al. [56] reported that the carbonation depth measured with phenolphthalein reflects the depletion of C-S-H gel with high calcium indicated by using thermodynamic modeling.

Cement	No curing	3 days of curing	28 days of curing
CEM I			
	1 cm	0.6 cm	0.3 cm
H3L0-15-0			
	0.8 cm	0.5 cm	0.4 cm
H3L10-10-0			
	0.8 cm	0.8 cm	0.5 cm
H3L15-0-0			
	1 cm	0.6 cm	0.5 cm
H3L20-0-0	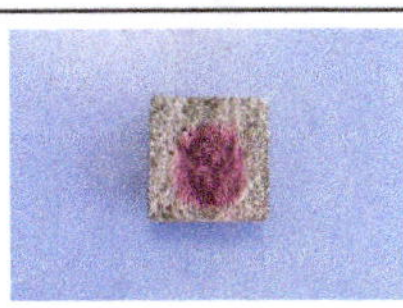		
	1.5 cm	1 cm	0.45 cm
H5L0-0-0			
	0.7 cm	0.4 cm	0.3 cm
H5L0-15-0	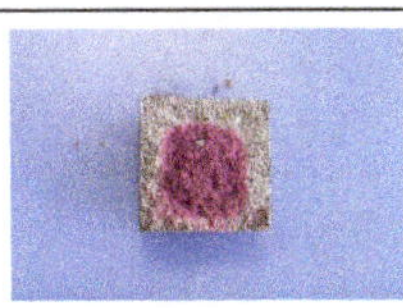		
	0.9 cm	0.65 cm	0.4 cm

Figure 6. *Cont.*

Cement	No curing	3 days of curing	28 days of curing
H5L15-0-0			
	0.9 cm	0.6 cm	0.4 cm
H7L0-15-0			
	0.6 cm	0.6 cm	0.4 cm
H7L10-10-0	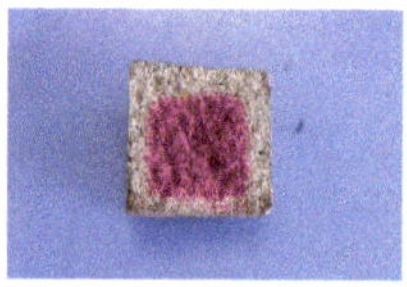		
	0.8 cm	0.7 cm	0.6 cm
H7L15-0-0	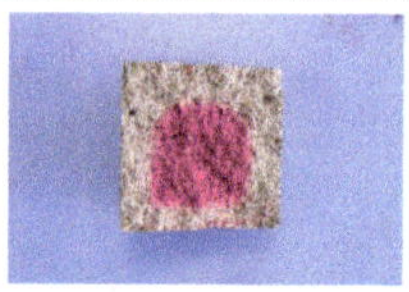		
	1 cm	0.7 cm	0.5 cm
H7L20-0-0			
	0.9 cm	0.6 cm	0.4 cm

Figure 6. Carbonation depth of samples after 365 days of natural carbonation.

The carbonation fronts shown in Figure 5 are sharp and reflect a gradual transition. By contrast, some authors [57,58] found non-pronounced carbonation fronts. It should be highlighted that non-cured mortars exhibited the highest level of dispersion in the carbonation depth. On the other hand, in both the 3-day curing condition and the 28-day curing condition, the carbonation rate increases with the higher silica fume content in the mortars. This fact can be attributed to the decrease in portlandite since it reacts with the reactive silicon present in the silica fume [9,21], which is an amorphous, highly reactive pozzolan, to form a new C-S-H gel with low Ca/Si ratio [21]. Therefore, the pH of the pore solution decreases. Furthermore, silica fume helps in accelerating the hydration of C_3S, C_2S, and C_4AF [34].

3.2. Carbonation Coefficient

Some models can be found in the literature for depicting the carbonation of cement-based materials [59], and most of them proceed by diffusion. Accordingly, the carbonation rate can be estimated from the carbonation depth measurements. Therefore, carbonation results have been modeled and analyzed according to Equation (1), where A (mm/$s^{0.5}$) is the carbonation coefficient [21,60]. This parameter is assumed to be constant [4,5,8]. Nevertheless, it depends on the environmental relative humidity, pore size distribution (PSD), hydration degree, carbon dioxide concentration, and binder composition, among others [61].

$$X = A\ \sqrt{t} \tag{1}$$

where X is the carbonation depth (mm), and t is the natural carbonation exposure time (year).

Figure 6 shows the carbonation coefficient, A, calculated from the carbonation depth results measured in mortars without wet curing (Figure 7a) or cured for 3 days (Figure 7b) or 28 days (Figure 7c), at 365 days and 730 days of natural carbonation. As expected, the longer the curing period, the better the carbonation resistance, i.e., lower carbonation coefficient [22–24].

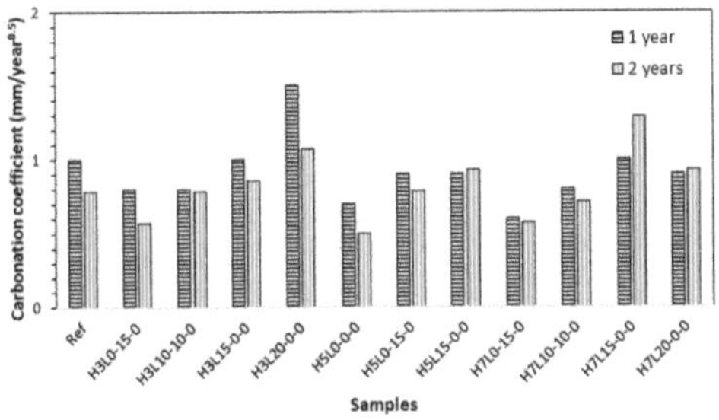

(a)

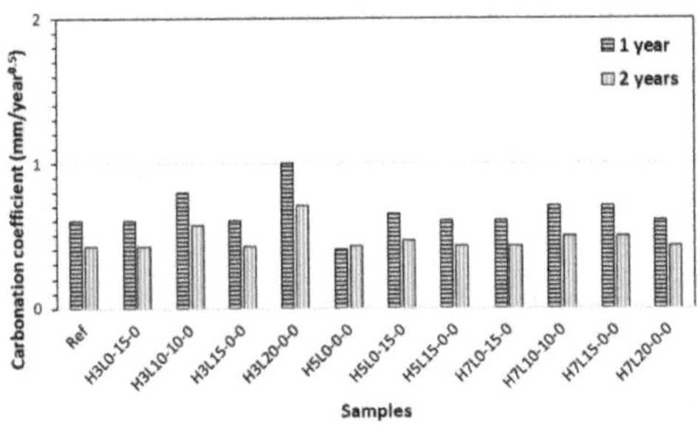

(b)

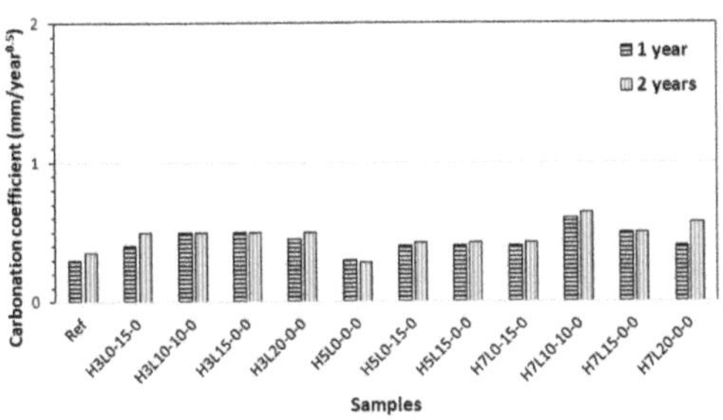

(c)

Figure 7. Carbonation coefficient calculated from the carbonation depth results measured at 365 and 730 days of natural carbonation of mortars: (**a**) without wet curing; (**b**) cured under water for 3 days; (**c**) cured under water for 28 days.

In most cases, the carbonation coefficient, A, calculated from the carbonation depth results measured in mortars after 365 days of exposure, was higher than the one calculated with the results taken after 730 days of exposure. This finding was especially pronounced in the mortars without curing or those that were cured for 3 days, while in the mortars cured for 28 days, we did not observe significant differences. This finding suggests that it would be necessary to constantly assess the carbonation coefficient if the cement-based material was not properly cured. As expected, there is a direct relationship between the carbonation depth and the carbonation coefficient at a certain age given by Equation (1).

3.3. Service Life Estimation

The Eurocodes are a series of 10 European standards coded from EN 1990 to EN 1999 [62], providing a common European approach for the design of civil engineering works, buildings and other construction products. Eurocode 2 (EN 1992) "Design of concrete structures" [62] specifies technical rules for the design of concrete, reinforced concrete, and prestressed concrete structures, using the limit state design philosophy. In particular, EN 1992-1-1 [62] deals with the rules and concepts for serviceability, safety, and durability of reinforced concrete structures [62]. This European standard considers the durability and cover to reinforcement requirements with regard to the carbon dioxide ingress rate and concrete cover. Chapter 6.4 defines the "exposure resistance classes, ERC" and classifies concrete with respect to resistance against corrosion induced by carbonation (class XRC). In addition, Annex P provides an alternative cover approach for durability without the use of ERC as defined in Chapter 6.4; in this case, the deemed-to-satisfy approach given in EN 206 is followed [63]. Table 3 shows the minimum concrete cover depth for carbon reinforcing steel required for corrosion induced by carbonation according to Eurocode 2, Annex P [64]. The recommended Structural Class is S4 for the standardized compressive concrete strengths (XC1: C20/25; XC2: C25/30; XC3, and XC4: C30/37) and design working life of 50 years.

Table 3. Minimum concrete cover depth (mm) for corrosion induced by carbonation for carbon reinforcing steel [64].

Environmental Requirement for Minimum Concrete Cover Depth (mm)	Structural Class					
Exposure Class: Corrosion Induced by Carbonation	**S1**	**S2**	**S3**	**S4**	**S5**	**S6**
XC1—Dry or permanently wet	10	10	10	15	20	25
XC2—Wet, rarely dry	10	15	20	25	30	35
XC3—Moderate humidity	10	15	20	25	30	35
XC4—Cyclic wet and dry	15	20	25	30	35	40

Natural carbonation results on ternary Portland cement mortars made with silica fume and limestone will be used to assess the potential for improvement of this new type of cement in comparison with the reference cement (CEM I). Furthermore, the minimum concrete cover required to prevent corrosion against carbonation specified by Eurocode 2, shown in Table 3, is taken as a benchmark figure. Since natural carbonation in the present study was performed at 60%RH and sheltered from rain, the corrosion exposure class induced by carbonation corresponds to the one coded as XC3 (Table 3). Accordingly, an S1 structural class requires a minimum cover thickness of 10 mm.

Table 4 shows that the carbonation depth estimated for 100 years of service life is lower than 10 mm for all the mixes when the mortars are cured for at least three days. By contrast, H3L20-0-0 and H7L15-0-0 mortars without curing exhibited carbonation depths of 10.7 mm and 12.8 mm, respectively. Therefore, these ternary cements comply with the requirements set out in the specification for S2—S6 structural classes (corrosion exposure class: XC3).

Table 4. Estimation of the carbonation depth (mm) for 100 years' service life (Exposure Class: Corrosion induced by carbonation).

	Curing Time/Carbonation Depth (mm)		
Mortar Code	**0**	**3**	**28**
Reference	7.8	4.3	3.6
H3L0-15-0	5.7	4.3	5.0
H3L10-10-0	7.8	5.7	5.0
H3L15-0-0	8.5	4.3	5.0
H3L20-0-0	10.7	7.1	5.0
H5L0-0-0	5.0	4.3	2.8
H5L0-15-0	7.8	4.6	4.3
H5L15-0-0	9.3	4.3	4.3
H7L0-15-0	5.7	4.3	4.3
H7L10-10-0	7.1	5.0	6.4
H7L15-0-0	12.8	5.0	5.0
H7L20-0-0	9.3	4.3	5.7

The choice of an adequate mix design for carbonation-induced reinforcement corrosion protection requires the consideration of several factors: composition (cement type), curing conditions, and exposure class (XC). In addition, the European standard EN 206 [63] specifies three strength classes: C20/25 for XC1, C25/30 for XC2, and C30/37 for XC3 and XC4. Consequently, the durability design of reinforced concrete structures commonly utilizes the deemed-to-satisfy rules concept, i.e., concrete mix design and concrete cover, which is mainly based on experience. This approach works well for traditional materials for which longtime experience is at the disposal of engineers. However, new Portland cement constituents and mix designs need assessment based on performance testing.

Nowadays, the exposure resistance classes (ERC) concept is proposed to classify concrete with regard to the resistance against corrosion induced by carbonation (XRC class). This system follows a performance-based concrete approach. Considering the exposure resistance classes given in EN 206 [63] and shown in Table 3 for corrosion induced by carbonation, the quality of the concrete given by the maximum carbonation coefficient (mm/year$^{0.5}$) and concrete cover (mm) are set up in the structural project.

Table 5 provides the maximum carbonation coefficient and concrete cover required for the exposure resistance classes (ERC) given for the XC3 exposure class and a design service life of 50 years defined in Chapter 6.4 of the EN 1992-1-1 [62].

Table 5. Maximum carbonation coefficient (mm/year$^{0.5}$) and minimum concrete cover (mm) for carbon reinforcing steel required for the exposure resistance classes (ERC) given for the XC3 exposure class and a design service life of 50 years [64].

Exposure Resistance Classes (ERC)	XRC 0.5	XRC 1	XRC 2	XRC 3	XRC 4	XRC 5	XRC 6	XRC 7
Cover (mm)	10	10	15	20	25	25	35	40
Maximum carbonation coefficient (mm/y$^{0.5}$)	0.6	1.2	2.4	2.7	3.6	4.5	5.4	6.4

All the mortar mixes cured for 28 days, with carbonation coefficients below 0.6 mm/year$^{0.5}$, can be used for all the exposure resistance classes (ERC). A similar conclusion can be reached for the mortars cured for three days, considering the carbonation coefficients obtained after two years of exposure. However, the carbonation coefficients calculated after one year of exposure are higher than the previous ones. Therefore, XRC 0.5 must be excluded. In addition, all the mortars without curing have carbonation coefficients over 0.6 mm/year$^{0.5}$, but only two over 1.2 mm/year$^{0.5}$, H3L20-0-0 and H7L15-0-0.

Therefore, XRC 0.5 is excluded for all the cases, and XRC 1 is only excluded in two cases (Table 5).

Finally, the carbonation of cement-based materials is considered by the cement sector as a lever to reach carbon neutrality by 2050 [1,2,6,8,16]. In addition, carbon dioxide uptake by mortars and concretes has recently been included in the *Global Carbon Budget* report [43] since natural carbonation stores permanently carbon dioxide. Furthermore, these blended cements usually absorb more carbon dioxide than CEM I [7,9,11,18]. Then, appropriate methodologies for future concrete mix design should consider durable and sustainable aspects, i.e., to ensure the reinforced concrete service life but also to account for carbon dioxide uptake [6].

4. Conclusions

In order to minimize climate change, the cement sector needs to develop new cements with a high addition content, such as ternary cements. For the first time, the carbonation resistance of new ternary cements made with silica fume and limestone was assessed by means of natural carbonation testing.

The results presented in this paper confirm that blended cements have a higher carbonation rate than CEM I. This finding suggests that appropriate methodologies for future concrete mix design should consider durable and sustainable aspects, i.e., to ensure the reinforced concrete service life, but also to account for carbon dioxide uptake within the climate change context.

Furthermore, the longer the curing period, the better the carbonation resistance. This finding is in agreement with the results reported in previous literature regarding other types of cements. It is well known that limestone and silica fume in blended cements led, separately, to pore refinement of the cement-based materials by the filling effect and the C-S-H gel formation, respectively. Both effects can justify that the carbonation resistance of these new ternary cements was less poor than expected.

In particular, it should be highlighted that the carbonation coefficient, A, calculated from the carbonation depth results measured in mortars after 365 days of exposure, was higher than the one calculated with the results obtained after 730 days in the mortars without curing or those that were cured for three days. In the mortar mixes cured for 28 days, there were hardly any differences found. Then, it is concluded that for studies conducted in bad curing conditions, the carbonation coefficient, A, should be considered from results at longer ages.

Finally, it should be pointed out that all the mortar mixes cured for 28 days present carbonation coefficients below 0.6 $mm/year^{0.5}$. Therefore, they can be used for all the exposure resistance classes (ERC) given in the European standard EN 206.

Author Contributions: Conceptualization, E.M., H.R. and M.Á.S.; methodology, E.M., H.R. and M.Á.S.; validation, E.M. and H.R.; investigation, E.M., H.R. and M.Á.S.; resources, E.M.; data curation, E.M., H.R. and M.Á.S.; writing—original draft preparation, E.M., H.R. and M.Á.S.; writing—review and editing, E.M., H.R. and M.Á.S.; formal analysis, H.R. and M.Á.S. All authors have read and agreed to the published version of the manuscript.

Funding: This research was funded by the CSIC (PIE 202060E176).

Institutional Review Board Statement: Not applicable.

Informed Consent Statement: Not applicable.

Data Availability Statement: Data are contained within the article.

Acknowledgments: Authors gratefully acknowledge the valuable contributions of Beatriz Aldea, Esther Puerto, and Carmen Barba.

Conflicts of Interest: The authors declare no conflicts of interest.

References

1. Zhang, C.Y.; Han, R.; Yu, B.; Wei, Y.M. Accounting process-related CO_2 emissions from global cement production under Shared Socioeconomic Pathways. *J. Clean. Prod.* **2018**, *184*, 451–465. [CrossRef]
2. Sanjuán, M.Á.; Andrade, C.; Mora, P.; Zaragoza, A. Carbon Dioxide Uptake by Cement-Based Materials: A Spanish Case Study. *Appl. Sci.* **2020**, *10*, 339. [CrossRef]
3. Bouarroudj, M.E.; Remond, S.; Bulteel, D. Use of grinded hardened cement pastes as mineral addition for mortars. *J. Build. Eng.* **2020**, *23*, 101863. [CrossRef]
4. Nie, S.; Zhou, J.; Yang, F.; Lan, M.; Li, J.; Zhang, Z.; Chen, Z.; Xu, M.; Li, H.; Sanjayan, J.G. Analysis of theoretical carbon dioxide emissions from cement production: Methodology and application. *J. Clean. Prod.* **2022**, *334*, 130270. [CrossRef]
5. Miller, S.A.; Horvath, A.; Monteiro, P.J.M. Readily implementable tech-niques can cut annual CO_2 emissions from the production of concrete by over 20%. *Environ. Res. Lett.* **2016**, *11*, 074029. [CrossRef]
6. Sanjuán, M.Á.; Argiz, C.; Mora, P.; Zaragoza, A. Carbon Dioxide Uptake in the Roadmap 2050 of the Spanish Cement Industry. *Energies* **2020**, *13*, 3452. [CrossRef]
7. Sanjuán, N.; Mora, P.; Sanjuán, M.Á.; Zaragoza, A. Carbon Dioxide Uptake Estimation for Spanish Cement-Based Materials. *Materials* **2024**, *17*, 326. [CrossRef]
8. Andersson, R.; Stripple, H.; Gustafsson, T.; Ljungkrantz, C. Carbonation as a method to improve climate performance for cement-based material. *Cement Concr. Res.* **2019**, *124*, 105819. [CrossRef]
9. Witkowski, H.; Koniorczyk, M. The influence of pozzolanic additives on the carbonation rate and Life Cycle Inventory of concrete. *Construct. Build. Mater.* **2020**, *254*, 119301. [CrossRef]
10. Jang, J.G.; Kim, G.M.; Kim, H.J.; Lee, H.K. Review on recent advances in CO_2 utilization and sequestration technologies in cement-based materials. *Construct. Build. Mater.* **2016**, *127*, 762–773. [CrossRef]
11. Kwon, S.-J.; Wang, X.-Y. CO_2 uptake model of lime-stone-powder-blended concrete due to carbonation. *J. Build. Eng.* **2021**, *38*, 102176. [CrossRef]
12. Marinković, S.; Carević, V.; Dragaš, J. The role of service life in Life Cycle Assessment of concrete structures. *J. Clean. Prod.* **2021**, *290*, 125610. [CrossRef]
13. *EN 16757*; Sustainability of Construction Works—Environmental Product Declarations—Product Category Rules for Concrete and Concrete Elements (Annex G). European Committee for Standardization (CEN): Brussels, Belgium, 2022.
14. Pade, C.; Guimaraes, M. The CO_2 uptake of concrete in a 100-year perspective. *Cement Concr. Res.* **2007**, *37*, 1348–1356. [CrossRef]
15. Yang, K.-H.; Seo, E.-A.; Tae, S.-H. Carbonation and CO_2 uptake of concrete. *Environ. Impact Assess. Rev.* **2014**, *46*, 43–52. [CrossRef]
16. Fitzpatrick, D.; Richardson, M.; Nolan, E. Sequestration of carbon dioxide by concrete infrastructure: A preliminary investigation in Ireland. *J. Sustain. Architect. Civ. Eng.* **2015**, *1*, 66–77. [CrossRef]
17. Xi, F.; Davis, S.J.; Ciais, P.; Crawford-Brown, D.; Guan, D.; Pade, C.; Shi, T.; Syddall, M.; Lv, J.; Ji, L.; et al. Substantial global carbon uptake by cement carbonation. *Nat. Geosci.* **2016**, *9*, 880–883. [CrossRef]
18. Andrade, C. Evaluation of the degree of carbonation of concretes in three environments. *Construct. Build. Mater.* **2020**, *230*, 116804. [CrossRef]
19. Habert, G.; Miller, S.A.; John, V.M.; Provis, J.L.; Favier, A.; Horvath, A.; Scrivener, K.L. Environmental impacts and decarbonization strategies in the cement and concrete industries. *Nat. Rev. Earth Environ.* **2020**, *1*, 559–573. [CrossRef]
20. AzariJafari, H.; Guo, F.; Gregory, J.; Kirchain, R. Carbon uptake of concrete in the US pavement network. *Resour. Conserv. Recycl* **2021**, *167*, 105397. [CrossRef]
21. Sanjuán, M.Á.; Andrade, C.; Cheyrezy, M. Concrete carbonation tests in natural and accelerated conditions. *Adv. Cement Res.* **2003**, *15*, 171–180. [CrossRef]
22. Sanjuán, M.Á.; Piñeiro, A.; Rodríguez, O. Ground granulated blast furnace slag efficiency coefficient (k-value) in concrete. Applications and limits. *Mater. Constr.* **2011**, *61*, 303–313. [CrossRef]
23. Gruyaert, E.; Heede, P.V.D.; Belie, N.D. Carbonation of slag concrete: Effect of the cement replacement level and curing on the carbonation coefficient–effect of carbonation on the pore structure. *Cement Concr. Compos.* **2013**, *35*, 39–48. [CrossRef]
24. Sanjuán, M.Á.; Estévez, E.; Argiz, C.; del Barrio, D. Effect of curing time on granulated blast-furnace slag cement mortars carbonation. *Cem. Concr. Compos.* **2018**, *90*, 257–265. [CrossRef]
25. Otieno, M.; Ikotun, J.; Ballim, Y. Experimental investigations on the effect of concrete quality, exposure conditions and duration of initial moist curing on carbona-tion rate in concretes exposed to urban, inland environment. *Construct. Build. Mater.* **2020**, *246*, 118443. [CrossRef]
26. Mäkikouri, S.; Vares, S.; Korpijärvi, K.; Papakonstantinou, N. The carbon dioxide emissions reduction potential of carbon-dioide-cured alternative binder concrete. *Recent Prog. Mater.* **2021**, *3*, 1–34. [CrossRef]
27. Benhelal, E.; Shamsaei, E.; Rashid, M.I. Challenges against CO_2 abatement strategies in cement industry: A review. *J. Environ. Sci.* **2021**, *104*, 84–101. [CrossRef] [PubMed]
28. Andrade, C.; Sanjuán, M.Á. Updating carbon storage capacity of Spanish cements. *Sustainability* **2018**, *10*, 4806. [CrossRef]
29. Sanjuán, M.Á.; Andrade, C.; Mora, P.; Zarogoza, A. Carbon dioxide uptake by mortars and concretes made with Portuguese cements. *Appl. Sci.* **2020**, *10*, 646. [CrossRef]
30. Andrade, C.; Sanjuán, M.Á. Carbon dioxide uptake by pure Portland and blended cement pastes. *Dev. Built Environ.* **2021**, *8*, 100063. [CrossRef]

31. Stripple, H.; Ljungkrantz, C.; Gustafsson, T. CO_2 Uptake in Cement-Containing Products. In *Background and Calculation Models for IPCC Implementation*, 2nd ed.; Report Number: B 2309; IVL Swedish Environmental Research Institute: Stockholm, Sweden, 2021; pp. 1–77. Available online: https://www.ivl.se/download/18.34244ba71728fcb3f3f8f9/1622457897161/B2309.pdf (accessed on 15 February 2024).
32. Younsi, A.; Turcry, P.; Aït-Mokhtar., A. Quantification of CO_2 uptake of concretes with mineral additions after 10-year natural carbonation. *J. Clean. Prod.* **2022**, *349*, 131362. [CrossRef]
33. Younsi, A. Long-term carbon dioxide sequestration by concretes with supplementary cementitious materials under indoor and outdoor exposure: Assessment as per a standardized model. *J. Build. Eng.* **2022**, *51*, 104306. [CrossRef]
34. Menéndez, E.; Sanjuán, M.Á.; Recino, H. Study of Microstructure, Crystallographic Phases and Setting Time Evolution over Time of Portland Cement, Coarse Silica Fume, and Limestone (PC-SF-LS) Ternary Portland Cements. *Crystals* **2023**, *13*, 1289. [CrossRef]
35. Zeraoui, A.; Maherzi, W.; Benzerzour, M.; Abriak, N.E.; Aouad, G. Development of Flash-Calcined Sediment and Blast Furnace Slag Ternary Binders. *Buildings* **2023**, *13*, 333. [CrossRef]
36. Srinivas, D.; Dey, D.; Panda, B.; Sitharam, T.G. Printability, Thermal and Compressive Strength Properties of Cementitious Materials: A Comparative Study with Silica Fume and Limestone. *Materials* **2022**, *15*, 8607. [CrossRef] [PubMed]
37. Zhao, Y.; Dong, X.; Zhou, Z.; Long, J.; Lu, G.; Lei, H. Investigation on Roles of Packing Density and Water Film Thickness in Synergistic Effects of Slag and Silica Fume. *Materials* **2022**, *15*, 8978. [CrossRef] [PubMed]
38. Li, P.; Wang, X.; Cao, H. Empirical Compression Model of Ul-tra-High-Performance Concrete Considering the Effect of Cement Hydration on Particle Packing Characteristics. *Materials* **2023**, *16*, 4585. [CrossRef] [PubMed]
39. Pinheiro, D.G.L.; Sousa, M.I.C.; Pelisser, F.; da Silva Rêgo, J.H.; Moragues Terrades, A.; Frías Rojas, M. Physical and Chemical Effects in Blended Cement Pastes Elaborated with Calcined Clay and Nanosilica. *Materials* **2023**, *16*, 1837. [CrossRef] [PubMed]
40. Papatzani, S.; Paine, K. A Step by Step Methodology for Building Sustainable Cementitious Matrices. *Appl. Sci.* **2020**, *10*, 2955. [CrossRef]
41. Kang, S.-H.; Hong, S.-G.; Moon, J. Performance Comparison between Densified and Undensified Silica Fume in Ultra-High Performance Fiber-Reinforced Concrete. *Materials* **2020**, *13*, 3901. [CrossRef]
42. Li, P.P.; Brouwers, H.J.H.; Chen, W.; Yu, Q.L. Optimization and characterization of high-volume limestone powder in sustainable ultra-high-performance concrete. *Constr. Build. Mater.* **2020**, *242*, 118112. [CrossRef]
43. Friedlingstein, P.; O'Sullivan, M.; Jones, M.W.; Andrew, R.M.; Bakker, D.C.E.; Hauck, J.; Landschützer, P.; Le Quéré, C.; Luijkx, I.T.; Peters, G.P.; et al. Global Carbon Budget 2023. *Earth Syst. Sci. Data* **2023**, *15*, 5301–5369. [CrossRef]
44. Sanjuán, M.Á.; Menéndez, E.; Recino, H. Mechanical Performance of Portland Cement, Coarse Silica Fume, and Limestone (PC-SF-LS) Ternary Portland Cements. *Materials* **2022**, *15*, 2933. [CrossRef]
45. Richard, P.; Cheyrezy, M. Composition of reactive powder concretes. *Cem. Concr. Res.* **1995**, *25*, 1501–1511. [CrossRef]
46. Sanjuán, M.Á.; Andrade, C. Reactive Powder Concrete: Durability and Applications. *Appl. Sci.* **2021**, *11*, 5629. [CrossRef]
47. Oertel, T.; Hutter, F.; Helbig, U.; Sextl, G. Amorphous silica in ultra-high performance concrete: First hour of hydration. *Cem. Concr. Res.* **2014**, *58*, 131–142. [CrossRef]
48. *EN 197-1*; Cement—Part 1: Composition, Specifications and Conformity Criteria for Common Cements. European Standard; The European Committee for Standardization (CEN): Brussels, Belgium, 2011.
49. *EN 196-2*; Method of Testing Cement—Part 2: Chemical analysis of cement. The European Committee for Standardization (CEN): Brussels, Belgium, 2013.
50. *EN 196-1*; Method of Testing Cement—Part 1: Determination of Strength. European Standard; The European Committee for Standardization (CEN): Brussels, Belgium, 2016.
51. *RILEM Committee CPC-18, Measurement of Hardened Concrete Carbonation Depth, (1988) TC14-CPC*; The International Union of Laboratories and Experts in Construction Materials, Systems and Structures (RILEM, from the Name in French): Champs-sur-Marne, France, 1988.
52. *CEN/TS 12390-10:2007*; Testing Hardened Concrete. Part 10: Determination of the Relative Carbonation Resistance of Concrete, CEN, 2007. The European Committee for Standardization (CEN): Brussels, Belgium, 2007.
53. Vogler, N.; Lindemann, M.; Drabetzki, P.; Kühne, H.C. Alternative pH-indicators for de-termination of carbonation depth on cement-based concretes. *Cem. Concr. Compos.* **2020**, *109*, 103565. [CrossRef]
54. Schultheiß, A.L.; Patel, R.A.; Vogel, M.; Dehn, F. Comparative Analysis of Engineering Carbonation Model Extensions to Account for Pre-Existing Cracks. *Materials* **2023**, *16*, 6177. [CrossRef] [PubMed]
55. Zhang, K.; Yio, M.; Wong, H.; Buenfeld, N. Development of more accurate methods for determining carbonation depth in cement-based materials. *Cem. Concr. Res.* **2024**, *175*, 107358. [CrossRef]
56. Shi, Z.; Lothenbach, B.; Geiker, M.R.; Kaufmann, J.; Leemann, A.; Ferreiro, S.; Skibsted, J. Experimental studies and thermodynamic modeling of the carbonation of Portland cement, metakaolin and limestone mortars. *Cem. Concr. Res.* **2016**, *88*, 60–72. [CrossRef]
57. Rahman, A.A.; Glasser, F.P. Comparative studies of the carbonation of hydrated cements. *Adv. Cem. Res.* **1989**, *2*, 49–54. [CrossRef]
58. Parrott, L.J.; Killoh, D.C. Carbonation in a 36-year-old, in-situ concrete. *Cem. Concr. Res.* **1989**, *19*, 649–656. [CrossRef]
59. Zhang, Q. Mathematical modeling and numerical study of carbonation in porous concrete materials. *Appl. Math. Comput.* **2016**, *281*, 16–27. [CrossRef]
60. Tuutti, K. *Corrosion of Steel in Concrete-CBI Research 4*, 1st ed.; Swedish Cement and Concrete Research Institute: Stockholm, Sweden, 1982; pp. 1–468.

61. Saetta, A.V.; Schrefler, B.A.; Vitaliani, R.V. The carbonation of concrete and the mechanism of moisture, heat and carbon-dioxide flow-through porous materials. *Cem. Concr. Res.* **1993**, *23*, 761–772. [CrossRef]
62. *EN 1992-1-1*; Eurocode 2: Design of Concrete Structures—Part 1-1: General Rules and Rules for Buildings. European Committee for Standardization (CEN): Brussels, Belgium, 2004.
63. *EN 206:2013+A1:2016*; Concrete—Specification, Performance, Production and Conformity. CEN: Brussels, Belgium, 2016.
64. Rivera, R.A.; Sanjuán, M.Á.; Martín, D.A.; Costafreda, J.L. Performance of Ground Granulated Blast-Furnace Slag and Coal Fly Ash Ternary Portland Cements Exposed to Natural Carbonation. *Materials* **2021**, *14*, 3239. [CrossRef]

materials

MDPI

Article

An Evaluation of the Radioactive Content of Ashes Obtained from the Use of Fuels from Recycled Materials by Co-Processing in the Cement Industry

José Antonio Suarez-Navarro [1], Miguel Ángel Sanjuán [2,*], Pedro Mora [3] and María del Mar Alonso [4]

1 Department for Environment, Environmental Radioactivity and Radiological Surveillance Unit, CIEMAT Research Center of Energy, Environmental and Technology, Av. Complutense, 22, 28040 Madrid, Spain; ja.suarez@ciemat.es

2 Department of Building Materials, Civil Engineering School, Technical University of Madrid, Profesor Aranguren, s/n, Ciudad Universitaria, 28040 Madrid, Spain

3 Department of Geological and Mines Engineering, Mine and Energy Engineering School, Technical University of Madrid (UPM), C/Ríos Rosas, 21, 28003 Madrid, Spain; pmora@oficemen.com

4 Eduardo Torroja Institute for Construction Sciences (IETcc-CSIC), 28033 Madrid, Spain; mmalonso@ietcc.csic.es

* Correspondence: masanjuan@ieca.es

Abstract: The co-processing of different wastes as fuels in the manufacture of cement clinker not only meets the objectives of a circular economy but also contributes to the reduction in CO_2 emissions in the manufacture of Portland cement. However, waste used as alternative fuels, such as sludge or organic-rich residues, may contain naturally occurring radionuclides that can be concentrated during the combustion process. In this study, the presence of natural radionuclides (radioactive series of uranium, thorium, and ^{40}K) and anthropogenic radionuclides (^{137}Cs) in these wastes has been investigated by gamma spectrometry. Possible relationships between the radioactive content and the obtained chemical composition, determined by X-ray fluorescence, have also been studied by applying a principal component analysis (PCA). The results showed that the wastes with the highest radioactive content were sewage sludge with activity concentrations of ^{238}U and ^{210}Pb of 321 ± 38 Bq kg^{-1} and 110 ± 14 Bq kg^{-1}, respectively. A correlation between radioactive content and Fe_2O_3 concentration was also observed. The annual effective dose rates to workers for the ashes estimated from the ash content ranged from 0.0033 mSv to 0.092 mSv and therefore do not pose a risk to workers as they are lower than the 1 mSv per year limit for the general public (DIRECTIVE 2013/59/EURATOM).

Keywords: recycled materials; co-processing; principal component analysis; cement; gamma spectrometry; XRF

1. Introduction

In the cement sector, as in other productive sectors, co-processing aims to promote sustainability by efficiently using waste materials, contributing to environmental protection, resource conservation, landfill reduction, and cost savings [1]. In fact, co-processing is determined when there is a dual material and energy benefit from waste in the same operation. Co-processing can be quantified by the recycling rate, defined as the proportion of waste materials recycled instead of being disposed of by usual routes such as landfilling or incineration [2]. The recycling rate plays an essential role in promoting sustainable waste management practices, reducing the environmental impact of waste, and moving towards the targets of the circular economy by encouraging the reuse and recycling of materials to minimize waste production.

In the manufacture of cement clinker, the percentage of recyclable materials from solid recovered fuels (SRFs) during co-processing in the cement industry can range from

76.8% to 100% [3]. This percentage represents the SRF ash that is considered recycled after incorporation into the clinker of Portland cements. SRF ashes containing non-combustible minerals, which are necessary for the manufacture of clinker, are integrated during the thermal recovery process. For this reason, SRF ash becomes a secondary raw material for the manufacture of cement clinker. The SRF is determined from the determination of the oxides in the SRF ash obtained by analytical techniques such as XRF or ICP-OES, which allows the recycling rate to be evaluated [4]. For this purpose, the R-index is used, which is given by the following expression [5]:

$$R\text{-}index = \frac{AC}{100} \cdot (\omega_1 + \omega_2 + \cdots + \omega_n) \quad (1)$$

where AC is the ash content (wt%$_{DT}$) and ω_n is the mass share of the oxides in the ash (wt%$_{DT}$). Although there are different criteria to determine the *R-index*, the most accurate value is the one determined from nine oxides, Al_2O_3, CaO, Fe_2O_3, K_2O, MgO, Na_2O, SO_3, SiO_2, and TiO_2, which will be indicated with the term R-index$_9$[(%d)] [1].

The materials used in co-processing in the cement industry can be from a number of different origins [2]: urban (UW) and industrial waste (IW), meat and bone meal (MBM), sewage sludge (STPS), end-of-life vehicle waste (ELVM), wood (WD), etc. The few existing studies on the potential hazardousness of these materials are based on the determination of heavy metals. Xu et al. [3] highlighted the need for control of this waste in China due to the presence and release of heavy metals, such as Cr and Pb. From a radiological protection point of view, urban wastes (UWs) have been studied in different incineration plants [4], and it has been determined that, in general, they do not represent a risk for workers or the general population. The presence of natural radionuclides and the radioactive content are increased by concentration in the ashes obtained after incineration. On the other hand, anthropogenic radionuclides would be present due to the possible existence of sources (^{137}Cs, ^{60}Co, or ^{241}Am) or the disposal of materials containing radionuclides from nuclear medicine (^{131}I, ^{99m}Tc, etc.). While some studies recommend radiation protection monitoring [5], others conclude that the presence of radionuclides in these wastes is at background levels [4]. Waste from residues with high organic content could concentrate 40K, as has been shown in biomass ashes in previous studies [6]. Additionally, studies on recycled metal waste from end-of-life vehicles (ELVM) showed the presence of natural and anthropogenic radionuclides (^{137}Cs) at ambient levels [7]. However, the wastes that have been most extensively studied have been sewage treatment plant sludge (STPS) [8]. Sewage sludge can contain both naturally occurring radionuclides [9] and anthropogenic radionuclides such as ^{137}Cs [10], although ^{131}I and other radionuclides from hospitals (^{99m}Tc, ^{123}I, ^{67}Ga, and ^{201}Tl) are the more common radionuclides and most studies focus mainly on them [11,12].

These radionuclides used in nuclear medicine are characterized by a half-life in the order of days, which would reduce their danger due to decay. For this reason, sludges would be one of the fuels to be investigated to assess their potential hazard from a radiological protection perspective. It is also important to try to correlate the possible presence of radionuclides with the chemical composition of the waste used as fuel, using the same oxides as those used to determine the R-index. These correlations could provide valuable information for predicting which wastes would require monitoring from a radiological protection perspective. There are different databases on natural radionuclides in building materials and NORM waste [13,14]. Previous studies have established a correlation between these radionuclides and chemical composition by applying a principal component analysis [15–17]. However, no studies have yet correlated the radiological content of materials used in co-processing with their chemical composition.

In view of the above, the objective was to determine the radiological content of different wastes used for co-processing in the cement industry and its relationship with the chemical composition in the form of a percentage of metal oxides. To this end, the following partial objectives were proposed: (i) to classify the wastes according to their chemical composition using a principal component analysis and cluster analysis; (ii) to determine the activity concentrations of the raw materials by gamma spectrometry, estimating their presence in the final ashes; and (iii) to evaluate the annual effective doses of workers interacting with the ashes produced in the co-processing process.

2. Experimental Section

2.1. Sample Preparation and Chemical Characterization

The number of samples studied was 63 from 6 different types of waste, chosen for their current and expected future consumption, as well as for their potential recycling rate [2]. They were supplied by LOEMCO. Figure 1 shows a picture of each waste type, indicating the nomenclature used in this study and the number of samples of each type.

Figure 1. Types of waste used, together with its nomenclature and the number of samples of each type.

Samples were labelled with the acronym of the residue and consecutive numbers up to the total number of samples, e.g., WD1, WD2, . . ., WD6.

A significant fraction of each sample was fractionated according to UNE-EN 15443:2011 [18]. The MBM, LD, and MD samples were quartered using groove splitters. CDRU, CDRI, and VFU samples were cut by cone stacking and subsequent quartering. The samples were subsequently weighed and oven-dried at 100 °C for 24 h to obtain 100–300 g aliquots. These samples were used to determine the activity concentration of the gamma emitters present (see Section 2.2). An equivalent amount was calcined in a muffle at 550 °C and the ashes were ground in a vibratory ring mill to a particle size of 100 μm. Finally, the samples were pelleted in a platinum crucible using a muffle (between 1200 and 1500 °C) with a flux ($Li_2B_4O_7$) containing an oxidizing agent ($LiNO_3$).

The chemical composition of the oxides in the samples was determined by X-ray fluorescence, following the recommendations of the UNE-EN 15410:2012 standard [19]. The equipment used was a Philips PW-1404 sequential spectrometer (LOEMCO, Madrid, Spain). The equipment had a rhodium anode X-ray tube, operated at a voltage of 30–50 kV and a current of 0.4–0.8 mA under vacuum conditions. Fluorescence spectra were acquired for 60–180 s per sample, and the equipment software allowed the quantification of the Al_2O_3, CaO, Fe_2O_3, K_2O, MgO, Na_2O, SO_3, SiO_2, and TiO_2 contents present in the different ash samples of the recovered solid fuels investigated. Samples were measured only once, as the laboratory's quality controls (based on ISO/IEC 17025:2017 [20]) ensure that a single measurement guarantees the reproducibility of the results.

The relationships between the oxides obtained and the types of samples were studied using the software RStudio version 2023.12.1, build 402. The principal component analysis (PCA) was performed using the libraries 'FactoMineR' and 'Factoextra'. The criteria used to perform the PCA were the same as in previous works by the same authors [15,16]. The vectors of the variables selected were those with Kaiser–Meyer–Olkin (KMO) indices greater than 0.6, ensuring sufficient correlation between these variables [21]. The 'Psych' library was used to determine these parameters. The results obtained were represented in an HJ-Biplot graph [16], which allows the simultaneous plotting of the vectors determined from the variables studied (percentages of oxides), defining the components, with the scores obtained for the samples according to these vectors. Finally, the samples were grouped by the cluster analysis using the PCA results.

2.2. Measurement by Gamma Spectrometry

Among the 63 samples used, 13 representative samples of each residue type were selected and measured by gamma spectrometry, following the procedure described in Section 2.1; they were measured on four coaxial detectors: two extended range (XtRa, with an operating potential of up to +4000 V), one broad energy (BEGE with an operating potential of up to −4500), and one reverse electrode (REGe with an operating potential of up to +5000). All three detectors were characterized by Canberra, which allowed the calculation of counting efficiencies as a function of energy using the mathematical code LabSOCS [22]. The detectors were connected to a compact Mirion-Canberra DSA-LX electronics chain that contains the integrated high-voltage source, amplifier, and analog-to-digital converter, and allows connection to a computer (Figure 1) [23]. The spectra were acquired and analyzed using Genie 2000 software [24]. The samples were measured for a minimum of 80,000 s, and the backgrounds for 600,000 s. The spectra were analyzed taking into account the following parameters: (i) the peaks were located with a significance threshold of 3, (ii) the energy tolerance was ±2 keV for both samples and the background, (iii) the baseline was set as a step function to establish the regions of interest, and (iv) the fit of the photopeaks was Gaussian. The lower limit of detection was determined using the Currie criterion [25].

The laboratory where the measurements were performed is accredited based on the UNE-EN ISO/IEC 17025:2017 standard [20], which establishes periodic quality controls, as well as participation in intercomparative exercises. The above laboratory criteria ensured the reproducibility of the results from a single measurement.

Figure 2 shows the photopeaks used for the determination of the activity concentration of the gamma-emitting radionuclides of the uranium, actinium, and thorium radioactive series, together with that of ^{40}K of a natural origin and ^{137}Cs of an anthropogenic origin [26].

The activity concentration of ^{226}Ra was determined by removing the ^{235}U interference from the 186 keV photopeak. Also, the interference produced by ^{228}Ac in the 1460 keV photopeak was removed using the algorithm described in [27]. The true coincidence summing effect was corrected by the Peak-to-Total method using the total efficiency curve of the characterized detectors [28].

Natural radioactive series	Radionuclide	Energy (keV)	Photons per 100 disint.
Uranium	^{210}Pb	46.539 (1)	4.252 (40)
	^{234}Th	63.20 (2)	3.75 (8)
	^{226}Ra	186.211 (13)	3.555 (19)
	^{214}Pb	351.932 (2)	35.60 (7)
	^{214}Bi	609.312 (7)	45.49 (19)
		1120.287 (10)	14.91 (3)
		1764.494 (14)	15.31 (5)
	^{234m}Pa	1001.441 (18)	0.856 (8)
Actinium	^{235}U	143.761 (3)	10.94 (6)
		163.356 (3)	5.08 (3)
		185.720 (4)	57.0 (3)
		205.316 (4)	5.02 (3)
Thorium	^{212}Pb	238.632 (2)	43.6 (5)
	^{208}Tl	583.187 (6)	85.0 (3)
	^{228}Ac	911.196 (6)	26.2 (8)
	^{137}Cs	661.655 (3)	85.01 (20)
	^{40}K	1460.822 (6)	10.55 (11)

Figure 2. The schematic of the measurement system (DSA-LX, HPGe gamma detector, dewar vessel filled with liquid nitrogen, lead shielding, Tien and Cooper shielding for Pb X-ray removal) and photopic of the gamma emitters of the three natural radioactive series, and ^{40}K and ^{137}Cs from the fallout. The values of energy and photons per 100 disintegrations were taken from the tables of the Laboratorie National Henri Becquerel [26].

2.3. *Calculation of the Natural Radionuclide Enrichment Factor of the Ashes Obtained from the Various Wastes after Use as Fuel*

The activity concentrations of the different wastes studied were determined as described in Section 2.2. However, to assess the true radiological hazard and determine the effective doses, which will be described in Section 2.4, it is necessary to replicate the final state of the waste after combustion. Therefore, the final activity concentration was calculated considering the percentages of ash generated during the combustion process (CA). The enrichment factor in natural radionuclides (F_e) with respect to the activity concentration of the uncalcined residues was determined by the following expression:

$$F_e = A_n \cdot \frac{1}{\frac{AC}{100}} \quad (2)$$

where A_n is the activity concentration of radionuclide n (Bq kg^{-1}), and AC is the ash content (%).

2.4. *Assessment of the Effective Doses of the Residues Used for Workers*

The doses from the final waste ash were assessed following the criteria set out in Radiation Protection 122 [29]. The doses were determined for the 'small heap' scenario for workers, where the total annual dose rate (E_A, µSv y^{-1}) is given by expression (3) [30].

$$E_A = E_{ext} + E_{inh} + E_{ing} \quad (3)$$

where E_{ext}, E_{inh}, and E_{ing} are the annual effective dose rates from external gamma radiation, inhalation, and ingestion, respectively. These effective doses are determined for each of the radionuclides contained in the sample by the following expressions:

$$E_{ext} = D_{ext} \cdot T_e \cdot A_r \tag{4}$$

$$E_{inh} = D_{inh} \cdot T_e \cdot B_r \cdot C_d \cdot A_r \tag{5}$$

$$E_{ing} = D_{ing} \cdot T_e \cdot R_i \cdot A_r \tag{6}$$

where D_{ext}, D_{inh}, and D_{ing} are the dose conversion factors for external radiation, inhalation, and ingestion for each radionuclide; T_e is the annual exposure time (1800 h y^{-1}); B_r is the respiratory rate (1.2 m^3 h^{-1}); C_d is the dust concentration (2.0 $\times$ 10^{-5} kg m^{-1}); R_i is the ingestion rate (1.0 $\times$ 10^{-5} kg h^{-1}); and A_r is the activity concentrations of the radionuclide for which the effective dose is to be estimated [31].

3. Results

3.1. Relationship between Chemical Composition and Type of Waste

Table 1 shows the percentage ranges of the oxides in each type of residue used in this study, along with the median values. Figure 3 shows the results from the principal component analysis (PCA) of the chemical composition of the residues in oxide form, as determined by XRF. Figure 3a presents the scree plot indicating that factors 1 and 2 account for 66.5% of the total variance. The PCA variable weight plot (Figure 3b) illustrates the correlations between the oxides of the various residues as obtained by XRF. The variables that met the criterion of having a Kaiser–Meyer–Olkin (KMO) value greater than 0.6 were CaO, Na_2O, SO_3, MgO, SiO_2, and Fe_2O_3. However, the variables K_2O, MgO, and TiO_2 did not meet the criterion and were thus excluded from the PCA. The variables that showed correlation with one another were the CaO and Na_2O oxides, MgO and SO_3, and SiO_2 and Fe_2O_3. Conversely, it was noted that the CaO and Na_2O oxides were inversely proportional to the SiO_2 and Fe_2O_3 oxides, forming a 180° angle between them. Additionally, these oxides were unrelated to the presence of SO_3 and MgO, as evidenced by the 90° angle they formed (with a cosine of 0). The percentage contributions of the variables to factors 1 and 2 are depicted in Figure 3c,d. CaO and SiO_2 had statistically significant contributions to factor 1, while SO_3, Fe_2O_3, and MgO contributed to factor 2.

Table 1. Percentage range of oxides of each of the residues used in this study, along with the mean obtained for each residue and oxide.

Oxide (%)	ELVM			STPS			MBM			UW			IW			WD		
	Min	Max	Med	Min	Max	Med	Min	Max	Med	Min	Max	Med	Min	Max	Med	Min	Max	Med
Al_2O_3	1.8	10.6	5.9	9.8	13.9	11.5	0.04	1.1	0.7	9.2	37.1	10.4	5.7	43.8	13.0	1.0	6.5	5.3
CaO	14.2	23.1	17.7	12.0	23.3	22.1	40.1	49.6	41.1	21.1	36.0	30.3	13.5	36.7	25.1	8.3	33.0	23.0
Fe_2O_3	1.7	19.4	7.8	9.6	18.9	11.0	0.2	3.8	0.8	1.9	7.6	2.5	1.0	20.1	3.7	0.7	6.2	3.8
K_2O	0.2	1.4	0.4	1.5	2.7	1.9	1.6	6.3	4.4	0.6	2.7	2.2	0.8	9.3	1.3	1.7	15.0	3.5
MgO	2.3	11.6	3.2	2.6	5.0	3.8	1.1	1.7	1.4	2.7	4.6	3.1	1.6	4.1	2.1	2.3	4.0	3.6
Na_2O	0.3	3.5	0.8	0.6	1.8	1.1	2.8	5.0	4.3	1.4	7.4	3.4	1.7	10.0	3.0	0.7	4.7	3.4
SO_3	1.0	5.5	2.2	1.1	5.4	3.6	0.4	1.2	0.7	2.0	4.3	3.3	0.8	4.1	1.8	1.0	2.8	2.4
SiO_2	11.6	39.3	32.4	15.1	25.0	21.8	1.0	7.3	5.7	15.4	31.6	18.4	7.3	40.1	23.9	6.2	38.9	27.2
TiO_2	0.9	2.6	1.3	0.7	1.0	1.0	0.01	0.4	0.06	1.9	16.9	6.1	0.7	22.9	3.4	0.07	8.5	5.5

Figure 4 shows the contributions of the individual samples to GWP factors 1 and 2. The grey bars represent the samples analyzed by gamma spectrometry. Factor 1 accounted for 75% of the ELVM residues, 100% of the MBM, 9% of the IW, and 71% of the STPS. Factor 2 accounted for 71% of the ELVM, 50% of the IW, 62% of the UW, and 14% of the STPS.

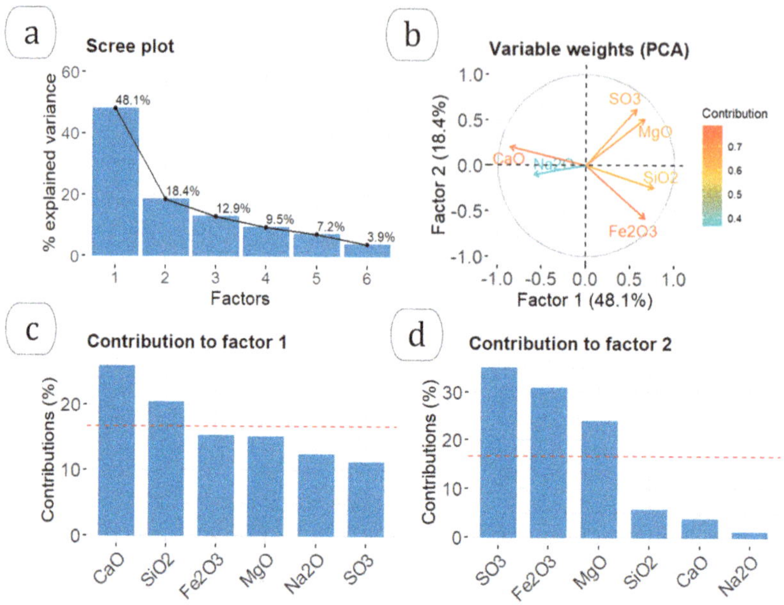

Figure 3. Results of the principal component analysis (PCA) of the residues according to the chemical composition obtained by XRF: (**a**) the scree plot with the % of variance explained according to the factors obtained, (**b**) correlations obtained between the different variables according to the 2 most important factors, (**c**) percentage contribution of the different variables to factor 1, and (**d**) percentage contribution of the different variables to factor 2.

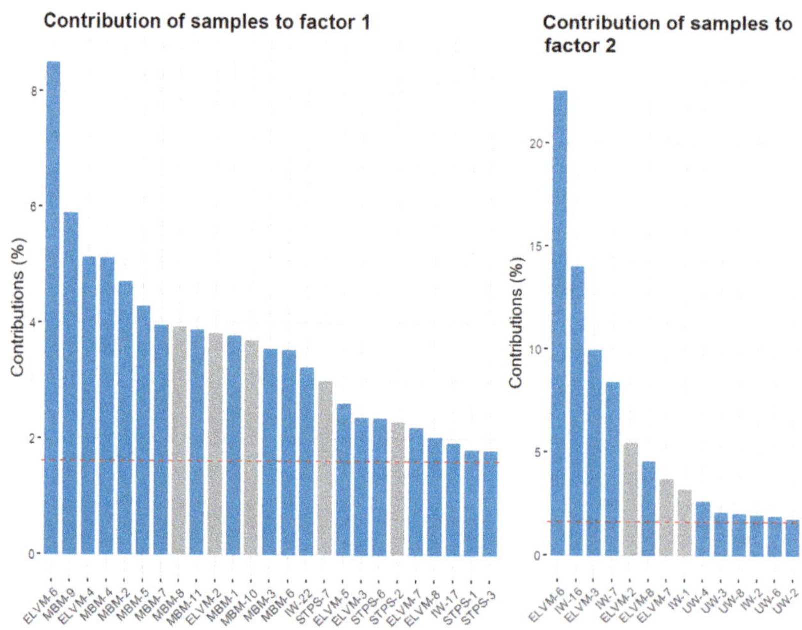

Figure 4. The contribution of the different samples to the factors with the highest GWP impact (48.1% and 18.4%, respectively). The dotted red line represents the expected average contribution (cutoff), with the columns shown for the two factors being those that exceeded this value.

Figure 5 shows the HJ-Biplot graph representing the results obtained from the PCA. The confidence ellipses for the different groups of residuals are also presented, as well as the scores of the different samples for the two factors represented. The results indicate two statistically significant groupings: meat and bone meal (MBM) waste represented in green and sewage sludge (STPS) waste represented in blue. The remaining wastes are centered on the graph, indicating a lack of substantial correlation with the two factors obtained. On the other hand, the ELVM wastes, grouped in the fuchsia sector, although not centered, exhibited a pronounced dispersion.

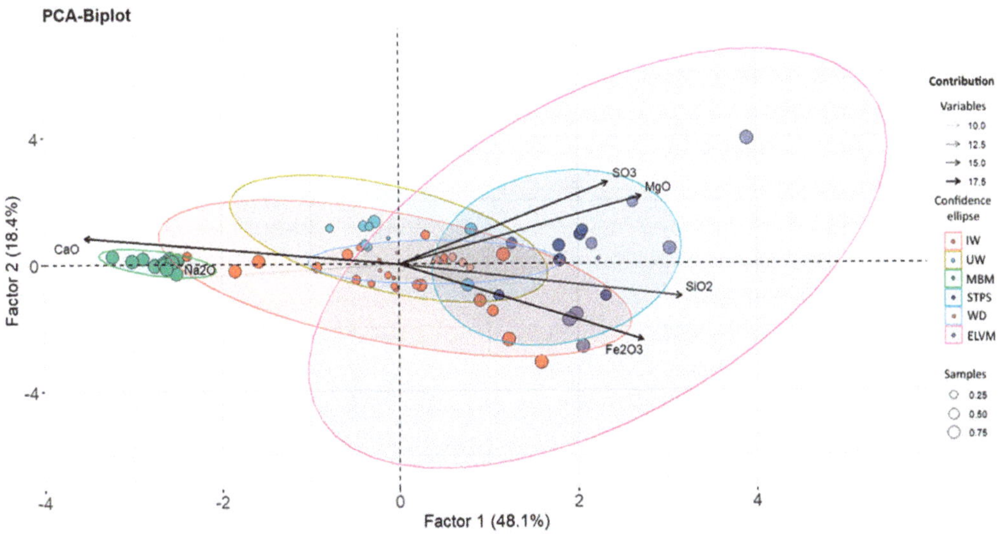

Figure 5. The HJ-Biplot graph presents the results of the principal component analysis along with the confidence ellipses for the different groups of residuals analyzed. The scores obtained for the various samples are shown in circular shapes whose sizes represent their contributions to the different components.

The confidence ellipses in Figure 5 were generated based on residue type and not on the PCA results. Therefore, Figure 6 displays the groupings derived from the PCA results through the cluster analysis. Residues analyzed by gamma spectrometry are highlighted with a red box. The six groups obtained demonstrate a clear clustering of the meat meals (MBM), showing their high dependence on factor 1 (Na_2O and CaO). The other residues exhibited greater heterogeneity and a lower correlation with the two factors identified, as they are centered in the graph. The wastes from sewage sludge (STP), although more dispersed than that from meat meal (MBM), are grouped together and show a relationship with both factors.

3.2. Activity Concentration of Natural Radionuclides in Different Wastes

Table 2 shows the activity concentrations of the natural radioactive series of uranium and thorium, ^{40}K, and ^{137}Cs in the 13 wastes selected and analyzed by gamma spectrometry. They were considered as reference levels for Portland cement as there are no world average levels for this type of waste. The world average level for radionuclides of the naturally occurring radioactive uranium and thorium series is 50 Bq kg^{-1}, while the world average level for ^{40}K is 500 Bq kg^{-1} [14,32]. Also, the value adopted for ^{137}Cs was the fallout value in soils, which is 4 Bq kg^{-1} [33]. The activity concentrations of the ELVM, MBM, IW, UW, and WD wastes were on the order of or lower than the selected reference value. STPS sewage sludge was the only waste with values above 50 Bq kg^{-1} for ^{234}Th and ^{210}Pb. Also, sample IW-6 had an activity concentration for ^{210}Pb also above 50 Bq kg^{-1}. The activity concentrations of the radionuclides of the thorium series were all below 50 Bq kg^{-1}. On the other hand, the activity concentrations of ^{40}K were all below 500 Bq kg^{-1}. Finally, the ^{137}Cs activity concentrations were all below the limit of detection (LOD) determined from the Currie criterion [25].

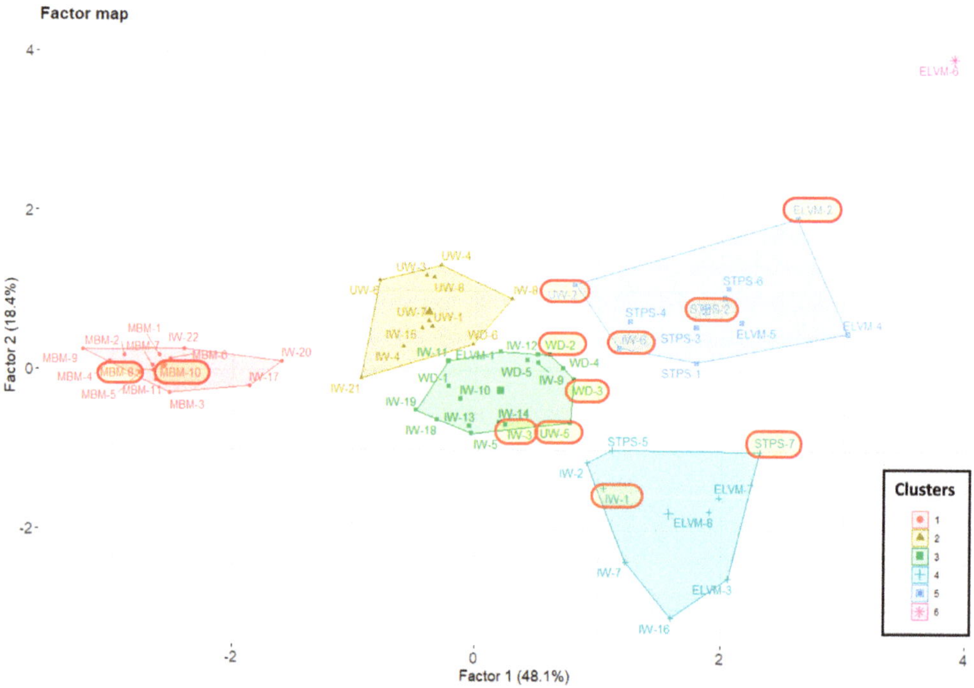

Figure 6. The cluster map created from the results of the principal component analysis. The total number of clusters resulting from the analysis was six.

Table 2. Activity concentrations of the natural radioactive series of uranium and thorium, ^{40}K, and ^{137}Cs for the municipal waste analyzed by gamma spectrometry.

Reference	Uranium Series					Thorium Series			^{40}K	^{137}Cs
	^{234}Th	^{226}Ra	^{214}Pb	^{214}Bi	^{210}Pb	^{228}Ac	^{212}Pb	^{208}Tl		
ELVM-2	<7.1	7.9 ± 5.9	<1.9	<2.4	<8.9	<3.3	2.43 ± 0.41	0.91 ± 0.31	<11.4	<0.8
ELVM-7	<12.8	<24.3	15.4 ± 4.6	16.8 ± 2.0	42 ± 12	11.0 ± 5.4	9.9 ± 2.5	3.8 ± 1.6	122 ± 24	<1.5
MBM-6	<5.6	<7.6	4.33 ± 0.72	4.26 ± 0.86	<6.5	<2.5	1.85 ± 0.29	0.73 ± 0.22	<17.5	<0.6
MBM-10	<11.2	<21.6	<3.0	<2.7	<11.3	<5.0	1.42 ± 0.54	<1.1	254 ± 26	<1.3
STPS-6	321 ± 38	<16.0	11.6 ± 2.1	10.5 ± 1.1	110 ± 14	16.9 ± 1.9	11.7 ± 1.7	4.2 ± 1.2	128 ± 18	<0.8
STPS-2	34.6 ± 9.2	28 ± 13	14.7 ± 2.1	14.2 ± 1.9	85 ± 10	17.4 ± 2.3	13.3 ± 1.4	5.07 ± 0.82	161 ± 19	<1.2
IW-1	<19.7	<33.9	<7.5	<6.7	<18.0	<10.7	<3.7	<3.0	51 ± 13	<2.7
IW-3	<17.9	<27.7	<4.8	<4.8	<14.7	<6.3	<1.7	<2.2	33 ± 23	<1.9
IW-6	27 ± 14	<39.0	<5.2	<5.8	69 ± 20	<8.7	3.6 ± 1.1	<2.5	66 ± 18	<2.5
UW-1	<20.5	<36.6	<6.3	<7.7	<20.2	<11.9	<2.2	<3.2	<39.0	<2.9
UW-5	<7.3	<16.4	3.7 ± 1.7	6.0 ± 1.2	14.1 ± 8.8	<2.4	2.22 ± 0.54	<1.0	46 ± 16	<0.7
WD-2	<14.7	<33.5	<4.9	<5.1	<14.3	<7.0	<2.3	<2.2	<26.7	<1.3
WD-3	<5.5	<8.7	4.43 ± 0.89	4.5 ± 1.1	7.9 ± 3.5	<2.9	2.58 ± 0.36	1.12 ± 0.27	<12.4	<0.7

The uncertainties are quoted for a coverage factor of k = 2.

3.3. *Enrichment in Natural Radionuclides of the Ashes Obtained from the Different Wastes after Use as Fuels*

Figure 7 shows the increase in the activity concentration of the natural radionuclides from the uranium series (^{234}Th, ^{226}Ra, ^{210}Pb), the thorium series (^{212}Pb), and ^{40}K when transforming the samples into ashes. The factors used to obtain the final activity concentration were calculated using expression 2, utilizing the ash percentages determined experimentally according to the procedure explained in Section 2.1. It can be seen that the ash percentages obtained were similar to the R-index$_9$[(%d)] determined. The activity concentrations of ^{226}Ra were estimated with ^{226}Ra (determined from the 186 keV peak) and by ^{214}Pb in cases where the ^{226}Ra activity was lower than the limit of detection (LOD). The sewage sludge (STPS-2 and STPS-6) showed the highest increase in the activity concentration of ^{234}Th, ^{226}Ra, and ^{210}Pb. IW-6 had the highest ^{210}Pb activity concentration in the ash (368 Bq kg^{-1}). On the other hand, MBM-10 obtained the highest concentration of ^{40}K activity in the ashes (1181 Bq kg^{-1}).

Reference	Ashes (%)	R-index$_9$[(%d)]
ELVM-2	16.7	13.5
ELVM-7	32.2	26.8
STPS-2	38.8	30.7
STPS-7	14.4	22.4
MBM-8	19.4	12.1
MBM-10	32.5	12.8
IW-1	21.5	10.7
IW-3	8.0	3.4
IW-6	4.1	14.7
UW-1	18.7	3.8
UW-5	3.9	12.0
WD-2	14.4	2.3
WD-3	2.8	12.0

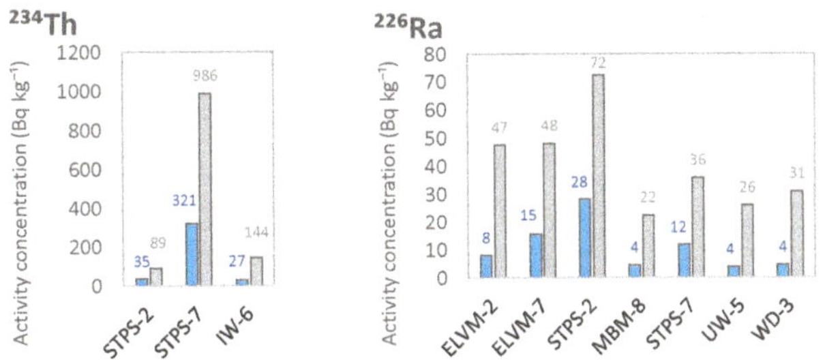

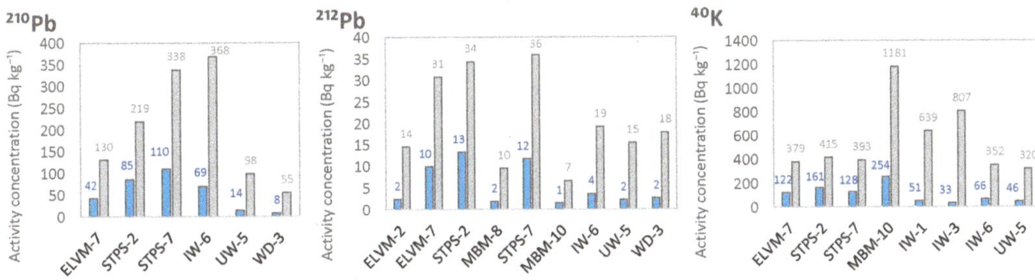

Figure 7. Increases in the activity concentration of radionuclides from the uranium (^{234}Th, ^{226}Ra, and ^{210}Pb) and thorium (^{212}Pb) series, along with ^{40}K.

3.4. *Effective Doses Received by Workers*

Table 3 presents the effective doses obtained with expressions 3, 4, 5, and 6 using the activity concentrations estimated for the different radionuclides in the ashes of each waste type. The maximum value of the total annual effective dose rate was for the sludge wastes (STPS-2 and STPS-7). The wastes IW-1 and IW-3 only contributed external dose rates from gamma radiation, as shown in Table 2, since only ^{40}K was detected. The annual effective dose rates for samples UW-1 and WD-2 could not be calculated because all activity concentrations of the studied radionuclides were below the limit of detection (L_D).

Table 3. Annual effective doses (mSv y^{-1}) obtained from the estimated activity concentrations for the different radionuclides in the studied wastes. E_A is the total annual effective dose, E_{ext} is the external dose from gamma radiation, E_{inh} is the annual effective dose from inhalation, and E_{ing} is the annual effective dose from ingestion.

Sample	E_{ext} (mSv y^{-1})	E_{inh} (mSv y^{-1})	E_{ing} (mSv y^{-1})	E_A (mSv y^{-1})
ELVM-2	0.028 ± 0.012	0.0035 ± 0.0015	0.00052 ± 0.00022	0.032 ± 0.013
ELVM-7	0.032 ± 0.013	0.0071 ± 0.0029	0.00142 ± 0.00058	0.040 ± 0.016
MBM-6	0.046 ± 0.015	0.0131 ± 0.0042	0.00208 ± 0.00066	0.061 ± 0.020
MBM-10	0.0134 ± 0.0029	0.00220 ± 0.00047	0.000294 ± 0.000063	0.0159 ± 0.0034
STPS-6	0.0260 ± 0.0062	0.063 ± 0.015	0.00309 ± 0.00073	0.092 ± 0.022

Table 3. *Cont.*

Sample	E_{ext} (mSv y^{-1})	E_{inh} (mSv y^{-1})	E_{ing} (mSv y^{-1})	E_A (mSv y^{-1})
STPS-2	0.0066 ± 0.0013	0.00137 ± 0.00027	0.000126 ± 0.000025	0.0081 ± 0.0016
IW-1	0.00329 ± 0.00042	-	-	0.00329 ± 0.00042
IW-3	0.0042 ± 0.0014	-	-	0.0042 ± 0.0014
IW-6	0.0036 ± 0.0013	0.0120 ± 0.0043	0.00063 ± 0.00023	0.0162 ± 0.0058
UW-1	-	-	-	-
UW-5	0.0175 ± 0.0077	0.0036 ± 0.0016	0.00074 ± 0.00033	0.0219 ± 0.0097
WD-2	-	-	-	-
WD-3	0.0189 ± 0.0058	0.0042 ± 0.0013	0.00088 ± 0.00027	0.0240 ± 0.0073

The uncertainties are quoted for a coverage factor of k = 1.

4. Discussion

The results obtained confirmed the initial objective, since the two sludges analyzed by gamma spectrometry were those with the highest activity concentrations and therefore the highest total annual effective dose rate calculated from the estimation of the radiological content of the ashes.

The GWP showed two statistically significant groupings that were strongly influenced by chemical composition, meat meals and sewage sludge (Figures 3 and 4). Meat meals were a priori together with woods the ones that could have the highest amount of ^{40}K as supported by previous studies [34,35]. However, only sample MBM-10 obtained a ^{40}K activity concentration of 254 ± 26 Bq kg^{-1}. This result is consistent with the observed relationship between these materials and the amount of Na_2O. Although K_2O and Na_2O performed equally well in the PCA, K_2O had to be removed as it did not meet the criterion of a KMO > 0.6. Therefore, the higher concentration of ^{40}K activity would be related to the presence of Na_2O and K_2O. The level of ^{40}K, together with the high concentration produced in its transformation into ash (Figure 5), would result in annual effective dose rates between 0.0081 and 0.024 mSv y^{-1}.

Sewage sludge had the highest activity concentrations of all the wastes analyzed. The activity concentrations were $_{238}U$ (^{234}Th) > ^{210}Pb > ^{226}Ra for the uranium series radionuclides. Additionally, the ratio between ^{226}Ra and ^{232}Th was 1. These results align with those of other studies [12]. The HJ-Biplot graph demonstrated that the sludges had a higher concentration of Fe_2O_3 (Figures 3 and 4). This suggests that the increased presence of uranium series radionuclides may be due to an association with iron compounds, as observed in other matrix types [15,16,30]. The ^{40}K activity concentrations were also consistent with other studies, ranging from 136 Bq kg^{-1} to 497 Bq kg^{-1} [4,8]. The present results did not indicate the presence of ^{137}Cs or ^{131}I, isotopes typically found in such residues [10,12]. The transformation into ash likely results in the loss of these isotopes due to their boiling points, 671 °C for ^{137}Cs and 184 °C for ^{131}I. Consequently, authors like Carvallo [5] advise controlling the calcination process of these residues to prevent high inhalation doses. The present results, obtained without calcination, suggest no hazard from ^{137}Cs. ^{131}I would not pose a radiological problem after 56 days of storage, given its half-life of 8.0233 (18) days. The estimated annual effective dose rates for the ashes, calculated using Radiation Protection 122 (RP 122) [29], were in line with other studies, with values between 0.02 mSv and 1 [9]. Similarly, the annual effective dose rates were consistent with those from probabilistic models like RESRAD, which estimated them to be 0.03 mSv y^{-1} [36]. These doses are significantly lower than the natural annual radioactive background of 2.4 mSv y^{-1} [37], indicating no radiological risk under the present study conditions.

Finally, the ^{210}Pb activity concentration results were 42 ± 12 Bq kg^{-1} and 69 ± 20 Bq kg^{-1} for ELVM-7 and IW-6, respectively. These activity concentrations were of the order of 50 Bq kg^{-1}, which was the average value for building materials adopted as a comparative criterion. These wastes are also characterized by a higher Fe_2O_3 content, suggesting a possible relationship between chemical composition and ^{210}Pb content, as observed for sewage sludge. However, the significant dispersion among the chemical compositions observed in the HJ-Biplot diagram necessitates further study of this type of waste. The ashes of these two materials would cause effective doses of 0.040 mSv y^{-1} and 0.016 mSv y^{-1}, respectively. These effective dose rates are consistent with those found by other researchers, who reported 0.06 mSv y^{-1} for metal smelting facilities from vehicles [7].

5. Conclusions

The use of waste materials for the co-processing process in cement production does not pose risks from a radiation protection perspective. The annual dose rates estimated in this study for workers ranged from 0.0033 mSv to 0.092 mSv. The highest activity concentrations were observed in sewage ash, with ^{238}U (^{234}Th) at 321 ± 38 Bq kg^{-1} and ^{210}Pb at 110 ± 14 Bq kg^{-1}. ^{210}Pb was quantified in samples from disused vehicle recycling and industrial waste. These radionuclides demonstrated a correlation with Fe_2O_3, which has also been observed in previous studies. The highest ^{40}K activity concentrations were found in wastes with a higher organic matter content, such as meat and bone meal. The radioactive content of this type of waste was of a natural origin, as anthropogenic radionuclides like ^{137}Cs were not detected in the wastes analyzed in this study. Future research could focus on the study of radionuclide leaching in ashes and also on those wastes with higher Fe_2O_3 content.

Author Contributions: Conceptualization, J.A.S.-N., M.Á.S., P.M. and M.d.M.A.; methodology, J.A.S.-N., M.Á.S., P.M. and M.d.M.A.; software, J.A.S.-N.; validation, J.A.S.-N., M.Á.S., P.M. and M.d.M.A.; formal analysis, J.A.S.-N.; investigation, J.A.S.-N., M.Á.S., P.M. and M.d.M.A.; data curation, J.A.S.-N.; writing—original draft preparation, J.A.S.-N.; writing—review and editing, J.A.S.-N., M.Á.S. and M.d.M.A.; supervision, M.Á.S. and M.d.M.A.; project administration, M.d.M.A.; funding acquisition, M.d.M.A. All authors have read and agreed to the published version of the manuscript.

Funding: This research has been conducted thanks to the support of Project PID2020-116002RB-100/AEI/10.13039/501100011033 (HORRADIONEX) (Spain).

Data Availability Statement: Data are contained within the article.

Conflicts of Interest: The authors declare no conflicts of interest. The funders had no role in the design of the study; in the collection, analyses, or interpretation of data; in the writing of the manuscript; or in the decision to publish the results.

References

1. Viczek, S.A.; Aldrian, A.; Pomberger, R.; Sarc, R. Determination of the material-recyclable share of SRF during co-processing in the cement industry. *Resour. Conserv. Recycl.* **2020**, *156*, 104696. [CrossRef]
2. Alarcón, A.; Mora, P.; Vallina, D.; Martin, M. Coprocesado: Recuperación material de la fracción mineral de los combustibles derivados de residuos en la industria cementera. *Cemento Hormigón* **2023**, *1019*, 14.
3. Xu, J.; Sun, S.F.; Karstensen, K.; Yan, D.H.; Peng, Z. Co-Processing Hazardous Waste in the Chinese Cement Industry—Status 2014. *Appl. Mech. Mater.* **2015**, *768*, 679–686. [CrossRef]
4. Kallio, A.; Virtanen, S.; Leikoski, N.; Iloniemi, E.; Kämäräinen, M.; Hildén, T.; Mattila, A. Radioactivity of residues from waste incineration facilities in Finland. *J. Radiol. Prot.* **2023**, *43*, 021502. [CrossRef] [PubMed]
5. Carvalho, F.P. Can the incineration of Municipal Solid Waste pose occupational and environmental radiation hazards? *Int. J. Occup. Environ. Saf.* **2017**, *1*, 1–10. [CrossRef]
6. Alonso, M.M.; Gascó, C.; Morales, M.M.; Suárez-Navarro, J.A.; Zamorano, M.; Puertas, F. Olive biomass ash as an alternative activator in geopolymer formation: A study of strength, radiology and leaching behaviour. *Cem. Concr. Compos.* **2019**, *104*, 103384. [CrossRef]
7. Ogundare, F.; Nwankwo, C. Radionuclide content of, and radiological hazards associated with samples from the different streams of metal recycling facilities. *Radioprotection* **2015**, *50*, 55–58. [CrossRef]
8. Camacho, A.; Montaña, M.; Vallés, I.; Devesa, R.; Céspedes-Sánchez, R.; Serrano, I. Temporal evolution of radionuclides in sludge from wastewater treatment plants. *J. Radioanal. Nucl. Chem.* **2013**, *295*, 297–306. [CrossRef]
9. Ozmen, S.F.; Topcuoğlu, B. *Chapter Determination of Natural Radioactivity Levels of Sludges Collected From Wastewater Treatment Plants of Antalya/Türkiye*; Firenze University Press: Florence, Italy, 2022.

10. Yui, K.; Kuramochi, H.; Osako, M. Understanding the Behavior of Radioactive Cesium during the Incineration of Contaminated Municipal Solid Waste and Sewage Sludge by Thermodynamic Equilibrium Calculation. *ACS Omega* **2018**, *3*, 15086–15099. [CrossRef]
11. Nakamura, A.; Hayabuchi, N.; Osaki, T.; Osaki, S. Output of radiopharmaceutical nuclides of known injected doses from a municipal sewage treatment system. *Health Phys.* **2005**, *88*, 163–168. [CrossRef]
12. Jiménez, F.; Debán, L.; Pardo, R.; López, R.; García-Talavera, M. Levels of ^{131}I and Six Natural Radionuclides in Sludge from the Sewage Treatment Plant of Valladolid, Spain. *Water Air Soil Pollut.* **2011**, *217*, 515–521. [CrossRef]
13. Alonso, M.M.; Suarez-Navarro, J.A.; Perez-Sanz, R.; Gasco, C.; Moreno de Los Reyes, A.M.; Lanzon, M.; Blanco-Varela, M.T.; Puertas, F. Data on natural radionuclide's activity concentration of cement-based materials. *Data Brief* **2020**, *33*, 106488. [CrossRef] [PubMed]
14. Trevisi, R.; Risica, S.; D'Alessandro, M.; Paradiso, D.; Nuccetelli, C. Natural radioactivity in building materials in the European Union: A database and an estimate of radiological significance. *J. Environ. Radioact.* **2012**, *105*, 11–20. [CrossRef] [PubMed]
15. Suárez-Navarro, J.A.; Alonso, M.M.; Gascó, C.; Pachón, A.; Carmona-Quiroga, P.M.; Argiz, C.; Sanjuán, M.Á.; Puertas, F. Effect of particle size and composition of granitic sands on the radiological behaviour of mortars. *Boletín Soc. Española Cerámica Vidr.* **2021**, *61*, 561–573. [CrossRef]
16. Cano, A.; Suarez-Navarro, J.A.; Puertas, F.; Fernandez-Jimenez, A.; Alonso, M.D.M. New Approach to Determine the Activity Concentration Index in Cements, Fly Ashes, and Slags on the Basis of Their Chemical Composition. *Materials* **2023**, *16*, 2677. [CrossRef]
17. Coletti, C.; Brattich, E.; Cinelli, G.; Cultrone, G.; Maritan, L.; Mazzoli, C.; Mostacci, D.; Tositti, L.; Sassi, R. Radionuclide concentration and radon exhalation in new mix design of bricks produced reusing NORM by-products: The influence of mineralogy and texture. *Constr. Build. Mater.* **2020**, *260*, 119820. [CrossRef]
18. *UNE-EN 15443:2011*; Combustibles Sólidos Recuperados. Métodos para la Preparación de la Muestra del Laboratorio. Asociación Española de Normalización (UNE): Madrid, Spain, 2011.
19. *UNE-EN 15410:2012*; Combustibles Sólidos Recuperados. Método para la Determinación del Contenido en Elementos Principales (Al, Ca, Fe, K, Mg, Na, S, Si, Ti). Asociación Española de Normalización (UNE): Madrid, Spain, 2012.
20. *UNE-EN ISO/IEC 17025:2017*; General Requirements for the Competence of Testing and Calibration Laboratories. Asociación Española de Normalización (UNE): Madrid, Spain, 2017.
21. Hill, B.D. *Sequential Kaiser-meyer-olkin Procedure as an Alternative for Determining the Number of Factors in Common-factor Analysis: A Monte Carlo Simulation*; Oklahoma State University: Stillwater, OK, USA, 2011.
22. Suárez-Navarro, J.A.; Moreno-Reyes, A.M.; Gascó, C.; Alonso, M.M.; Puertas, F. Gamma spectrometry and LabSOCS-calculated efficiency in the radiological characterisation of quadrangular and cubic specimens of hardened portland cement paste. *Radiat. Phys. Chem.* **2020**, *171*, 108709. [CrossRef]
23. MIRION. *DSA-LX Digital Signal Analyzer User's Manual*; Mirion Technologies (Canberra): Meriden, CT, USA, 2017.
24. CANBERRA. *Genie 2000 Spectrometry Software Customization Tools*; Canberra Industries: Meriden, CT, USA, 2009.
25. Currie, L.A. Limits for Qualitative Detection and Quantitative Determination. *Anal. Chem.* **1968**, *40*, 586–593. [CrossRef]
26. Be, M.; Chisté, V.; Dulieu, C.; Kellett, M.; Mougeot, X.; Arinc, A.; Chechev, V.; Kuzmenko, N.; Kibédi, T.; Luca, A. *Table of radionuclides (Vol. 8-A= 41 to 198)*; Bureau International Des Poids et Mesures (BIPM): Sèvres, France, 2016.
27. Suárez-Navarro, J.A.; Gascó, C.; Alonso, M.M.; Blanco-Varela, M.T.; Lanzon, M.; Puertas, F. Use of Genie 2000 and Excel VBA to correct for γ-ray interference in the determination of NORM building material activity concentrations. *Appl. Radiat. Isot.* **2018**, *142*, 1–7. [CrossRef]
28. Barba-Lobo, A.; Expósito-Suárez, V.M.; Suárez-Navarro, J.A.; Bolívar, J.P. Robustness of LabSOCS calculating Ge detector efficiency for the measurement of radionuclides. *Radiat. Phys. Chem.* **2023**, *205*, 110734. [CrossRef]
29. European Commission. *Radiation Protection 122. Practical Use of the Concepts of Clearance and Exemption Part II: Application of the Concepts of Exemption and Clearance to Natural Radiation Sources*; Office for Official Publications of the European Communities: Luxembourg, 2001.
30. Suárez-Navarro, J.A.; Lanzón, M.; Moreno-Reyes, A.M.; Gascó, C.; Alonso, M.M.; Blanco-Varela, M.T.; Puertas, F. Radiological behaviour of pigments and water repellents in cement-based mortars. *Constr. Build. Mater.* **2019**, *225*, 879–885. [CrossRef]
31. Iwaoka, K.; Tabe, H.; Yonehara, H. Natural radioactivity of bedrock bath instruments and hot spring instruments in Japan. *J. Radioanal. Nucl. Chem.* **2013**, *295*, 817–821. [CrossRef]
32. Janković, M.M.; Todorović, D.J. Concentrations of natural radionuclides in imported zirconium minerals. *Nucl. Technol. Radiat. Prot.* **2011**, *26*, 110–114. [CrossRef]
33. Ritchie, J.C.; McHenry, J.R. Application of Radioactive Fallout Cesium-137 for Measuring Soil Erosion and Sediment Accumulation Rates and Patterns: A Review. *J. Environ. Qual.* **1990**, *19*, 215–233. [CrossRef]
34. Wang, S.Y.; Xiao, X.; Wang, X.Q.; Dong, C.Q.; Li, W.Y.; Lu, Q.; Wang, T.P. Potassium recovery from the fly ash from a grate boiler firing agro-residues: Effects of unburnt carbon and calcination pretreatment. *J. Chem. Technol. Biotechnol.* **2017**, *92*, 801–807. [CrossRef]
35. Maeda, N.; Fukasawa, T.; Katakura, T.; Ito, M.; Ishigami, T.; Huang, A.-N.; Fukui, K. Existence Form of Potassium Components in Woody Biomass Combustion Ashes and Estimation Method of Its Enrichment Degree. *Energy Fuels* **2018**, *32*, 517–524. [CrossRef]

36. Wolbarst, A.B.; Chiu, W.A.; Yu, C.; Aiello, K.; Bachmaier, J.T.; Bastian, R.K.; Cheng, J.-J.; Goodman, J.; Hogan, R.; Jones, A.R.; et al. Radioactive materials in biosoils: Dose modeling. *Health Phys.* **2006**, *90*, 16–30. [CrossRef]
37. Cinelli, G.; De Cort, M.; Tollefsen, T. *European Atlas of Natural Radiation*; Publication Office of the European Union: Luxembourg, 2019.

materials

Article

Enhancing Mechanical Properties and Microstructures of Mass-Manufactured Sand Concrete by Incorporating Granite Powder

Jian Huang [1], Guangfeng Xu [2], Shujie Chen [2], Demei Yu [2,3], Tengfei Fu [2,*], Chao Feng [2,*] and Yulin Wang [4,5]

1 The Fifth Construction Co.Ltd. of CCCC Fourth Harbor Engineering Co., Ltd., Fuzhou 350008, China; swordy-hj@163.com
2 College of Transportation and Civil Engineering, Fujian Agriculture and Forestry University, Fuzhou 350108, China; yudemei0826@fafu.edu.cn (D.Y.)
3 CSCEC Strait Construction and Development Co., Ltd., Fuzhou 350015, China
4 School of Architecture and Civil Engineering, Wuyi University, Wuyishan 354300, China
5 Engineering Research Center of Prevention and Control of Geological Disasters in the Mountainous Areas of Northern Fujian, Fujian Province University, Wuyishan 354300, China
* Correspondence: futengfei@fafu.edu.cn (T.F.); fchao1227@fafu.edu.cn (C.F.)

Abstract: The production of manufactured sand and stone processing can cause dust pollution due to the generation of a significant amount of stone powder. This dust (mainly granite powder) was collected and incorporated as a cement replacement into mass-manufactured sand concrete in order to enhance the mechanical properties and microstructures. The heat of the hydration was measured by adding the granite powder into the cementitious material system. The mechanical properties, autogenous shrinkage, and pore structures of the concrete were tested. The results showed that the mechanical strength of the concrete increased first and then decreased with the increase in granite powder content. By replacing the 5% cement with the granite powder, the 28 d compressive and flexural strength increased by 17.6% and 20.9%, respectively. The autogenous shrinkage was mitigated by the incorporation of the 10% granite powder and decreased by 19.7%. The mechanism of the granite powder in the concrete was studied by X-ray diffraction (XRD), scanning electron microscopy (SEM), and mercury intrusion porosimetry (MIP). The porosity decreased significantly within the 10% granite powder. A microstructure analysis did not reveal a change in the type of hydration products but rather that the granite powder played a role in the microcrystalline nucleation during the hydration process.

Keywords: manufactured sand concrete; mechanical properties; heat of hydration; pore distribution; microstructure

1. Introduction

Manufactured sand has become the focus of research and applications due to the shortage of natural sand in order to achieve low-carbon, green, and sustainable development in the construction industry. A large amount of stone powder is produced as a consequence of this process, thus polluting the air and rivers. The resource utilization of stone powder is an effective way to solve this problem [1,2]. Stone powder is used as a mineral admixture to replace part of the cement, and it will affect the performance of the concrete through a filling effect in cement-based materials [3,4]. The stone powder will accelerate the cement hydration in cement paste under the condition of a constant water–cement ratio through the dilution effect of the cement paste. This is beneficial to improve the early strength of cement concrete [5,6].

The use of granite powder to replace part of the cement can significantly improve the carbonization and durability of the cement mortar [7]. Jain et al. found that replacing sand and cement with 30% granite powder and 15% waste glass powder can improve the durability of the concrete [8]. Singh et al. found that the mechanical strength of concrete was significantly improved by adding 10% marble and 25% granite powder [9]. In addition, Singh et al. pointed out that 30% fine sand replaced by fine granite solid waste can produce concrete with the optimum compressive strength [10]. Shwetha et al. considered the feasibility of using granite dust and debris to improve the tensile and bending strength of concrete [11]. Other research found that the introduction of stone powder into Portland cement systems was beneficial in improving the volume stability of the mortar and concrete [12]. Chouhan et al. [13] provided a comprehensive review on the influence of various shapes and sizes of stone waste as partial replacements of cement and sand regarding the workability, mechanical, and durability properties of mortar mixes and concrete. These studies suggest that the dimensional stone waste enhanced the bond properties and the durability characteristics [13]. Additional details can also be found in the literature [13]. Mashaly demonstrated that up to a 20% granite sludge replacement of cement could enhance the freeze and thaw resistance, abrasion, and sulfate attack, with negligible decreases in the physical and mechanical properties [14]. In other studies, the addition of stone powder was found to reduce the early shrinkage of cement-based materials but increase the later shrinkage [15,16].

The granite powder mainly improved the microstructure and properties of the concrete via the filling effect (referring to the physical presence of fine mineral additions in cement to enable denser packing) and nucleation effect (referring to the surface of the mineral filler providing nucleation sites for C–S–H formation). The improvement generally increased with the replacement of the granite powder. Xiong et al. found that tobermorite was generated in the slurry of granite–stone powder composite cementitious material by studying the effects of different-lithology stone powders on the hydration products of cement-based materials [17]. Li et al. found that the concrete pores first increased and then decreased with the increase in the granite powder content [18].

Granite powder can improve the mechanical strength and elastic modulus of concrete with manufactured sand in later ages as a supplementary cementitious material replacing fly ash with different proportions [19]. The granite powder was beneficial regarding the high-strength concrete durability with manufactured sand within a certain range of replacements. However, too-high values of fine powder could inhibit the hydration of the cement and then adversely affect the durability of the resulting concrete [20]. Shen et al. found that the particle shape and surface texture of manufactured sand had less influence on the performance of the concrete than that of stone powder [21].

At present, there are few studies on the effect of granite powder on the performance and mechanism of mass concrete composed of manufactured sand. This work incorporated granite powder into mass concrete in the hope of reducing the heat generation due to hydration and increasing the utilization of solid waste materials. The working mechanism of the granite powder was analyzed via XRD and SEM. The hydration behavior of the cementitious material was studied through isothermal calorimetry. The mechanical properties and early shrinkage properties of the granite powder mass concrete were also evaluated.

2. Materials and Methods

2.1. Materials and Mixture Proportion

The local P.O 42.5 cement (Lianshi Cement, Fuzhou, China) was used in this work. Granite powder (GP) was obtained from a local stone processing mill. Then, it was sieved through a 0.075 mm sieve. The specific surface area of the resulting powder is 3.075 m^2/g with a density of 2.18 g/cm^3. The chemical composition of the cement and granite powder from XRF analysis are shown in Table 1, and the SEM image and XRD diffractogram of the granite powder are presented in Figures 1 and 2.

Table 1. Chemical compositions of cement and granite powder (%).

Items	CaO	SiO_2	Al_2O_3	Fe_2O_3	MgO	K_2O	Na_2O
Cement	59.3	21.28	5.99	3.31	2.16	0.13	0.49
Granite powder	2.18	70.36	14.56	2.41	1.22	5.37	2.51

Figure 1. SEM image of granite powder (mag. ×10,000).

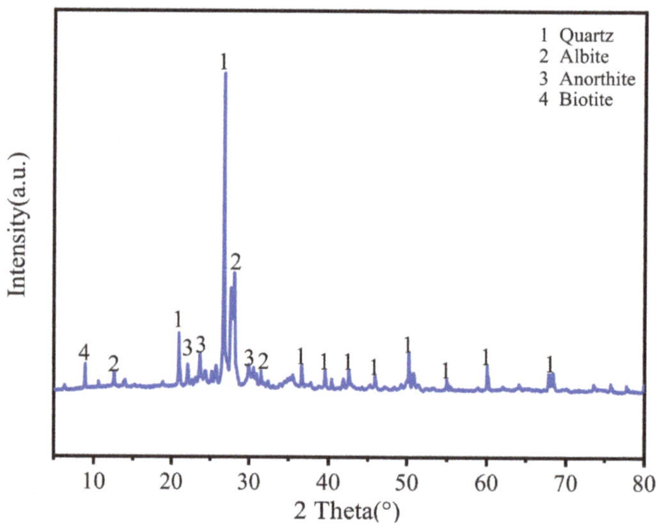

Figure 2. XRD pattern of granite powder.

The granite powder particles have rough surfaces showing irregular granular shapes (Figure 1), and it is observed that the granite powder is in the form of quartz, albite, anorthite, and biotite (Figure 2). Figure 3 shows the particle distributions of cement and granite powder, showing that the sieved granite powder is finer than cement. The fineness modulus of the river sand was 2.82. The coarse aggregate of basalt was used with a gradation of 5–25 mm, and the apparent density was 2999 kg/m^3. The mixing water was clean tap water. The fine aggregate used was combined using manufactured sand (MS) and natural fine sand (FS) according to a ratio of 0.64:0.36. The apparent density was 2596 kg/m^3. Gradation of the fine aggregate is shown in Table 2.

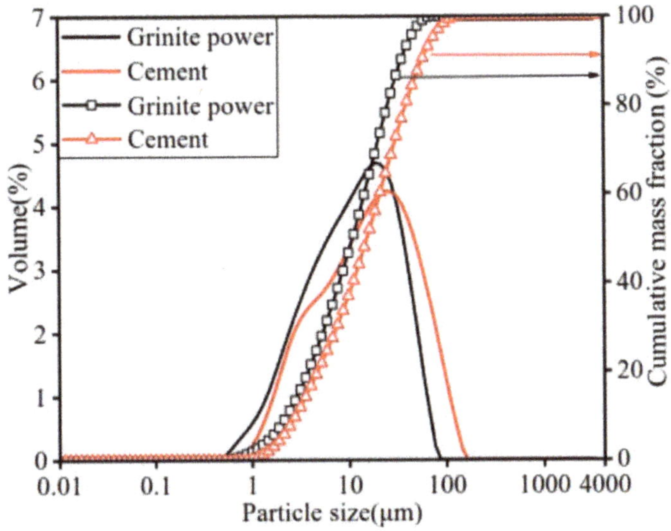

Figure 3. Particle distributions of cement and granite powder. (Cumulative mass fraction follows arrow to the right axis).

Table 2. Sieve analysis of fine aggregate (passing, %).

Sieve Opening (mm)	9.5	4.75	2.36	1.18	0.6	0.3	0.15	0.075	Fineness Modulus
MS	100	95.1	65.2	49.8	31.2	14.2	8.2	3.6	3.28
FS	100	100	100	100	98.1	54.0	12.6	1.3	1.35

Mixture proportions of mass-manufactured sand concrete with granite powder and fly ash are presented in Table 3. This particular mixture design was used for the Fuzhou subway roof, which is a 2 m deep reinforced concrete slab, to support the superstructures.

Table 3. Concrete mixture proportions (kg/m^3).

Items	Cementitious Materials			Coarse Aggregate	Fine Aggregate	Water	Superplasticizer
	Cement	Granite Powder	Fly Ash				
AGP0	338.40	0	59.4	1056	704	151.2	0.75
AGP5	321.48	16.92	59.4	1056	704	151.2	0.75
AGP10	304.56	33.84	59.4	1056	704	151.2	0.75
AGP15	287.64	50.76	59.4	1056	704	151.2	0.75

2.2. *Experimental Methods*

2.2.1. Mechanical Properties Testing

Compressive strength of the concrete specimens was tested by compression testing machine (YAW4206, Shanghai, China). Flexural strength of the specimens was tested by an electro-hydraulic servo universal testing machine (HUT605A, Wan Testing Equipment, Shenzhen, China). Three specimens of 100 mm × 100 mm × 100 mm were tested based on China National Standard GB/T 50081-2019 [22]. Three 100 mm × 100 mm × 400 mm specimens were evaluated according to ASTM C78 [23]. All samples were cured in a standard condition of 20 ± 3 °C and 95 ± 5% RH.

2.2.2. Heat of Hydration Measurement

The hydration heat test was performed on a TAM Air isothermal calorimeter (TAM Air, TA Instruments, New Castle, DE, USA). The hydration exothermic rate and cumulative exothermic amount of different cementitious systems were measured according to ASTM C186-2017 [24]. The water–binder ratio was 0.38. For each mixture, the result was averaged between two specimens.

2.2.3. Autogenous Shrinkage Measurement

The autogenous shrinkage of concrete was carried out according to ASTM C1698-2019 [25]. The prepared concrete was poured into a 420 mm (Φ70 mm) corrugated tube, placed horizontally on a steel frame. The room temperature was controlled at (20 ± 2) °C, and the relative humidity was above 50%. The results were averaged between three specimens.

2.2.4. Pore Structure Analysis

Three specimens of 20 mm × 20 mm × 20 mm were prepared. The samples were soaked in anhydrous ethanol to stop the hydration of cement and dried in a 60 ± 5 °C vacuum oven for 5 h, then cooled to room temperature. The pore structure of concrete was measured by Auto Pore IV 9500 mercury intrusion porosimeter (Norcross, GA, USA).

2.2.5. Microstructure Characterization

Samples of approximately 5 mm × 5 mm were taken from the AGP0 and AGP10 28 d specimens and dried and sputtered with gold. Microscopic morphology was observed using an electron microscope of model Verios G4 UC (Thermo Fisher Scientific, Waltham, MA, USA).

Mineral composition of concrete was analyzed using a D8 Advance type X-ray diffractometer (Bruker, Billerica, MA, USA). The 28 d concrete specimens were ground until all the pieces broke down and sieved through 0.075 mm sieve. The samples were made into 10 mm × 10 mm filament specimens. The testing condition was at 2θ of 5–90° and the step of 12°/min.

3. Results

3.1. Mechanical Properties

3.1.1. Compressive Strength

The compressive strength of the manufactured sand concrete with different granite powder content is shown in Figure 4. The compressive strength of 7 d and 28 d peaked at the 5% granite powder content. Compared with the control group (AGP0), the compressive strength in the AGP5 and AGP10 groups at 7 d and 28 d increased significantly. AGP5 increased by 18.9% and 17.6%, and AGP10 increased by 12.5% and 8.8%, respectively. When the replacement exceeded 10%, the AGP15 group was 8.1% and 6.7% lower than the AGP0 group at 7 d and 28 d, respectively. Therefore, the high replacement of the granite powder could adversely affect the compressive strength.

3.1.2. Flexural Strength

The flexural strength of manufactured sand concrete with different granite powder content is shown in Figure 5. The flexural strength is similar to the compressive strength, and it increases first and then decreases with the increase in powder content. The 7 d and 28 d flexural strength of the concrete in AGP5 increased by 18.1% and 20.9% compared with the controls (AGP0), and AGP10 increased by 12.6% and 10.3%, respectively. However, a further increase in the replacement of the granite powder would not result in a significant increase in the flexural strength.

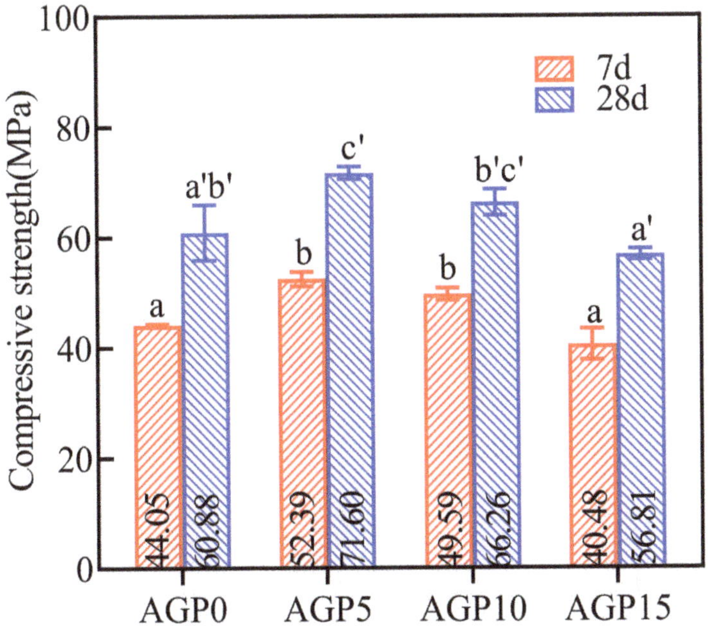

Figure 4. Compressive strength values of manufactured sand concrete with different stone powder content. Note: data were analyzed with one-way ANOVA. Significant differences exist between any two groups when they do not share a common letter over the columns (also applies for Figure 5).

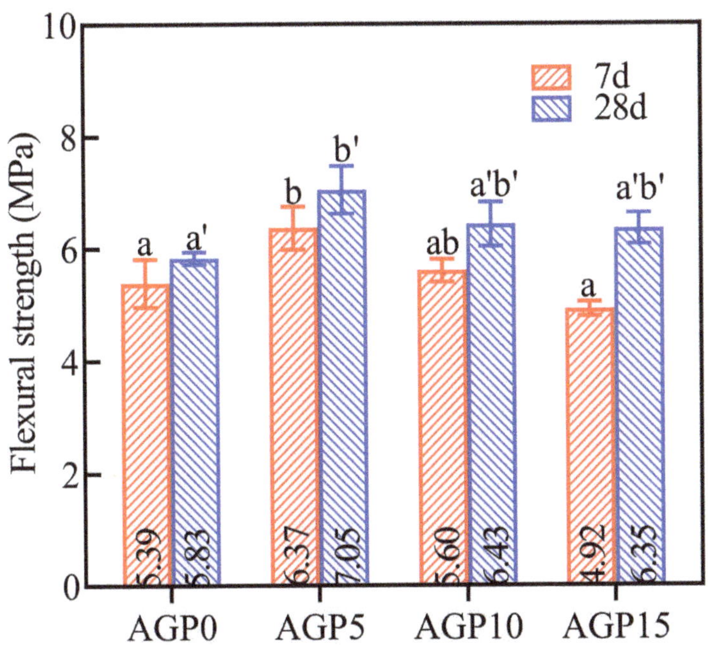

Figure 5. The flexural strength of manufactured sand concrete with different granite powder replacements.

3.2. Heat of Hydration

The effect of the granite powder on the heat release rate of the cement hydration is shown in Figure 6. The effect on the cumulative heat is shown in Table 4 and Figure 7. From Figure 6a, it can be seen that the peak of the heat evolution curve shifted to the right for the cement pastes blended with granite powder when compared to AGP0. Therefore, the granite powder has a slight retardation effect on the hydration of the cement. Table 4 and Figure 7 show that the granite powder can also decrease the hydration heat. The cumulative hydration heat values at 150 h for AGP5, AGP10, and AGP15 are 2.03%, 4.22%, and 7.23% lower than for AGP0. This is due to granite powder being considered inert material. Partially replacing the cement would inevitably reduce the total heat of the hydration in the system, which can be beneficial regarding the mass concrete casting by reducing the risk of early cracking due to the temperature gradient.

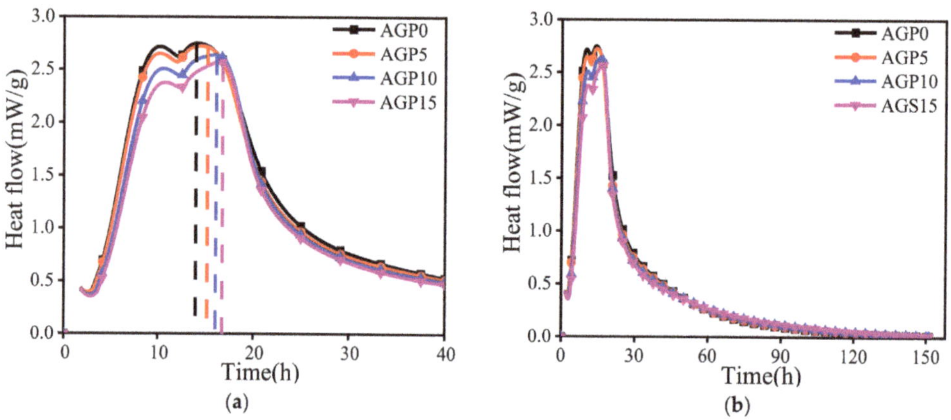

Figure 6. Effect of granite powder on hydration heat release rate of cement: (**a**) heat flow in the first 40 h (dashed line indicating corresponding time of heat flow peak); (**b**) heat flow within 150 h.

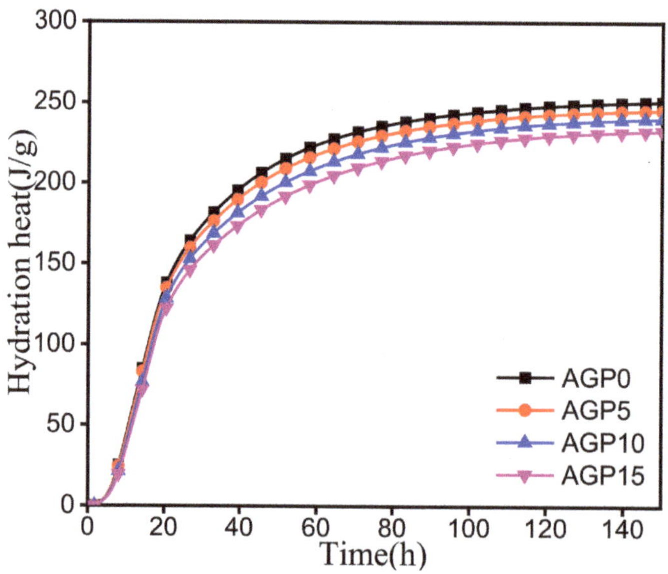

Figure 7. Cumulative heat release over time.

Table 4. Cumulative heat release of cement at 150 h.

Mixture	Hydration Heat (J/g)	Relative Increase (%)
AGP0	250.98	0
AGP5	245.88	−2.03
AGP10	240.38	−4.22
AGP15	232.84	−7.23

3.3. Autogenous Shrinkage

The effect of different granite powder content on the autogenous shrinkage is shown in Figure 8. It can be seen that the autogenous shrinkage of the concrete occurred mainly on the first day and did not change significantly in the next few days. The increase in the granite powder content had a greater effect on the autogenous shrinkage. The autogenous shrinkage rates of the AGP5 and AGP10 concrete groups were comparatively small. Compared with the AGP0 group, there were significant decreases of 10.4% and 19.7%, respectively. When the replacement reached 15%, the autogenous shrinkage rate of the concrete increased significantly. However, it should be noted that high powder content would create smaller pores (50 nm or less) so as to significantly increase the autogenous shrinkage.

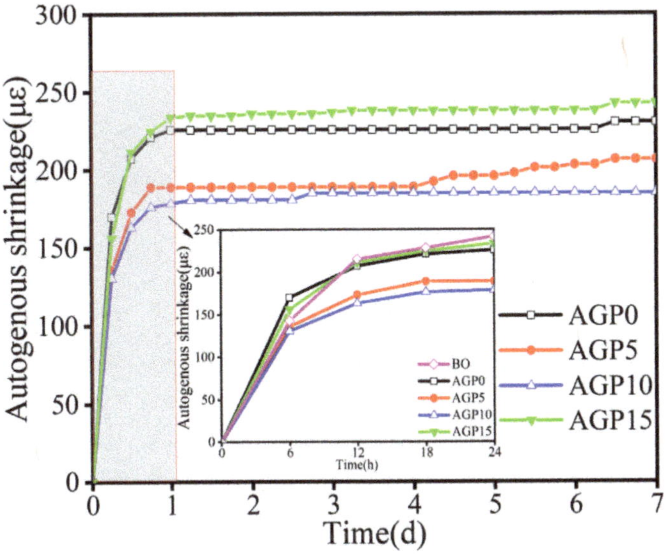

Figure 8. Effects of different amounts of stone powder substitution on autogenous shrinkage.

3.4. Microstructure Analysis

3.4.1. Pore Structure

The results of the MIP test of the concrete pore structure are shown in Table 5 and Figure 9a. Based on the volume proportions of the pore structure distributions shown in Figure 9b, it can be seen that, when the granite powder replaces the cement in a certain range, the filling effect of the granite powder on the concrete is significant. With the increase in the granite powder content, the larger pores (≥50 nm) in the concrete are significantly reduced, the compactness of the concrete is improved, and the porosity is reduced. When the replacement amount of the granite powder continues to increase, the filling effect diminishes because the granite powder cannot fill the pores close to or smaller than its own size. This can also help to explain the test results of the mechanical properties.

Table 5. Test results of concrete pore structure.

Mixture	Total Pore Volume (mL/g)	Total Hole Area (m^2/g)	Average Pore Size (nm)	Porosity (%)
AGP0	0.0535	6.009	15.35	11.9
AGP5	0.055	8.623	21.51	9.58
AGP10	0.036	7.734	15.68	6.50
AGP15	0.0694	13.74	19.51	11.9

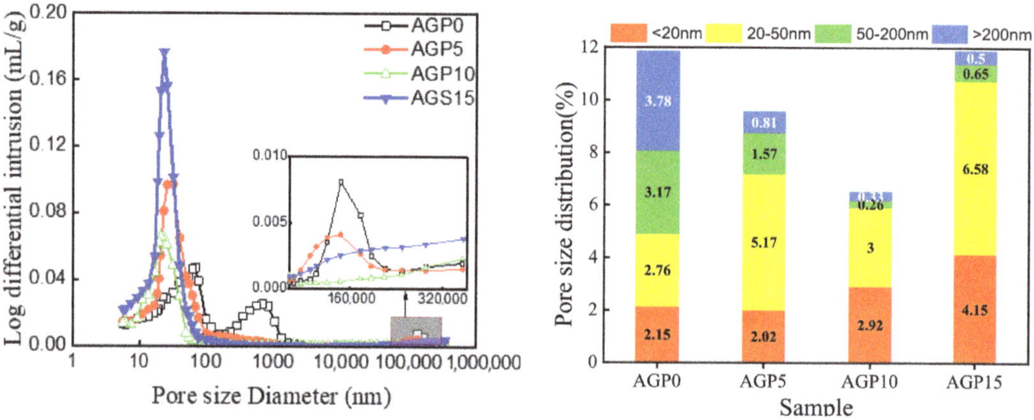

(**a**) (**b**)

Figure 9. Pore-size distributions of concrete with different granite powder content: (**a**) differential intrusion; (**b**) pore-size distribution.

3.4.2. SEM

The SEM images of the 28 d hydration products of the two groups of AGP0 and AGP10 are shown in Figure 10. There was a significant difference with the addition of the granite powder. There were a large number of pores and micro-cracks between the hydration products, and the structure was not as dense. After the incorporation of the granite powder, it can be observed that the granite powder and the hydration products in AGP10 are cemented together (Figure 10b) with significantly decreased porosity compared to AGP0. Also, the number of micro-cracks is significantly reduced. This is likely due to the nucleation effect of the granite powder, forming a denser structure by combining the hydration products with stronger integrity.

3.4.3. XRD

The XRD results of the 28 d hydration products of granite powder-enhanced concrete are shown in Figure 11. Compared with the control group (AGP0), there were significantly higher proportions of quartz and albite phases in the AGP10 and AGP15 groups. The characteristic peak of calcium hydroxide (CH) in the AGP10 group is higher than that in the AGP0 group, indicating a higher degree of hydration. Due to the presence of certain nucleation in granite powder, CH crystallization is promoted. The CH characteristic peak of the AGP15 group is lower, which is due to the high content and the excessive reduction in the total amount of cement, which leads to the reduction in the amount of hydration products, and the dilution of granite powder leads to a reduction in the hydration degree.

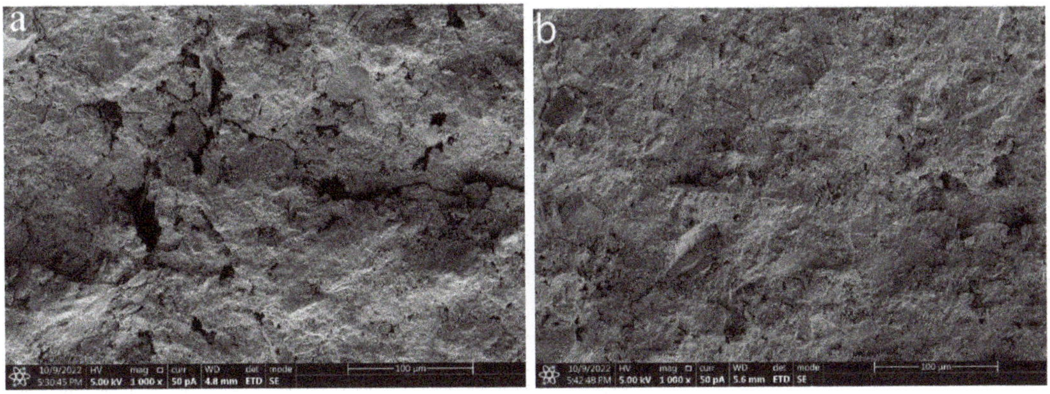

Figure 10. SEM images of 28 d hydration products (**a**) AGP0; (**b**) AGP10.

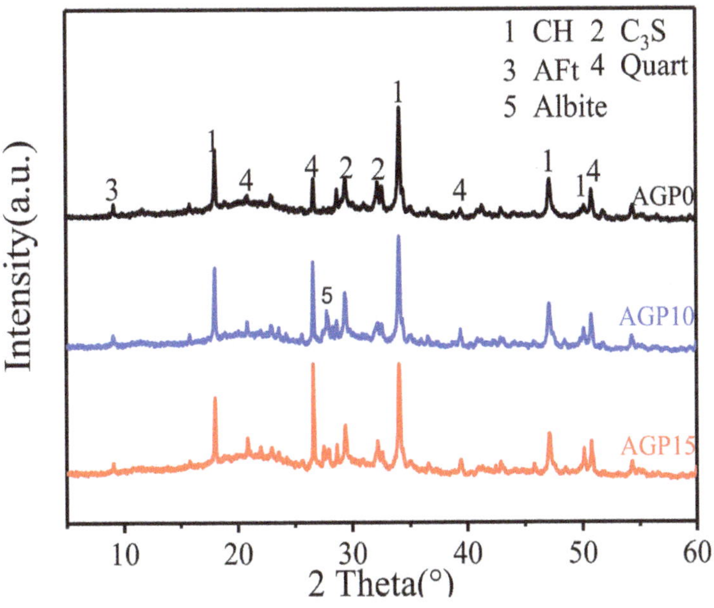

Figure 11. XRD patterns of 28 d samples.

4. Conclusions

Granite powder was incorporated to improve the mechanical properties of mass-manufactured sand concrete. The mechanical strength of mass machine-made sand concrete increases first and then decreases with the increase in granite powder. The 28 d compressive strength with the 5% and 10% replacements increased by 17.6% and 8.8%, and the 28 d flexural strength increased by 20.9% and 10.3%, respectively. Granite powder with an appropriate replacement can reduce the autogenous shrinkage. Compared with the control group, the autogenous shrinkage rate of the concrete with the 10% granite powder replacement decreased significantly by 19.7%. The heat of hydration of the cementitious material was reduced by incorporating the granite powder. The porosity was decreased significantly within a certain amount of granite powder. The granite powder did not change the type of the hydration products but played a role in the microcrystalline nucleation during the hydration process.

Author Contributions: Conceptualization, D.Y. and T.F.; validation, T.F.; formal analysis, C.F.; investigation, J.H., G.X. and S.C.; data curation, C.F.; writing—original draft preparation, J.H.; writing—review and editing, T.F. and Y.W.; supervision, T.F.; project administration and funding acquisition, D.Y. All authors have read and agreed to the published version of the manuscript.

Funding: This research was funded by the Fifth Construction Co., Ltd. of CCCC Fourth Harbor Engineering Co., Ltd., China (grant No. ZJSHW-FZDTLZ-2021-E3-048), China Construction Shares Technology Research and Development Project China (grant No. CSCEC-2023-Z-07), Engineering Research Center of Prevention and Control of Geological Disasters in the Mountainous Areas of Northern Fujian, Fujian Province University, China (grant No. WYERC2024-3), and Science Foundation Projects of Fujian Province (2023J01476).

Institutional Review Board Statement: Not applicable.

Informed Consent Statement: Not applicable.

Data Availability Statement: The original contributions presented in the study are included in the article, further inquiries can be directed to the corresponding author.

Conflicts of Interest: Author Jian Huang was employed by the company The Fifth Construction Co., Ltd. of CCCC Fourth Harbor Engineering Co., Ltd., and author Demei Yu was employed by the company CSCEC Strait Construction and Development Co., Ltd. The remaining authors declare that the research was conducted in the absence of any commercial or financial relationships that could be construed as a potential conflict of interest.

References

1. Luo, Y.; Bao, S.; Zhang, Y. Recycling of granite powder and waste marble produced from stone processing for the preparation of architectural glass–ceramic. *Constr. Build. Mater.* **2022**, *346*, 128408. [CrossRef]
2. Jia, L.; Huang, M.; Huang, M.; Luo, Y.; Zhang, X.; Liao, X.; Bao, S. Preparation of granite powder–based geopolymer by synergistic action of calcination and phosphoric acid. *J. Am. Ceram. Soc.* **2024**, *107*, 501–513. [CrossRef]
3. Wang, D.; Shi, C.; Farzadnia, N.; Shi, Z.; Jia, H.; Ou, Z. A review on use of limestone powder in cement-based materials: Mechanism, hydration and microstructures. *Constr. Build. Mater.* **2018**, *181*, 659–672. [CrossRef]
4. Campos, H.; Klein, N.; Marques Filho, J. Proposed mix design method for sustainable high-strength concrete using particle packing optimization. *J. Clean. Prod.* **2020**, *265*, 121907. [CrossRef]
5. Ma, J.; Yu, Z.; Ni, C.; Shi, H.; Shen, X. Effects of limestone powder on the hydration and microstructure development of calcium sulphoaluminate cement under long-term curing. *Constr. Build. Mater.* **2019**, *199*, 688–695. [CrossRef]
6. Moon, G.; Oh, S.; Jung, S.; Choi, Y. Effects of the fineness of limestone powder and cement on the hydration and strength development of PLC concrete. *Constr. Build. Mater.* **2017**, *135*, 129–136. [CrossRef]
7. Li, L.; Wang, Y.; Tan, Y.; Kwan, A.; Li, L. Adding granite dust as paste replacement to improve durability and dimensional stability of mortar. *Powder Technol.* **2018**, *333*, 269–276. [CrossRef]
8. Jain, K.; Sancheti, G.; Gupta, L. Durability performance of waste granite and glass powder added concrete. *Constr. Build. Mater.* **2020**, *252*, 119075. [CrossRef]
9. Singh, C.; Aggarwal, V. Experimental investigation of concrete strength properties by partial replacement of cement-sand with marble-granite powder. *Mater. Today Proc.* **2022**, *62*, 3734–3737. [CrossRef]
10. Singh, S.; Khan, S.; Khandelwal, R.; Chugh, A.; Nagar, R. Performance of sustainable concrete containing granite cutting waste. *J. Clean. Prod.* **2016**, *119*, 86–98. [CrossRef]
11. Shwetha, K.; Mahesh, K.; Dalawai, V.; Anadinni, S.; Sowjanya, G. Comparative study on strengthening of concrete using granite waste. *Mater. Today Proc.* **2022**, *62*, 5317–5322. [CrossRef]
12. Wu, L.; Farzadnia, N.; Shi, C.; Zhang, Z.; Wang, H. Autogenous shrinkage of high-performance concrete: A review. *Constr. Build. Mater.* **2017**, *149*, 62–75. [CrossRef]
13. Chouhan, H.; Kalla, P.; Nagar, R.; Gautam, P. Influence of dimensional stone waste on mechanical and durability properties of mortar: A review. *Constr. Build. Mater.* **2019**, *227*, 116662. [CrossRef]
14. Mashaly, A.; Shalaby, B.; Rashwan, M.; Pradeep, K. Performance of mortar and concrete incorporating granite sludge as cement replacement. *Constr. Build. Mater.* **2018**, *169*, 800–818. [CrossRef]
15. Danish, A.; Mosaberpanah, M.; Salim, M.; Fediuk, R.; Waqas, R. Reusing marble and granite dust as cement replacement in cementitious composites: A review on sustainability benefits and critical challenges. *J. Build. Eng.* **2021**, *44*, 102600. [CrossRef]
16. Medina, G.; Bosque, I.; Frías, M.; Rojas, M.; Medina, C. Granite quarry waste as a future eco-efficient supplementary cementitious material (SCM): Scientific and technical considerations. *J. Clean. Prod.* **2017**, *148*, 467–476. [CrossRef]
17. Xiong, Z.; Wang, P.; Wang, Y. Hydration behaviors of portland cement with different lithologic stone powders. *Int. J. Concr. Struct. Mater.* **2015**, *9*, 55–60. [CrossRef]

18. Li, T.; Tier, L. Microscopic mechanism analysis of the influence of stone powder with different replacement ratio on concrete performance. *Arab. J. Geosci.* **2022**, *16*, 1132–1142. [CrossRef]
19. Li, H.; Huang, F.; Cheng, G.; Xie, Y.; Tan, Y.; Li, L.; Yi, Z. Effect of granite dust on mechanical and some durability properties of manufactured sand concrete. *Constr. Build. Mater.* **2016**, *109*, 41–46. [CrossRef]
20. Zheng, S.; Chen, J.; Wang, W. Effects of Fines Content on Durability of High-Strength Manufactured Sand Concrete. *Materials* **2023**, *16*, 522. [CrossRef]
21. Shen, W.; Yang, Z.; Cao, L.; Cao, L.; Liu, Y.; Yang, H.; Liu, Z.; Bai, J. Characterization of manufactured sand: Particle shape, surface texture and behavior in concrete. *Constr. Build. Mater.* **2016**, *114*, 595–601. [CrossRef]
22. *GB/T 50081-2019*; Standard for Test Methods of Concrete Physical and Mechanical Properties. Ministry of Housing and Construction, People's Republic of China: Beijing, China, 2019.
23. *ASTM C78*; Standard Test Method for Flexural Strength of Concrete (Using Simple Beam with Third-Point Loading). ASTM International: West Conshohocken, PA, USA, 2010.
24. *ASTM C186*; Standard Test Method for Heat of Hydration of Hydraulic Cement. ASTM International: West Conshohocken, PA, USA, 2017.
25. *ASTM C1698*; Standard Test Method for Autogenous Strain of Cement Paste and Mortar. ASTM International: West Conshohocken, PA, USA, 2019.

materials

MDPI

Article

Mechanical and Microstructural Investigation of Geopolymer Concrete Incorporating Recycled Waste Plastic Aggregate

Blessing O. Adeleke [1,*], John M. Kinuthia [1], Jonathan Oti [1], Duncan Pirrie [1] and Matthew Power [2]

1 Faculty of Computing, Engineering and Science, University of South Wales, Pontypridd CF37 1DL, UK; john.kinuthia@southwales.ac.uk (J.M.K.); jonathan.oti@southwales.ac.uk (J.O.); duncan.pirrie@southwales.ac.uk (D.P.)

2 Vidence Inc., 4288 Lozells Avenue, Suite 213L, Burnaby, BC V5A 0C7, Canada; matthew.power@videnceinc.com

* Correspondence: blessing.adeleke@southwales.ac.uk; Tel.: +44-1443482167

Abstract: The effective use of waste materials is one of the key drivers in ensuring sustainability within the construction industry. This paper investigates the viability and efficacy of sustainably incorporating a polylactic acid-type plastic (WP) as a 10 mm natural coarse aggregate (NA) replacement in geopolymer concrete. Two types of concrete (ordinary Portland cement—OPC and geopolymer) were produced for completeness using a concrete formulation ratio of 1:2:3. The ordinary concrete binder control was prepared using 100% OPC at a water/binder ratio of 0.55, while the geopolymer concrete control used an optimum alkaline activator/precursor—A/P ratio (0.5) and sodium silicate to sodium hydroxide—SS/SH volume ratio (1.2/0.8). Using the same binder quantity as the control, four concrete batches were developed by replacing 10 mm NA with WP at 30 and 70 wt% for ordinary and geopolymer concrete. The mechanical performance of the developed concrete was assessed according to their appropriate standards, while a microstructural investigation was employed after 28 days of curing to identify any morphological changes and hydrated phases. The results illustrate the viability of incorporating WP in geopolymer concrete production at up to 70 wt% replacement despite some negative impacts on concrete performance. From a mechanical perspective, geopolymer concrete indicated a 46.7–58.3% strength development superiority over ordinary concrete with or without WP. The sample composition and texture quantified using automated scanning electron microscopy indicated that adding WP reduced the presence of pores within the microstructure of both concrete types. However, this was detrimental to the ordinary concrete due to the low interfacial zone (ITZ) between calcium silicate hydrate (CSH) gel and WP, resulting in the formation of cracks.

Keywords: alkali–alkaline activator; GGBS; silica fume; consistency; compressive strength; scanning electron microscopy; waste plastic; aluminosilicate; geopolymer

1. Introduction

Since the 20th century, the global production of plastics has significantly increased from approximately 240 million tonnes in 2002 to 299 million tonnes in 2013 and is continuously rising [1]. For instance, 359 million tons of plastic were produced in 2018, with 51% coming from Asia and 17% from Europe [2]. According to the Association of Plastic Manufacturers—PlasticsEurope [3] and Brahney et al. [4], the current rate of plastic production was estimated at 370 million metric tons per year; at that rate, about 11 billion metric tons of plastic will have accumulated in the environment by 2025. In India alone, the annual waste plastic production is currently at 40 million tonnes, increasing at a rate of 1.5% to 2% yearly [1].

The construction industry has also resulted in waste plastic (WP) generation with the constant need for new infrastructure worldwide [5]. Unfortunately, this has resulted in a high demand for construction materials which have contributed significantly to waste production, and this was identified as one key factor in the construction industry contributing to pollution worldwide [6]. The negative environmental implications of WP, as corroborated by Zhang et al. [7] and Xiang et al. [8], cannot be overemphasised as it disintegrates into micro-plastics (MPs, 1 μm–5 mm) and nano-plastics (NPs, 1 nm–1000 nm) in nature due to ultraviolet (UV) radiation, physical abrasion, biological metabolism, mechanical, and temperature changes, and the resultant micro-plastics impact on water and soil quality. Furthermore, micro- and nano-plastics can be transported through aeolian and aqueous processes. They possess an enhanced capability to adsorb and transfer other environmental pollutants, such as heavy metals and organic contaminants [9]. Despite the high contamination potential of WP at the finest particle level, Oliveira et al. [10] and Bridson et al. [11] further described WP as a worrying environmental problem resulting from the leaching of hazardous substances used in producing polymer products, such as lubricants, surfactants, solvents, flame-retardant catalysts, antioxidants, plasticizers, and colourants. Hence, all these industrial products potentially pose significant risks to animal and human health in the form of undesirable immune responses, acute reactions, cytotoxicity, and hypersensitivity [12–14].

The increased production of WP does, therefore, require effective strategies to be adopted and implemented to reduce the potential negative health and environmental impacts. These strategies include recycling/reusing using mechanical, biological, or thermochemical techniques, incineration for energy recovery, and landfilling [1,6]. However, only 9% of waste plastic is recycled worldwide, while 80% is disposed in landfill or dispersed in both terrestrial and aquatic ecosystems [6,7,15]. Additionally, the incineration approach is not sustainable as it involves a combustion process that emits carbon dioxide (CO_2) gas along with potentially toxic elements into the atmosphere. Geyer et al. [16] and Yang et al. [17] also agreed that incineration does not entirely destroy WP since the bottom ash can contain remnants of partly burnt waste (micro-plastics), which is known to be a potential pollution source.

The application of waste in the manufacturing of construction materials is broadly acknowledged as an effective approach to limit the harmful consequences of WP and potential arsenic toxicity [18]. Past studies have shown that reusing waste materials has become a widespread strategy in developing and developed countries to reduce the environmental impacts of waste materials [19,20]. As such, applying WP as an aggregate in construction materials is a reasonable reuse/recycling strategy that promotes environmentally friendly and green construction by substituting these wastes, as aggregates and fibres, for natural raw materials [1,2]. This approach involves incorporating waste materials (including arsenic-containing wastes) into the construction process, thus preventing their dispersal and to mobilise potential risks to health and the environment. This effectively mobilises the toxic substances within a solid matrix and reduces the chances of releasing them into the environment [18,21]. Previous studies investigated the use of WP in ordinary concrete that used plastic in two forms: plastic fibres or aggregates [1,2,22,23]. In those instances, a complete or partial natural aggregate's replacement with WP resulted in natural resource savings. After the initial grinding and sizing of WP, granules below the size of 4 mm are employed as fine aggregates, while the fibres are used as additives. Also, roughly crushed WP above 5 mm was used to replace coarse aggregates [6,24]. In terms of lower density, unit weight and thermal conductivity, building materials incorporating WP have demonstrated suitable material properties compared to building materials without WP replacement. However, there is deterioration of the mechanical properties of building materials as the WP dosage increases [24]. Examples of such applications include their incorporation in building brick, paver block, concrete, bitumen and asphalt mixtures, etc. [6].

Compared with traditional ordinary Portland cement (OPC) concrete, the high-strength geopolymer concrete relies on the use of sodium silicate and possesses the additional benefits of reduced CO_2 emissions and energy consumption [25]. Geopolymers are inorganic alumino-silicate polymers with a three-dimensional network, which are obtained using the reaction of aluminosilicate-rich materials obtained from industrial, municipal, and agricultural wastes and by-products such as fly ash, ground granulated blast-furnace slag (GGBS), rice husk ash, etc., with a highly alkaline aqueous solution [26,27]. This reaction is an alkali activation process where the final products of geopolymerisation are usually manipulated by the chemical composition of the alkaline activators and source aluminosilicate materials that acts as a precursor [28]. Provis [29] also described *"Geopolymers as a form of a subset within the broader alkali-activated binder family. These low-calcium binding materials result from the reaction of an aluminosilicate powder (precursor) with an alkaline activator, typically composed of hydroxide, silicate, carbonate, or sulfate. This reaction leads to the formation of an amorphous aluminosilicate gel and secondary nanocrystalline zeolite-like structures."* Hama and Hilal [30] investigated the effect of partially replacing sand with plastic waste in self-compacting OPC concrete. The range of compressive strength values was about 65 MPa to 37 MPa and that a decrease in compressive strength was observed as the plastic content increased in comparison with the control mix without WP. The reason for this observation was that the plastic waste was a soft material compared with natural aggregate. Also, Pacheco-Torgal [31] investigated the impact of using polyethylene terephthalate (PET) from plastic bottles as a partial substitute for sand within the OPC concrete with up to 50% plastic replacement. The research concluded that the recycled PET could be substituted into eco-friendly concrete at specific replacement rates. A reason for this conclusion was that by replacing sand with PET, the self-weight of concrete is decreased, making the material more suitable for non-structural elements that do not require high compressive strength. Shaikh [32] reinforced OPC, OPC-fly ash-based composites, and an ambient-cured geopolymer using recycled PET (e.g., polypropylene fibre of length 12–19 mm). This improved the compressive strength of geopolymer composites better than other cementitious composites to ~57 MPa and ~47 MPa when 1.0 vol% and 1.5 vol% PET fibres were added, respectively.

Geopolymer concretes composed of fly ash, blast-furnace slag, and PET waste (bottles) decreased the compressive and tensile strength compared with the control mix without plastics from 20.5 MPa to 13.8 MPa and from 2.7 MPa to 1.2 MPa, respectively. In turn, the 10% replacement of sand with plastic granules as fine aggregate contributed to a slight improvement in the mechanical properties [33]. Lazorenko et al. [6] employed the use of WP as fillers by assessing sustainable construction using novel geopolymer composites by integrating waste plastic of different sizes and shapes. The results indicated that geopolymer composites with fine PET particles showed high strength, while increasing the size of PET particles decreased the workability of the geopolymer. Another example of an existing outcome obtained from replacing certain components in concrete composition with plastic is from Ahmed et al. [2], who investigated geopolymer concrete composites incorporated with recycled plastic aggregates and modified with nano-silica. The authors suggested an optimal use of 50% replacement levels of recycled plastics as a replacement for fine aggregates in geopolymer concrete.

Notably, little or no research has been conducted to explore the application of WP as a coarse aggregate in geopolymer concrete using either a commercial or laboratory-synthesised silica fume (SF)-derived sodium silicate (SS) solution. Thus, this paper aims to evaluate the viability and efficacy of sustainably incorporating WP as a 10 mm NA replacement in geopolymer concrete production while assessing the mechanical and microstructural performance of the produced geopolymer. The current research findings aim to add to the existing knowledge on geopolymer concrete research, encourage more eco-conscious practices in geopolymer concrete production, whilst promoting a greener and more sustainable future in the construction sector.

2. Materials

This study utilised ordinary Portland cement (CEM-II/B-V 32.5R), silica fumes (SFs), sodium hydroxide (NaOH), ground granulated blast-furnace slag (GGBS), aggregates, and waste plastics (WP). OPC was manufactured and supplied by Lafarge Cement UK in accordance with BS EN 197—1:2011 [34]. GGBS was exploited as an aluminosilicate precursor material in this study and was manufactured in accordance with BS EN 15167-1:2006 [35] by Civil and Marine Slag Cement Ltd., Llanwern, Newport, UK.

The SF was a light-grey amorphous reactive micro-silica powder that was manufactured in Norway by Elkem Silicon Materials and supplied under the trading name of Elkem Un-densified Micro-silica 971 by Tarmac Cement and Lime Company, Buxton Lime and Powders, Derbyshire, Derby, UK. NaOH was used in the form of white laboratory-grade pellets, with a molecular weight of 40 g/mol, pH value of 14, and was commercially sourced from Fisher Scientific Ltd., Loughborough, Leicestershire, UK. The pellets were to produce a sodium hydroxide solution of 10 M [5]. The solution was then used as an activator to dissolve SF to create a sodium silicate solution with SiO_2/Na_2O molar ratio of 2. The coarse aggregates are limestone aggregates (20 mm and 10 mm), while the fine aggregate was a natural river sand dredged from the Bristol Channel, UK. A local supplier supplied both coarse and fine aggregates according to the requirements of BS EN 12620:2002+A1:2008 [36]. The WP is a polylactic acid-type plastic used in construction as insulation, furnishing and fibre within carpets, food packaging, and 3D printing [37]. The plastic material used for this research was waste 3D printer filaments broken down using a plastic shredder with the capacity to produce plastic granules between 0–10 mm (see Figure 1). In addition, a particle size distribution curve for the shredded WP between a range of 1–4 mm and sieved samples are shown in Figure 2 according to BS EN ISO 17892-4 [38]. Table 1 shows the chemical oxide compositions for OPC, GGBS, and SF, while Tables 2 and 3 present their physical properties. Also, Table 3 presents some physical properties of the aggregates (coarse and fine) and WP according to the relevant standards.

Table 1. Oxide compositions of OPC, GGBS, and SF [5].

Oxides	Compositions (%)		
	Cement	**GGBS**	**SF**
CaO	61.49	37.99	0.2
MgO	3.54	8.78	0.1
SiO_2	18.84	35.54	97.1
Al_2O_3	4.77	11.46	0.2
Na_2O	0.02	0.37	-
P_2O_5	0.1	0.02	0.03
Fe_2O_3	2.87	0.42	0.01
Mn_2O_3	0.05	0.43	-
K_2O	0.57	0.43	0.2
TiO_2	0.26	0.7	-
V_2O_5	0.06	0.04	-
BaO	0.05	0.09	-
SO_3	3.12	1.54	0.1
Loss on ignition	4.3	2	0.5

Table 2. Physical compositions of OPC, GGBS, and SF [5].

Other Properties	Cement	GGBS	SF
Insoluble residue	0.5	0.3	-
Bulk density (kg/m^3)	1400	1200	300
Specific gravity	3.15	2.9	3.15
Glass content (%)	-	90	-
Blaine fineness (m^2/kg)	365	450	-
Alkalinity value (pH)	13.41	10.4	7
Colour	Grey	Off-white	Grey
Physical form	Fine powder	Fine powder	Powder

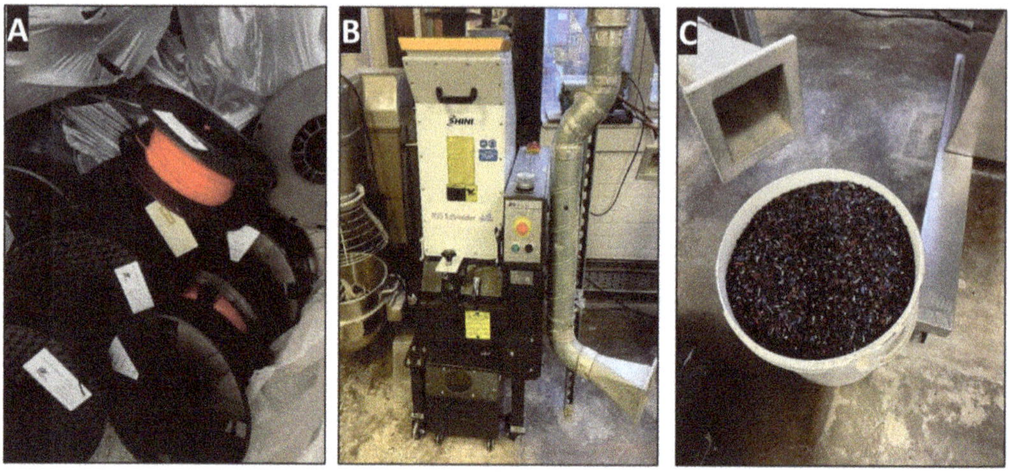

Figure 1. Production of the plastic waste material: (**A**) plastic waste; (**B**) plastic shredder; and (**C**) shredded plastic granules.

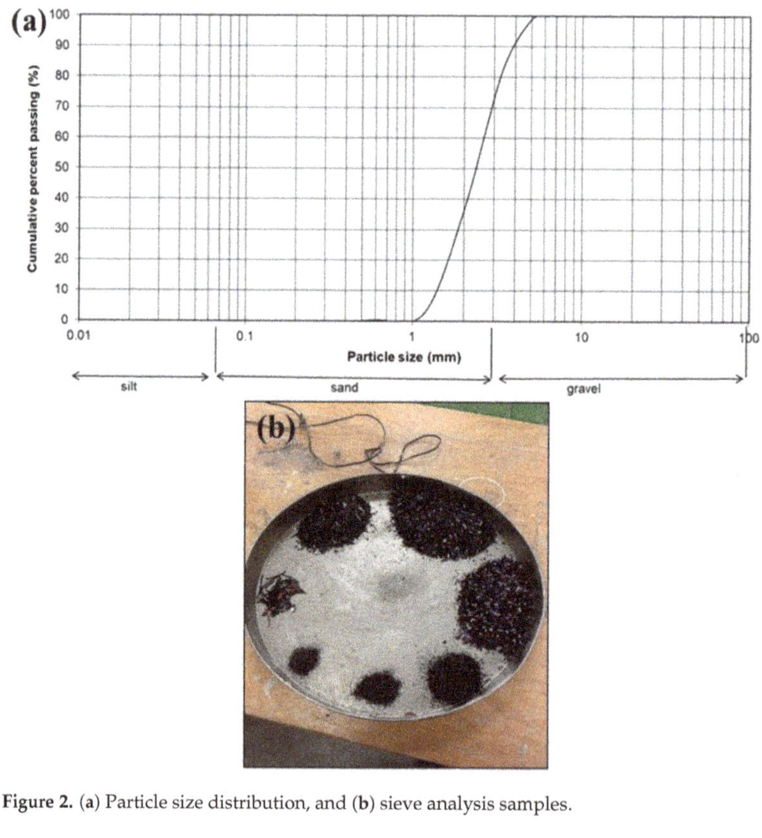

Figure 2. (**a**) Particle size distribution, and (**b**) sieve analysis samples.

Table 3. Some physical properties of the coarse and fine aggregates.

Physical Properties	Coarse Aggregates		Fine Aggregates (Sand)	Waste Plastics (WP)
	20 mm	10 mm		
Uniformity coefficient (CU)	1.3	3.3	0.11	-
Curvature coefficient (CC)	7.5	1.5	1.75	-
Flakiness index (%)	23	30–35	-	-
Elongation index (%)	12	17–22	-	-
Shape index (%)	7	12	-	-
Impact value	15	23	-	-
Fineness modulus (mm)	-	4	1.54	-
Uncompacted bulk density (kg/m^3)	2570	1350	1500	-
Pre-dried particle density (kg/m^3)	-	2690	2600	-
Water absorption (%)	1.1	2	21	-
Heat Deflection Temperature—Hdt (°C)	-	-	-	52
Density (g/cm^3)	-	-	-	1.24
Tensile strength (MPa)	-	-	-	50
Flexural strength (MPa)				80
Impact strength (Unnotched) Izod (J/m)	-	-	-	96.1
Shrink rate %(in/in)	-	-	-	0.37–0.41 (0.0037–0.0041)

3. Methodology

3.1. Mix Design

The viability and efficacy of sustainably incorporating WP in geopolymer concrete production were assessed in this study. Two controls (C0 and GC0) for ordinary and geopolymer concrete mixes were adopted for completeness as described in the mix design (see Table 4). Mix C0 was prepared with 100% OPC using a water/binder ratio of 0.55, while GC0 was developed in accordance with Adeleke et al. [5] to produce an optimum geopolymer mix by activating GGBS with a binary activator blend of sodium silicate (SS) and sodium hydroxide (SH) solution (1.2/0.8) at an activator/precursor (A/P) mass ratio of 0.5. Using the same binder quantity as the control, other concrete batches were developed by replacing 10 mm NA with WP at 30 and 70 wt% for both ordinary (CWP1 and CWP2) and geopolymer (GCWP1 and GWPC2) concrete using a correction factor of 0.38. The correction factor is the ratio of plastic to aggregate used in the concrete mix according to each percentage replacement level (30 and 70 wt%), allowing the aggregate volume to be replaced instead of its mass. It is worth noting that the concrete mix ratio for all the mix codes were fixed as a binder–sand–aggregates ratio of 1:2:3, and they only vary in terms of the water content, type of binder, and WP replacement levels.

Table 4. Mix composition of ordinary concrete (C0) and geopolymer concretes (GC).

Mix Code	Elaborated Abbreviation	Concrete Binder						W (L)	Aggregates (kg)			
		OPC (kg)	Geopolymer Binder						NFA	WP	Natural Coarse Aggregate	
			GGBS (kg)	A/P Ratio	SS:SH	Activator (mL) SS	Activator (mL) SH				10 mm	20 mm
C0	C0-OPC-0%WP-100%NA (Control 1)	3.9	-	-	-	-	-	2.2	7.9	-	3.9	7.9
CWP1	CWP1-OPC-30%WP-70%NA	3.9	-	-	-	-	-	2.2	7.9	1.2	2.7	7.9
CWP2	CWP2-OPC-70%WP-30%NA	3.9	-	-	-	-	-	2.2	7.9	2.7	1.2	7.9
GC0	GC0–0%WP-100%NA (Control 2)	-	2.6	0.5	1.2:0.8	540	360	1.1	7.9	-	3.9	7.9
GCWP1	GCWP1–30%WP-70%NA	-	2.6	0.5	1.2:0.8	540	360	1.1	7.9	1.2	2.7	7.9
GCWP2	GCWP2–70%WP-30%NA	-	2.6	0.5	1.2:0.8	540	360	1.1	7.9	2.7	1.2	7.9

OPC—ordinary Portland cement; GGBS—ground granulated blast-furnace slag; A/P—activator/precursor ratio; SS:SH—sodium silicate to sodium hydroxide ratio; natural fine aggregate—NFA; water—W; waste plastic—WP.

3.2. Concrete Specimen Preparation and Testing Methods

The same alkali–alkaline activator (10 M), test sample preparation (geopolymer concrete), and testing methods as in Adeleke et al. [5] were used in this study to produce fresh concrete. A range of tests was conducted, including a slump test (S), unconfined compressive strength (UCS), and automated scanning electron microscopy with linked energy dispersive spectrometers (SEM-EDS). The consistency of the fresh concrete was measured using the slump test (S) in accordance with BS EN 12350-2:2019 [39]. For each batch mix, nine cube (100 mm × 100 mm × 100 mm) test specimens were developed according to BS EN 206:2013+A2:2021 [40]. The geopolymer concrete specimens were placed in a sealed environment (plastic film) to prevent evaporation and kept for 24 h at 20 ± 2 °C. After that, all hardened test specimens were de-moulded after 24 h, while the geopolymer specimens were cured in a sealed container and the ordinary concrete specimens were soaked in water at 20 ± 2 °C and a humidity of 90% [41].

The mechanical performance of the hardened concrete was evaluated in terms of UCS according to BS EN 12390-3:2019 [42] at 7, 28, and 56 days of curing. Lastly, a compositional and microstructural investigation was employed on some of the prepared test specimens after 28 days of curing using the SEM-EDS [43,44]. Representative subsamples were prepared as polished 30 mm diameter blocks which were carbon coated prior to the analysis. A Hitachi SU3900 scanning electron microscope manufactured by Hitachi High-Tech Corporation, Maidenhead, UK, fitted with two large-area (60 mm^2) Bruker SDD energy dispersive spectrometers and running the Bruker AMICS automated mineralogy package was used to undertake the analysis at an accelerating voltage of 20 kV coupled with a beam current of approximately 10 nA which was used. ED spectra were acquired at a 10 µm interval across the area analysed, and each EDS spectra was compared with a spectral library of known minerals and compositions allowing each analysis spot to be assigned to a compositional grouping. Data outputs from automated SEM-EDS mineral analysis include quantitative modal mineralogy along with false-colour compositional maps and full-area SEM-BSE images. Although plastics are not routinely measured during automated mineralogy analysis, post-processing of the images allowed the plastics to be delineated and quantified. The morphology of representative samples was also interactively examined using an MIRA 3 TESCAN scanning electron microscope.

4. Results and Discussions

4.1. Consistency of Fresh Concrete

Figure 3 illustrates the consistency results obtained for the slump. The general trend indicates a lower slump for the geopolymer concrete (GC0, GCWP1, and GCWP2) compared to the ordinary concrete (C0, CWP1, and CWP2). Furthermore, the ordinary concrete has the highest slump values within a 60–90 mm range, while the geopolymer mixes experienced a slump within the 40–55 mm range. The disparity in slump (geopolymer and traditional concrete) could be attributed to the viscosity/quantity of the liquid employed in the concrete mixes and the varying kinetic reactions in each system [5,22]. In addition, GGBS decreases water demand in ordinary concrete mixes as a result of its smooth round particles and lower water absorption; therefore, its use in large quantities within the geopolymer mixes could have negatively impacted the slump values [45].

Furthermore, observation showed that for both the geopolymer and ordinary concrete mixes, there is a decline in slump values for every increase in the quantity of WP. For instance, there was an 11% and 25% decline in slump with a WP quantity of 30 and 70% for the ordinary concrete, while a 9% and 20% decline was observed in the geopolymer concrete at 30 and 70% WP dosage. This shows that any WP replacement of more than 70% could impact the overall workability of the concrete produced. The justification for this slump reduction could be attributed to the non-uniformity in the shape of WP, which could have improved the cohesiveness within each concrete batch mix [23]. In contrast, the 10 mm NA with a more spherical/regular shape enhances the homogeneity and fluidity of the freshly produced concrete mixes.

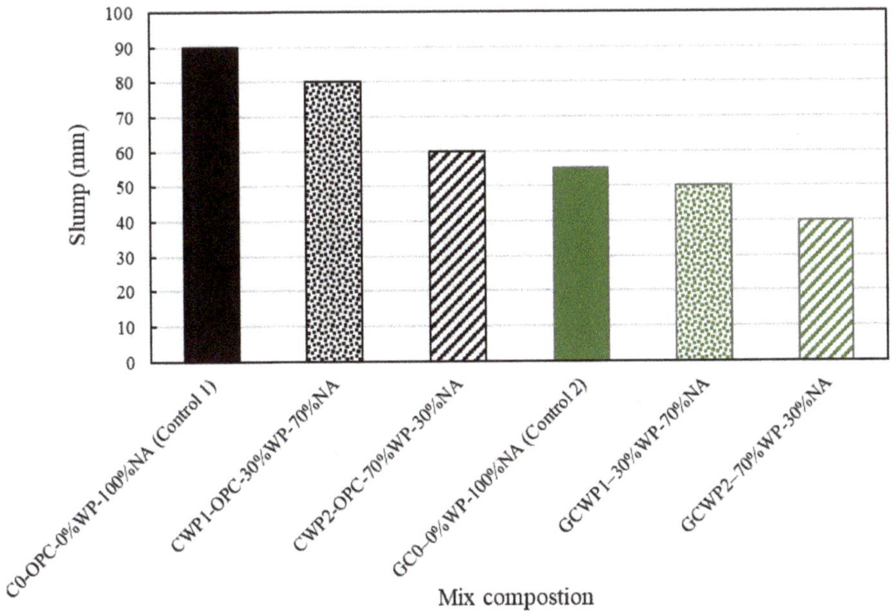

Figure 3. Consistency of concrete mixes measured—slump test.

A similar finding of decreasing workability of geopolymer mixtures with the inclusion of waste plastic aggregate as a replacement for fine sand was reported by Lazorenko et al. [6], where the authors demonstrated that the inclusion of waste plastic aggregates reduced the workability of the fly ash-based geopolymer mortars when they introduced shredded flakes and strips and ground particles of different sizes to the geopolymer system. Ahmed et al. [2] also indicated that due to the lower specific gravity of recycled plastics than the other elements within a geopolymer mix, they form a structure that resembles a mesh, keeping them together (coagulation) and preventing the concrete mixture from flowing, thus impacting on the overall consistency. Overall, it can be suggested from the consistency results and BS EN 206:2013+A2:2021 [40] that all the concrete formulations can be classified as S2 standard mix, except GCWP2, which is classified as an S1 dry mix.

4.2. Concrete Strength Development

Figure 4 illustrates the unconfined compressive strength (UCS) development for all the concrete mixes (ordinary—C0, CWP1, CWP2, and geopolymer—GC0, GCWP1, GCWP2) at 7, 28, and 56 days of the ambient curing period. Observation showed a steady increase in strength development for all of the formulated geopolymer and ordinary concrete mixes over the moist curing periods. In addition, mix C0 and GC0 achieved the highest UCS of 25.5 MPa and 55.4 MPa, respectively, while mix CWP2 and GWP2 with 70% WP replacement levels, performed the lowest with a UCS of 21.5 and 40.4 MPa at 28 days. The strength gain by the ordinary concrete mix can be attributed to the initial hydration of the cement components with water to produce cementing gels (C-S-H), which are required for strength development. In contrast, the trend of strength development for the geopolymer concrete can be attributed to a polymerisation process that involves the chemical reaction of aluminosilicate minerals (GGBS) under alkaline conditions (SS/SH solution), creating a three-dimensional amorphous aluminosilicate matrix that displays strength comparable or superior to ordinary concrete [27,28,46].

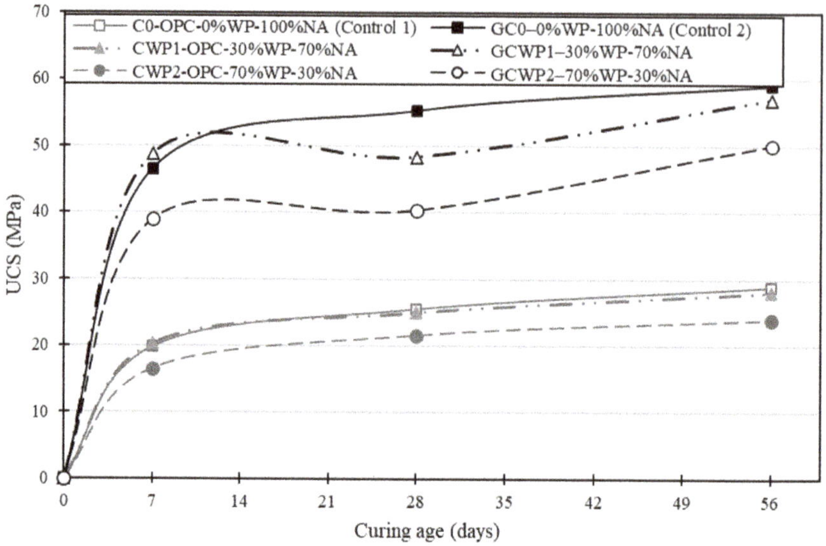

Figure 4. UCS development of the ordinary and geopolymer concrete mixes made with different AP and SS/SH ratios.

Figure 5 shows the percentage change in the UCS of each type of concrete formulation (ordinary and geopolymer) compared with their respective controls. Observations indicated that replacing 10 mm coarse aggregates with WP at a 30% dosage improved the UCS for both ordinary (2.8%) and geopolymer concrete (4.8%). In contrast, a strength reduction of 16.7% and 16.3% for both ordinary and geopolymer concrete was identified at 7 days of moist curing. A similar trend of declining UCS was observed for all the concrete formulations after 28 and 56 days of moist curing. For instance, the mix of GCWP2 with 70% WP experienced the largest reduction in UCS, while the mix of CWP1 with 30% WP displayed the lowest UCS reduction compared with the control at both 28 and 56 days.

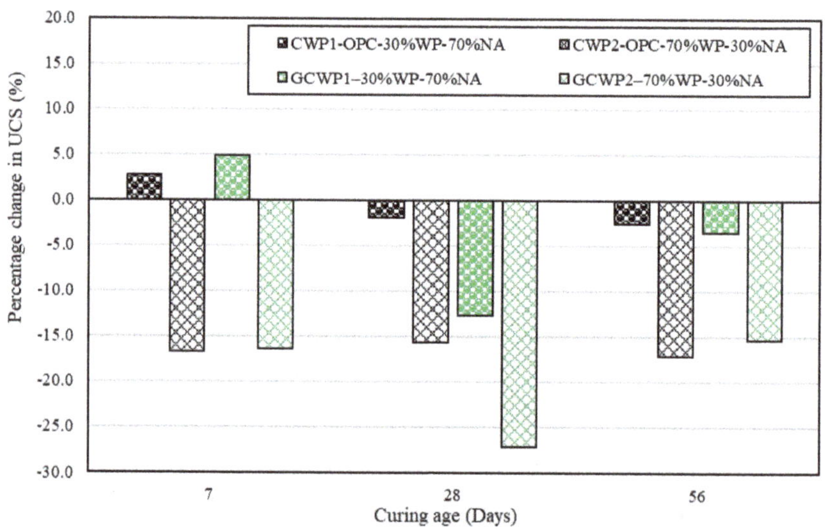

Figure 5. Percentage change in the UCS of each type of concrete formulation (ordinary and geopolymer) compared with their respective controls.

Overall, the UCS reduction for both concrete formulations with the addition of different dosages of WP could be attributed to the lower elastic modulus of WP compared with natural 10 mm coarse aggregate. In addition, it is worth remembering that plastic is considered a hydrophobic material, meaning less water is required for hydration during curing operations [47]. Therefore, the observed decrease in UCS can be linked with the hydrophobic nature and surface smoothness of plastic aggregates, which adds to weak interfacial bonding between the plastic filler (relatively less-rough surfaces of the WPs) and the geopolymer binder matrix. This behaviour is also typical for OPC-based composites incorporated with plastic aggregates [6,24].

Despite some reduction in UCS for the geopolymer concrete when compared with the control, a continuous trend of UCS strength increase can be seen in Figure 4 over the curing age of 7, 28, and 56 days of moist curing. Ahmed et al. [2] attributed this phenomenon to the continuation of the polymerisation process, which enhances the entire geopolymer structures, and, therefore, zones of interfacial transition between WP surfaces and the geopolymer matrix system improve. This behaviour is also typical for ordinary concrete with continuous cement hydration over the curing period.

Furthermore, a UCS comparison between geopolymer and ordinary concrete as illustrated in Figure 6 showed that despite the inclusion of WP, the geopolymer concrete indicated significant strength development at each curing age duration, indicating the robustness of geopolymer concrete. Moreover, the variation in UCS for the observed concrete mix formulation at each curing age could be due to the varied mix composition incorporating WP at 30 and 70% WP, which impacts the alkali activation/polymerisation process and cement hydration of the developed concrete systems.

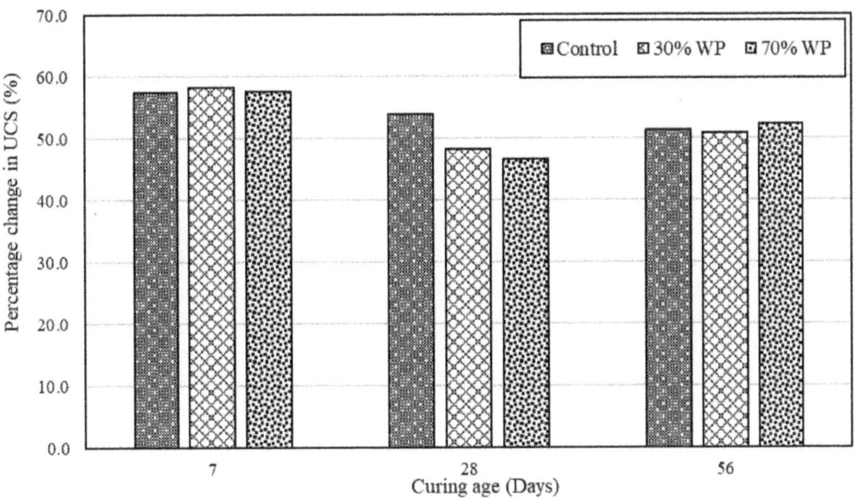

Figure 6. UCS development comparison between geopolymer and ordinary concrete.

4.3. Compositional and Microstructural Analysis

The area % quantitative composition of the ordinary and geopolymer concretes with 0, 30, and 70% waste plastics are shown in Tables 5–7 for 7-, 28-, and 56-day curing periods, respectively. The quantitative SEM-EDS analysis reports the relative abundance of both the natural aggregate phases (as their constituent minerals), plastics, and the cementitious phases reported as chemical compositional groups. It should be noted that these modal abundance data relate to the 30 mm diameter areas analysed and as such may not be representative of the modal abundance by volume especially as there are coarse aggregate grains present. The OPC reports to the Ca silicate categories, whilst the geopolymer binding reports to the Ca silicate Na, Mg, Al compositional group.

Table 5. Modal (area %) for ordinary and geopolymer concretes at 7 days of curing based on SEM-EDS automated mineral analysis.

Minerals	**Area (%)**					
	OPC-0%WP-100%NA	**OPC-30%WP-70%NA**	**OPC-70%WP-30%NA**	**GC-0%WP100%NA**	**GC-30%WP70%NA**	**GC-70%WP30%NA**
Plastic	1.58	3.52	16.28	0.23	0.19	1.91
Quartz	23.28	29.92	29.11	28.64	30.37	12.67
K Feldspar	1.04	1.30	1.05	0.95	1.54	0.53
Plagioclase	0.57	0.60	0.64	0.68	0.64	0.14
Muscovite	0.13	0.20	0.29	0.12	0.13	0.25
Biotite	0.19	0.29	0.28	0.28	0.34	0.05
Chlorite	0.26	0.28	0.29	0.27	0.34	0.06
Kaolinite	0.09	0.09	0.07	0.04	0.06	0.03
Illitic Clays	0.42	0.49	0.55	0.33	0.39	0.31
Calcite	12.50	13.76	28.42	6.22	19.59	66.97
Dolomite	34.97	17.50	4.03	33.88	16.57	8.18
Fe Oxides and Siderite	0.21	0.25	0.30	0.37	0.30	0.08
Ti Oxides	0.14	0.04	0.03	0.02	0.04	0.03
Apatite	0.09	0.01	0.00	0.00	0.08	0.01
Pyrite	0.09	0.00	0.02	0.00	0.01	0.01
Clinopyroxene and Amphibole	0.17	0.03	0.05	0.03	0.05	0.03
Zircon	0.01	0.00	0.00	0.00	0.00	0.00
Ca Silicate High Ca	2.23	2.21	1.76	0.02	0.01	0.00
Ca Silicate Mid Ca	16.87	24.16	23.47	0.52	0.75	1.03
Ca Silicate Low Ca	1.37	1.74	2.30	0.25	0.20	0.14
Ca Silicate	1.00	1.28	2.50	2.08	2.12	0.93
Ca Al Silicate High Ca	0.63	0.87	0.88	0.04	0.10	0.32
Ca Silicate Na Mg Al	0.01	0.01	0.01	21.43	23.05	6.64
Ca Fe Silicate	0.49	0.53	0.58	0.29	0.31	0.06
Ca Mg Silicate	2.12	3.04	2.26	3.26	2.70	1.42
Fe Silicate	0.14	0.14	0.13	0.19	0.20	0.08
Ca Al Oxide	0.38	0.46	0.55	0.00	0.01	0.00
Ca Sulphates	0.49	0.71	0.32	0.02	0.01	0.00
Undifferentiated	0.10	0.07	0.09	0.06	0.10	0.04

Table 6. Modal (area %) for ordinary and geopolymer concretes at 28 days of curing based on SEM-EDS automated mineral analysis.

Minerals	**Area (%)**					
	OPC-0%WP-100%NA	**OPC-30%WP-70%NA**	**OPC-70%WP-30%NA**	**GC-0%WP100%NA**	**GC-30%WP70%NA**	**GC-70%WP30%NA**
Plastic	2.00	2.80	10.31	0.06	3.66	6.82
Quartz	28.86	25.64	26.56	25.21	21.88	26.11
K Feldspar	1.12	1.06	1.19	0.55	0.82	1.16
Plagioclase	0.46	0.51	0.71	0.57	0.57	0.51
Muscovite	0.09	0.21	0.09	0.12	0.09	0.14
Biotite	0.25	0.26	0.14	0.29	0.18	0.30
Chlorite	0.19	0.25	0.23	0.29	0.23	0.29
Kaolinite	0.03	0.05	0.04	0.05	0.03	0.03
Illitic Clays	0.32	0.42	0.36	0.38	0.24	0.38
Calcite	17.92	38.85	7.99	8.01	32.24	28.46
Dolomite	16.32	2.64	36.35	42.66	20.08	18.72
Fe Oxides and Siderite	0.32	0.27	0.24	0.35	0.20	0.18
Ti Oxides	0.02	0.02	0.01	0.02	0.01	0.02
Apatite	0.06	0.03	0.00	0.05	0.04	0.04
Pyrite	0.00	0.02	0.00	0.00	0.02	0.01
Clinopyroxene and Amphibole	0.02	0.02	0.01	0.05	0.03	0.04
Zircon	0.00	0.00	0.00	0.00	0.00	0.00
Ca Silicate High Ca	1.94	1.78	1.21	0.01	0.01	0.01
Ca Silicate Mid Ca	24.05	22.60	19.22	0.41	0.54	0.83
Ca Silicate Low Ca	1.88	0.59	0.92	0.16	0.18	0.20
Ca Silicate	1.16	0.86	0.72	1.49	1.98	1.85
Ca Al Silicate High Ca	0.56	0.62	0.62	0.05	0.08	0.16
Ca Silicate Na Mg Al	0.02	0.01	0.01	16.50	18.20	18.11
Ca Fe Silicate	0.50	0.43	0.43	0.28	0.18	0.30
Ca Mg Silicate	2.50	1.55	1.80	2.24	2.01	1.93
Fe Silicate	0.16	0.18	0.17	0.16	0.09	0.15
Ca Al Oxide	0.45	0.40	0.49	0.00	0.00	0.00
Ca Sulphates	0.27	0.31	0.40	0.02	0.01	0.01
Undifferentiated	0.52	0.41	0.05	0.07	0.05	0.06

Table 7. Modal (area %) for ordinary and geopolymer concretes at 56 days of curing based on SEM-EDS automated mineral analysis.

Minerals	Area (%)					
	OPC-0%WP-100%NA	OPC-30%WP-70%NA	OPC-70%WP-30%NA	GC-0%WP100%NA	GC-30%WP70%NA	GC-70%WP30%NA
Plastic	0.10	9.89	10.10	0.19	2.47	11.30
Quartz	14.76	20.37	27.43	13.07	38.80	19.24
K Feldspar	0.57	0.76	1.07	0.53	1.54	0.76
Plagioclase	0.27	0.32	0.63	0.40	0.86	0.50
Muscovite	0.07	0.15	0.14	0.09	0.20	0.15
Biotite	0.16	0.12	0.23	0.06	0.36	0.20
Chlorite	0.15	0.11	0.21	0.11	0.32	0.24
Kaolinite	0.05	0.04	0.04	0.03	0.05	0.03
Illitic Clays	0.19	0.32	0.32	0.19	0.47	0.36
Calcite	17.06	54.96	39.58	69.77	15.90	53.76
Dolomite	50.33	0.54	0.79	2.72	2.62	1.42
Fe Oxides and Siderite	0.22	0.22	0.30	0.08	0.36	0.19
Ti Oxides	0.02	0.03	0.02	0.02	0.03	0.02
Apatite	0.01	0.00	0.01	0.01	0.02	0.01
Pyrite	0.02	0.12	0.00	0.10	0.01	0.01
Clinopyroxene and Amphibole	0.01	0.01	0.01	0.02	0.05	0.03
Zircon	0.00	0.00	0.00	0.00	0.00	0.00
Ca Silicate High Ca	0.59	1.20	1.66	0.01	0.03	0.01
Ca Silicate Mid Ca	12.03	15.72	21.44	0.86	0.82	0.97
Ca Silicate Low Ca	0.37	0.49	0.92	0.27	0.42	0.35
Ca Silicate	0.66	1.18	1.11	1.31	2.95	1.95
Ca Al Silicate High Ca	0.49	0.81	0.80	0.13	0.09	0.22
Ca Silicate Na Mg Al	0.01	0.00	0.01	8.67	29.82	17.06
Ca Fe Silicate	0.30	0.28	0.32	0.07	0.29	0.19
Ca Mg Silicate	1.04	1.31	2.00	1.34	3.61	2.09
Fe Silicate	0.07	0.09	0.10	0.06	0.27	0.12
Ca Al Oxide	0.34	0.33	0.40	0.00	0.00	0.00
Ca Sulphates	0.12	0.20	0.35	0.01	0.02	0.00
Undifferentiated	0.08	0.32	0.13	0.07	0.10	0.13

Figures 7 and 8 show the AMICS automated SEM-EDS images from sections of the ordinary and geopolymer concrete cube specimen at 0%, 30%, and 70%WP at age 7 days of ambient curing. The images show that the ordinary and geopolymer concrete morphology is visibly different, confirming a different type of microstructure after the initial 7-day curing period. Furthermore, visual inspection indicates the presence of pores (see Figures 7 and 8) which are more evident in the control mixes (C0 and GC0) but are gradually reduced with each replacement level of WP in both the ordinary and geopolymer concrete. It is possible that the added WP acted as fillers due to their non-uniformity in shape and size (5–10 mm). However, the ordinary concrete mix of CWP1 and CWP2 matrixes were loose due to cracks between the boundaries of the added WP and the CSH gel formation (Figure 9). Shilar et al. [48] described these boundaries as the interfacial zone (ITZ) between the cementitious hydrate paste and aggregate and are widely known as the weakest link in ordinary concrete, where microcracks typically appear first under stress. The presence of the cracks could be due to the low bonding of the cementitious hydrate gel (CSH) with the WP, which explains the low compressive strength depicted in Figures 4 and 5.

In contrast to the ordinary concrete mix, the presence of pores in the geopolymer mix of GC0, GCWP2, and GCWP2 did not significantly impact the UCS due to the robustness of the geopolymer gel (Na-bearing aluminosilicates) bonding with both the WP and the natural aggregates. Ahmed et al. [2] supported this claim by stating that the use of NaOH-activated sodium silicate enhances the creation of an aluminium-enriched aluminosilicate surface on the aggregates (WP and natural) by speeding up the Si-preferential dissolution of kaolin and albite, thus, increasing the ITZ bonding strength between the aggregates and geopolymer gel.

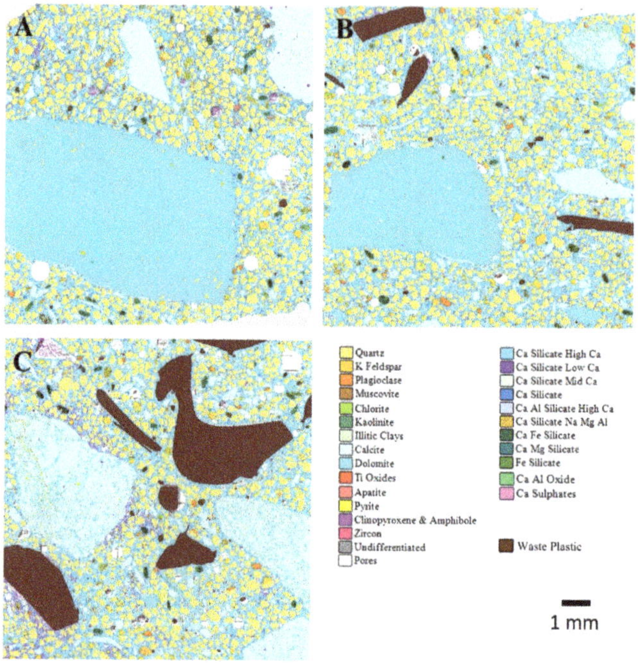

Figure 7. AMICS automated mineralogy false-colour images of 7-day-cured ordinary concrete mixes for (**A**) C0—control, (**B**) CWP1 with 30% WP, and (**C**) CWP2 with 70% WP.

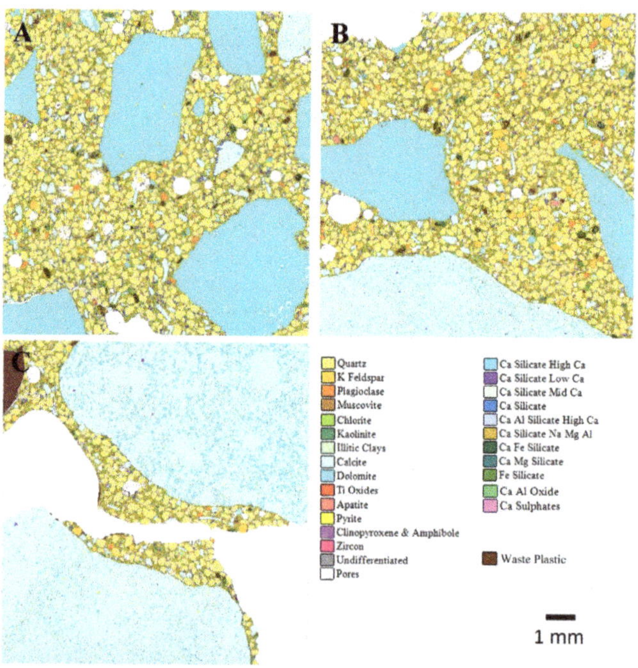

Figure 8. AMICS automated mineralogy false-colour images of 7-day-cured geopolymer concrete mixes for (**A**) GC0—control, (**B**) GCWP1 with 30% WP, and (**C**) GCWP2 with 70% WP.

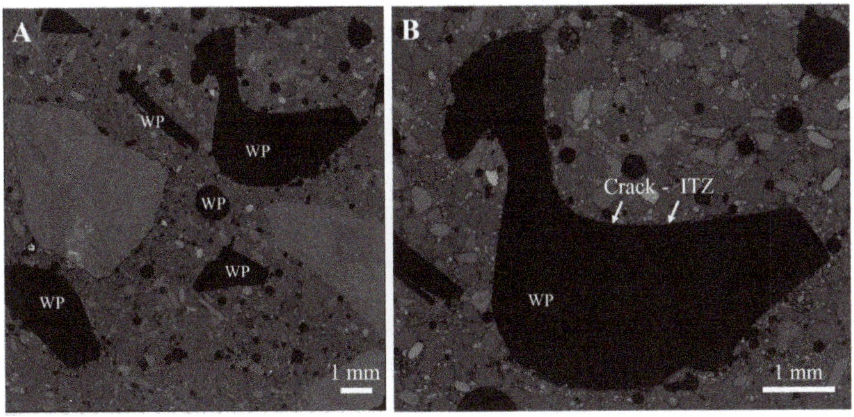

Figure 9. SEM-BSE images of 7-day-cured ordinary concrete mixes for CWP2 with 70% WP. (**A**) Whole area measured; (**B**) enlarged area showing a crack at the margin of the waste plastic (WP).

The automated SEM-EDS (AMICS) images from slices of the dried ordinary and geopolymer concrete cube specimens with 0%, 30%, and 70% WP at 28 days of curing are shown in Figures 10 and 11. Generally, observations showed a reduction in the pores for both ordinary and geopolymer concrete after 28 days of curing age. The reduction in pore sizes could be attributed to the production of the relevant cementitious hydrate (CSH and geopolymer gel) that fills up the pores to form a denser structure. The importance of the microscopic inspection of pores has been investigated by some researchers who concluded that the porosity and density of the structure are directly proportionate to the composite's durability and mechanical performance [2,48].

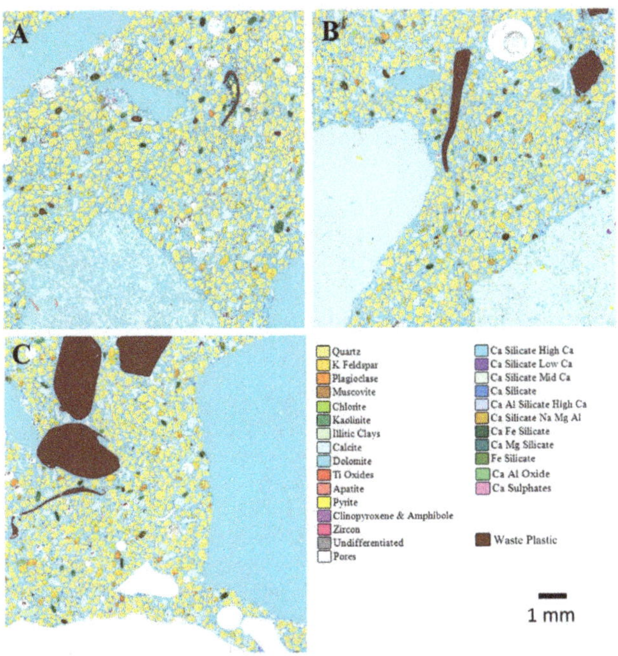

Figure 10. AMICS automated mineralogy false-colour images of 28-day-cured ordinary concrete mixes for (**A**) C0—control, (**B**) CWP1 with 30% WP, and (**C**) CWP2 with 70% WP.

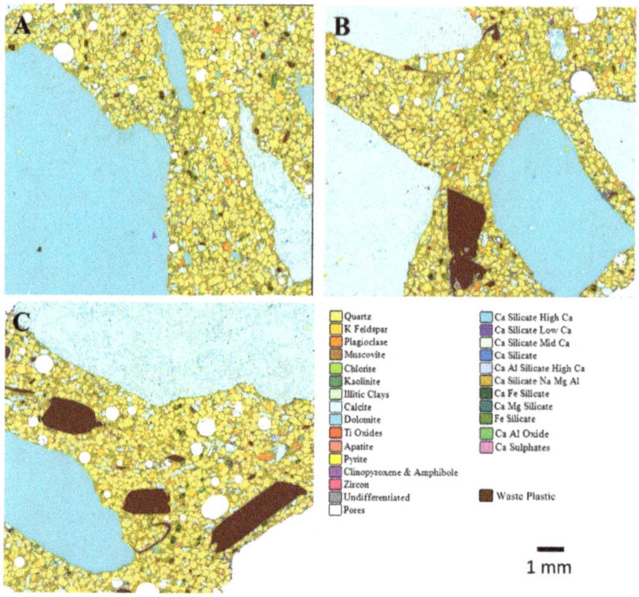

Figure 11. AMICS automated mineralogy false-colour images of 28-day-cured geopolymer concrete mixes for (**A**) GC0—control, (**B**) GCWP1 with 30% WP, and (**C**) GCWP2 with 70% WP.

Although some pores were still evident at 28 days of curing, geopolymer mixes GC0, GCWP2, and GCWP2 depict a denser and more uniform microstructure at 0%, 30%, and 70% WP and experienced higher unconfined compressive strength compared with ordinary concrete mixes.

5. Conclusions

The results from this study indicate the practicality and efficacy of incorporating waste plastic (polylactic acid-type) in geopolymer concrete production by replacing 10 mm of natural coarse aggregates (NA) with up to 70 wt%. The following conclusions can be drawn:

1. The consistency results indicated a gradual decline in slump values for every increase in the quantity of WP for both geopolymer and ordinary concrete mixes. This behaviour is more pronounced in geopolymer formulations due to the high viscosity of the developed geopolymer paste and the non-uniformity in the shape of WP. Therefore, any WP replacement of more than 70 wt% could impact the overall workability of the concrete produced. However, appropriate admixtures (superplasticiser) could be applied to improve the workability further.
2. A UCS comparison between geopolymer and ordinary concrete with and without the inclusion of WP, indicated superior strength development for the geopolymer concrete within the range of 46.7–58.3% at each curing age duration, indicating the robustness of the geopolymerisation reaction in concrete. Mix C0 and GC0 without WP achieved the highest UCS of 25.5 MPa and 55.4 MPa, respectively, while mix CWP2 and GWP2 with 70% WP replacement levels, performed the lowest with a UCS of 21.5 and 40.4 MPa at 28 days. In addition, the compressive strength of the ordinary and geopolymer concrete formulations declined with increasing dosages of WP (30% and 70%) due to the weak interfacial bonds between WP and the binder matrix. Moreover, this was more pronounced in ordinary concrete mixes, while the binder matrix in geopolymer concrete showed more interaction with WP.

3. An SEM analysis indicates that adding WP in both ordinary and geopolymer concrete reduced the presence of pores within the microstructure. However, this was detrimental to the ordinary concrete due to the low ITZ between the CSH gel and WP, resulting in the formation of cracks. The observation justifies the significant reduction in strength for ordinary concrete compared with geopolymer concrete. In contrast to the ordinary concrete mix, the presence of pores in the geopolymer mixes (GC0, GCWP2, and GCWP2), which reduced over the duration of the curing period, did not significantly impact the UCS due to the robustness of the geopolymer gel (Na-bearing aluminosilicates) and improved ITZ with both the WP and the natural aggregates.
4. Automated SEM-EDS quantifies the natural aggregates, waste plastics, and cementitious hydrates such as CSH and Na-bearing aluminosilicates (geopolymer) gel present.
5. Although a concrete mix with 70% WP showed a significant reduction in UCS in both concrete formulations, geopolymer concrete can still be applied to construction practice depending on the required design strength.
6. Further research can be carried out to improve the consistency and strength of geopolymer concrete incorporated with other types of plastic for its use in structural applications.

Author Contributions: Conceptualization, methodology, B.O.A. and J.M.K.; data curation, writing—original draft preparation, B.O.A.; visualization, investigation, B.O.A.; supervision, J.M.K. and J.O.; software, validation, B.O.A. and M.P.; writing—reviewing and editing, B.O.A., J.M.K., J.O. and D.P. All authors have read and agreed to the published version of the manuscript.

Funding: This research was funded by the Faculty of Computing Engineering and Science of the University of South Wales through the Research Investment Strategy (RIS) project on the development of capacity for inter-disciplinary engineering research into advanced materials.

Institutional Review Board Statement: Not applicable.

Informed Consent Statement: Not applicable.

Data Availability Statement: Data are contained within the article.

Acknowledgments: The authors would like to acknowledge the Advanced Materials Testing Centre (AMTeC), within the Faculty of Computing Engineering and Science at the University of South Wales, UK, for the continuous support during the implementation of the laboratory experimentations.

Conflicts of Interest: Author Matthew Power was employed by the company Vidence Inc. The remaining authors declare that the research was conducted in the absence of any commercial or financial relationships that could be construed as a potential conflict of interest.

References

1. Ponmalar, S.; Revathi, P. Waste recycled plastic granules substitute for aggregate in concrete—Review. *Mater. Today Proc.* **2022**, *65*, 1441–1448. [CrossRef]
2. Ahmed, H.U.; Mohammed, A.S.; Mohammed, A.A. Engineering properties of geopolymer concrete composites incorporated recycled plastic aggregates modified with nano-silica. *J. Build. Eng.* **2023**, *75*, 106942. [CrossRef]
3. PlasticsEurope. Plastics—The Facts 2020. An Analysis of European Plastics Production, Demand and Waste Data. 2020. Available online: https://www.plasticseurope.org/application/files/5716/0752/4286/AF_Plastics_the_facts-WEB-2020-ING_FINAL.pdf (accessed on 16 January 2024).
4. Brahney, J.; Hallerud, M.; Heim, E.; Hahnenberger, M.; Sukumaran, S. Plastic rain in protected areas of the United States. *Science* **2020**, *368*, 1257. [CrossRef] [PubMed]
5. Adeleke, B.O.; Kinuthia, J.M.; Oti, J.; Ebailila, M. Physico-Mechanical Evaluation of Geopolymer Concrete Activated by Sodium Hydroxide and Silica Fume-Synthesised Sodium Silicate Solution. *Materials* **2023**, *16*, 2400. [CrossRef]
6. Lazorenko, G.; Kasprzhitskii, A.; Fini, E.H. Sustainable construction via novel geopolymer composites incorporating waste plastic of different sizes and shapes. *Constr. Build. Mater.* **2022**, *324*, 126697. [CrossRef]
7. Zhang, K.; Hamidian, A.H.; Tubic, A.; Zhang, Y.; Fang, J.K.H.; Wu, C.; Lam, P.K.S. Understanding plastic degradation and microplastic formation in the environment: A review. *Environ. Pollut.* **2021**, *274*, 116554. [CrossRef] [PubMed]
8. Xiang, Y.; Jiang, L.; Zhou, Y.; Luo, Z.; Zhi, D.; Yang, J.; Lam, S.S. Microplastics and environmental pollutants: Key interaction and toxicology in aquatic and soil environments. *J. Hazard. Mater.* **2022**, *422*, 126843. [CrossRef]

9. Sridharan, S.; Kumar, M.; Singh, L.; Bolan, N.S.; Saha, M. Microplastics as an emerging source of particulate air pollution: A critical review. *J. Hazard. Mater.* **2021**, *418*, 126245. [CrossRef] [PubMed]
10. Oliveira, M.; Almeida, M.; Miguel, I. A micro(nano)plastic boomerang tale: A never ending story? *TrAC Trends Anal. Chem.* **2019**, *112*, 196–200. [CrossRef]
11. Bridson, J.H.; Gaugler, E.C.; Smith, D.A.; Northcott, G.L.; Gaw, S. Leaching and extraction of additives from plastic pollution to inform environmental risk: A multidisciplinary review of analytical approaches. *J. Hazard. Mater.* **2021**, *414*, 125571. [CrossRef]
12. Wang, L.; Wu, W.M.; Bolan, N.S.; Tsang, D.C.W.; Qin YLi, M.; Hou, D. Environmental fate, toxicity and risk management strategies of nanoplastics in the environment: Current status and future perspectives. *J. Hazard. Mater.* **2021**, *401*, 123415. [CrossRef]
13. Horton, A.A. Plastic pollution: When do we know enough? *J. Hazard. Mater.* **2022**, *422*, 126885. [CrossRef]
14. Mofijur, M.; Ahmed, S.F.; Rahman, S.M.A.; Arafat Siddiki, S.Y.; Islam, A.; Shahabuddin, M.; Ong, H.C.; Mahlia TM, I.; Djavanroodi, F.; Show, P.L. Source, distribution and emerging threat of micro- and nanoplastics to marine organism and human health: Socio-economic impact and management strategies. *Environ. Res.* **2021**, *195*, 110857. [CrossRef]
15. Borrelle, S.B.; Ringma, J.; Law, K.L.; Monnahan, C.C.; Lebreton, L.; McGivern, A.; Murphy, E.; Jambeck, J.; Leonard, G.H.; Hilleary, M.A.; et al. Predicted growth in plastic waste exceeds efforts to mitigate plastic pollution. *Science* **2020**, *369*, 1515–1518. [CrossRef] [PubMed]
16. Geyer, R.; Jambeck, J.R.; Law, K.L. Production, use, and fate of all plastics ever made. *Sci. Adv.* **2017**, *3*, e1700782. [CrossRef]
17. Yang, Z.; Lu, F.; Zhang, H.; Wang, W.; Shao, L.; Ye, J.; He, P. Is incineration the terminator of plastics and microplastics? *J. Hazard. Mater.* **2021**, *401*, 123429. [CrossRef]
18. Koley, S. Sustainability appraisal of arsenic mitigation policy innovations in West Bengal, India. *Infrastruct. Asset Manag.* **2023**, *10*, 17–37. [CrossRef]
19. Bayiha, B.N.; Yamb, E.; Metekong, J.S.; Billong, N.; Nzengwa, R. Optimizing the Choice of Limestone Deposits for the Production of Portland Cement in Cameroon. *Int. J. Mater. Sci. Appl.* **2018**, *7*, 174–185. [CrossRef]
20. Billong, N.; Oti, J.; Kinuthia, J. Using silica fume based activator in sustainable geopolymer binder for building application. *Constr. Build. Mater.* **2021**, *275*, 122177. [CrossRef]
21. Adeleke, B.O.; Kinuthia, J.M.; Oti, J.E. Impacts of MgO waste:GGBS formulations on the performance of a stabilised natural high sulphate bearing soil. *Constr. Build. Mater.* **2022**, *15*, 125745. [CrossRef]
22. Thorneycroft, J.; Orr, J.; Savoikar, P.; Ball, R.J. Performance of structural concrete with recycled plastic waste as a partial replacement for sand. *Constr. Build. Mater.* **2018**, *161*, 63–69. [CrossRef]
23. Tota-Maharaj, K.; Adeleke, B.O.; Nounu, G. Effects of waste plastics as partial fine-aggregate replacement for reinforced low-carbon concrete pavements. *Int. J. Sustain. Eng.* **2022**, *15*, 192–207. [CrossRef]
24. Lazorenko, G.; Kasprzhitskii, A. Geopolymer additive manufacturing: A review. *Addit. Manuf.* **2022**, *55*, 102782. [CrossRef]
25. Sun, K.; Peng, X.; Wang, S.; Zeng, L.; Ran, P.; Ji, G. Effect of nano-SiO_2 on the efflorescence of an alkali-activated metakaolin mortar. *Constr. Build. Mater.* **2020**, *253*, 118952. [CrossRef]
26. Simão, L.; Fernandes, E.; Hotza, D.; Ribeiro, M.J.; Montedo, O.R.K.; Raupp-Pereira, F. Controlling efflorescence in geopolymers: A new approach. *Case Stud. Constr. Mater.* **2021**, *15*, e00740. [CrossRef]
27. Skariah Thomas, B.; Yang, J.; Bahurudeen, A.; Chinnu, S.N.; Abdalla, J.A.; Hawileh, R.A.; Hamada, H.M. Geopolymer concrete incorporating recycled aggregates: A comprehensive review. *Clean. Mater.* **2022**, *3*, 100056. [CrossRef]
28. Nath, P.; Sarker, P.K. Effect of GGBFS on setting, workability and early strength properties of fly ash geopolymer concrete cured in ambient condition. *Constr. Build. Mater.* **2014**, *66*, 163–171. [CrossRef]
29. Provis, J.L. Geopolymers and other alkali activated materials: Why, how, and what? *Mater. Struct.* **2014**, *47*, 11–25. [CrossRef]
30. Hama, S.M.; Hilal, N.N. Fresh properties of self-compacting concrete with plastic waste as partial replacement of sand. *Int. J. Sustain. Built Environ.* **2017**, *6*, 299–308. [CrossRef]
31. Pacheco-Torgal, F. Introduction to the use of recycled plastics in eco-efficient concrete. In *Use of Recycled Plastics in Eco-Efficient Concrete*; Woodhead: Sawston, UK, 2019; pp. 1–8.
32. Shaikh, F.U.A. Tensile and flexural behaviour of recycled polyethylene terephthalate (PET) fibre reinforced geopolymer composites. *Constr. Build. Mater.* **2020**, *245*, 118438. [CrossRef]
33. Chithambar Ganesh, A.; Deepak, N.; Deepak, V.; Ajay, S.; Pandian, A.; Karthik. Utilization of PET bottles and plastic granules in geopolymer concrete. *Mater. Today Proc.* **2021**, *42*, 444–449. [CrossRef]
34. *BS EN 197-1:2011*; "Cement", Part 1: Composition, Specifications and Conformity Criteria for Common Cements. European Committee for Standardisation: London, UK, 2011.
35. *BS EN 15167-1:2006*; "Ground Granulated Blast Furnace slag for Use in Concrete, Mortar and Grout", Part 1: Definitions, Specifications and Conformity Criteria. European Committee for Standardisation: London, UK, 2006.
36. *BS EN 12620:2002+A1:2008*; Aggregates for Concrete. ASRO: Bucharest, Romania, 2008.
37. Gutierrez, R. All You Need to Know About PLA Plastic/Material. 2020. Available online: https://www.wjoffice.com/blog/all-you-need-know-about-pla-plasticmaterial (accessed on 25 November 2023).
38. *BS EN ISO 17892-4:2016*; "Geotechnical Investigation and Testing—Laboratory Testing of Soil", Part 4: Determination of Particle Size Distribution. European Committee for Standardisation: London, UK, 2016.
39. *BS EN 12350-2:2019*; "Testing Fresh Concrete", Part 1: Slump Test. European Committee for Standardisation: London, UK, 2019.

40. *BS EN 206:2013+A2:2021*; "Concrete" Specification, Performance, Production and Conformity. European Committee for Standardisation: London, UK, 2021.
41. *BS EN 12390-2:2019*; "Testing Hardened Concrete", Part 2: Making and Curing Specimens for Strength Tests. European Committee for Standardisation: London, UK, 2019.
42. *BS EN 12390-3:2019*; "Testing Hardened Concrete", Part 3: Compressive Strength of Test Specimens. European Committee for Standardisation: London, UK, 2019.
43. Pirrie, D.; Rollinson, G.K. Unlocking the applications of automated mineral analysis. *Geol. Today* **2011**, *27*, 226–235. [CrossRef]
44. Schulz, B.; Sandmann, D.; Gilbricht, S. SEM-Based Automated Mineralogy and Its Application in Geo- and Material Sciences. *Minerals* **2020**, *10*, 1004. [CrossRef]
45. Özbay, E.; Erdemir, M.; Durmuş, H.İ. Utilization and efficiency of ground granulated blast furnace slag on concrete properties—A review. *Constr. Build. Mater.* **2016**, *105*, 423–434. [CrossRef]
46. Swathi, B.; Vidjeapriya, R. Influence of precursor materials and molar ratios on normal, high, and ultra-high performance geopolymer concrete—A state of art review. *Constr. Build. Mater.* **2023**, *392*, 132006. [CrossRef]
47. Belmokaddem, M.; Mahi, A.; Senhadji, Y.; Pekmezci, B.Y. Mechanical and physical properties and morphology of concrete containing plastic waste as aggregate. *Constr. Build. Mater.* **2020**, *257*, 119559. [CrossRef]
48. Shilar, F.A.; Ganachari, S.V.; Patil, V.B.; Khan, T.M.Y.; Dawood Abdul Khadar, S. Molarity activity effect on mechanical and microstructure properties of geopolymer concrete: A review. *Case Stud. Constr. Mater.* **2022**, *16*, e01014. [CrossRef]

materials

MDPI

Article

Wet–Dry Cycles and Microstructural Characteristics of Expansive Subgrade Treated with Sustainable Cementitious Waste Materials

Samuel J. Abbey [1,2,*], Samuel Y. O. Amakye [1], Eyo U. Eyo [1], Colin A. Booth [2] and Jeremiah J. Jeremiah [1]

1 School of Engineering, College of Arts, Technology and Environment, University of the West of England, Bristol BS16 1QY, UK; samuel_amakye@outlook.com (S.Y.O.A.); eyo.eyo@uwe.ac.uk (E.U.E.); jeremiah.jeremiah@uwe.ac.uk (J.J.J.)

2 Centre for Architecture and Built Environment Research (CABER), College of Arts, Technology and Environment, University of the West of England, Bristol BS16 1QY, UK; colin.booth@uwe.ac.uk

* Correspondence: samuel.abbey@uwe.ac.uk

Abstract: This work presents an experimental study on the physico-mechanical and microstructural characteristics of stabilised soils and the effect of wetting and drying cycles on their durability as road subgrade materials. The durability of expansive road subgrade with a high plasticity index treated with different ratios of ground granulated blast furnace slag (GGBS) and brick dust waste (BDW) was investigated. Treated and cured samples of the expansive subgrade were subjected to wetting–drying cycles, California bearing ratio (CBR) tests, and microstructural analysis. The results show a gradual reduction in the California bearing ratio (CBR), mass, and the resilient modulus of samples for all subgrade types as the number of cycles increases. The treated subgrades containing 23.5% GGBS recorded the highest CBR value of 230% under dry conditions while the lowest CBR value of 15% (wetting cycle) was recorded for the subgrade treated with 11.75% GGBS and 11.75% BDW at the end of the wetting–drying cycles, both of which find useful application in road pavement construction as calcium silicate hydrate (CSH) gel was formed in all stabilised subgrade materials. However, the increase in alumina and silica content upon the inclusion of BDW initiated the formation of more cementitious products due to the increased availability of Si and Al species as indicated by EDX analysis. This study concluded that subgrade materials treated with a combination of GGBS and BDW are durable, sustainable and suitable for use in road construction.

Keywords: durability; wetting–drying; microstructural; California bearing ratio; expansive subgrade; brick dust waste; sustainable soil stabilisation

1. Introduction

Road subgrade materials are subjected to various weather conditions such as rain and extreme heat during spring and summer, especially in tropical areas. Subgrade materials found in shallow depths are vulnerable to continuous wetting and drying cycles in these regions [1]. The durability of the mechanical properties and the deformation behaviour of subgrade materials under these changing environmental conditions have been a concern and have been investigated using the wetting and drying cycle test [2,3]. These investigations have shown that subgrade strength decreases with an increase in wetting and drying cycles according to [4], as a result of the breakdown of interlocking bonds underpinned by constant swelling and shrinking from wetting and drying, respectively [5–7]. This further shows that various thresholds exist in terms of "cycles to failure" for different subgrade materials treated with different binders. The wetting and drying cycles can significantly affect the hydro-mechanical behaviour of unsaturated soils [4,8,9].

Expansive subgrades exhibit a significant increase in irreversible bulk volume upon wetting [2]. These subgrades are weak when wet and exhibit high shrink–swell behaviours, with changes in moisture content leading to movement within the subgrade capable of inducing large differential volume changes leading to cracking and other associated defects in road pavements [10–13]. Subgrades with these characteristics would fail quickly during wetting–drying cycles and hence are not suitable for use in road construction [14].

Consequently, the strength and durability of expansive subgrade materials are usually improved before construction, using chemical stabilisation processes entailing the addition of binders to the soil to improve their mechanical properties such as the California bearing ratio (CBR), the unconfined compressive strength (UCS), and the resilient modulus, and to reduce their water adsorption and, eventually, swell potential [12]. Studies have used different types of binders including cement, lime and other wastes and industrial by-products such as Silica fume, geo-textiles, brick and plastic wastes to improve on the performance of such problematic soils [15–28]. The application of these alumina and silica-rich by-product materials are well known for their strength-giving pozzolanic reaction in the presence of an activator, traditionally calcium-based. This reaction results in the formation of cementitious products, namely calcium silicate hydrate (C-H-S) gels which harden under proper curing conditions and periods to glue soil particles and fill-up pore spaces. The reduced pores and overall increase in density contribute to improving the strength and reducing the moisture absorption of the treated materials [13,29,30].

However, the durability of such stabilised material continues to be of immense concern with the current extremes in terms of weather. Researchers have continuously focused on sustainability based on durability. Several studies have moved beyond the success of strength enhancement to consider durability of such enhancement and resilience in the face of harsh climate changes. A study by Wassermann [31] on the effects of wetting–drying cycles on the performance of cement treated soils reported a higher mass loss in samples with lower binder contents with increasing cycles. Treated samples are reported to continuously get eroded with each wet–dry cycle. Durability studies on untreated and lime-treated soils showed that the higher void ratio and the increased pore size distribution were more severe in the untreated soil samples compared to lime-treated counterparts after subjecting both samples to wetting–drying cycles [32]. The increase in pore size distribution modified the compression curve of the untreated samples from a convex shape into a linear profile, indicating larger deformation under similar loading conditions after several cycles, due to a loss in interparticle bond with increasing cycles as expected [32].

In this study, the engineering properties of expansive subgrade materials were improved using sustainable waste materials by enhancing their durability to withstand large cycles of wetting–drying. Microstructural analysis via scanning electron microscopy (SEM) and mineralogical analysis via energy dispersive X-ray (EDX) were conducted on the treated subgrade materials to identify the changes in the morphology, elemental composition, and other formations in the stabilised subgrade materials. Artificially synthesised subgrade materials from a mixture of bentonite and kaolinite at various proportions were constituted to form high-plasticity-index subgrades. The subgrades were then treated using waste materials such as ground granulated blast furnace slag (GGBS) and brick dust waste (BDW) as partial replacement for cement and lime in the mix. After the subgrade materials were treated, a wetting–drying cycle test was conducted in accordance with [33,34] by recording their CBR values after each wetting and drying cycle. The result of this study provides insight into the viability of BDW in combination with GGBS as an eco-friendly binder. This is a vital step in the light of current efforts in transitioning from cement and lime-based binders to more sustainable materials to cut-down on the current carbon footprint and expand the knowledgebase in terms of alternative admixtures for sustainable infrastructural development.

2. Materials and Methods

2.1. Materials

The materials used in this study include ordinary Portland cement, lime, GGBS, BDW, bentonite, and kaolinite mixed at varying proportions to form two subgrade materials named subgrade 1 (25% bentonite + 75% kaolinite) and subgrade 2 (75% bentonite + 25% kaolinite). The choice of the composition of the subgrades was to simulate a lower and upper band in terms of expansivity of subgrades. Whereas subgrade 2 containing 75% bentonite is an extremely highly expansive soil with liquid limit of 294.07%, plastic limit of 45.38%, and a plasticity index of 248.69%. Subgrade 1 with 25% bentonite qualifies as a high plasticity soil with liquid limit of 131.26%, plastic limit of 28.24%, and a plasticity index of 102.52%. The bentonite and kaolinite used were supplied by Potclays Ltd. (Stoke-on-Trent, UK), while the cement (CEM I) used was supplied by Dragon Alfa Cement Ltd. (Bristol, UK), and in compliance with [35]. The lime used was supplied by Singleton Birch Ltd. (Barnetby, UK) and was in compliance with [36] to be classified as hydrated lime, while the BDW was supplied by Celtic Sustainable Ltd. (Ceredigion, UK) and in compliance with [37]. Furthermore, the GGBS used was supplied by Francis Flower in compliance with [38]. The sustainable waste materials used consisted of GGBS and BDW used at proportions of 23.5% GGBS and 0% BDW, and 11.75% GGBS and 11.75% BDW for both subgrade types. Compaction and Atterberg limit tests were conducted in compliance with [39–41] to determine the Optimum Moisture Content (OMC) and Maximum Dry Density (MDD) of untreated subgrade materials. The procedure and equipment used to conduct microstructural analysis in this study are as reported in the authors' previous study [12,42]. The study also conducted wetting–drying cycle on the subgrade materials formulated in this study in accordance with the wetting and drying procedure described in [33]. After each cycle of wetting–drying, the CBR samples were tested in accordance with [34]. The oxide compositions of the materials are presented in Table 1.

Table 1. Chemical Composition of Materials [43].

Oxide (%)	SiO_2	Al_2O_3	Fe_2O_3	FeO	MgO	CaO	K_2O	SO_3	TiO_2	Na_2O	Trace	L.O.I
Bentonite	63.02	21.08	3.25	0.35	2.67	0.65	-	-	-	2.57	0.72	5.64
Kaolinite	48.5	36.0	1.00	-	0.30	0.05	2.15	-	0.06	0.15	-	11.7
Cement	20	6.0	3.0	-	4.21	63	-	2.30	-	-	-	0.80
GGBS	35.35	11.59	0.35	-	8.04	41.99	-	0.23	-	-	-	-
Lime	3.25	0.19	0.16	-	0.45	89.2	0.01	2.05	-	-	-	-
BDW	52	41	0.7	-	0.12	4.32	0.53	0.33	0.65	0.05	-	2.01

2.2. Sample Preparation and Experimental Design

Wetting–drying cycle test was conducted in this study following the wetting and drying procedure outlined in [33]. A total of 21 cycles of wetting and drying were performed. Wetting–drying cycles conducted in this study were to simulate the effect of repeated wetting and drying cycles on treated road subgrade materials during the wet (flooding) and dry (drought) seasons. CBR samples were prepared in a 152 × 178 mm steel mould from 3 layers of stabilised soil each receiving 62 blows of a 2.5 kg rammer, using an automatic mechanical compaction machine. The samples were compacted at 1.2 of OMC to allow for adequate moisture for hydration and pozzolanic reaction during curing and stored at a temperature of 22 ± 2 °C in compliance with [34]. The stabilised samples were wrapped in clingfilm and cured for 28 days before subjecting to wetting–drying cycle test. After curing the samples, the clingfilm was removed and the samples were wrapped using duct tape leaving the top and bottom of the sample. This was done to hold the sample together in one piece during the wetting–drying cycles test. All CBR samples were submerged in water for 5 h at a room temperature of 20 ± 2 °C to simulate flooding during heavy rainfall. After 5 h of soaking, the samples were removed and weighed, and their mass was recorded to monitor the mass change in the sample in the course of the cycles. After the soaking period, one sample was set aside and tested for CBR (wetting cycle No 1) and the CBR value was recorded. The remaining soaked samples were then placed in an oven at a temperature of 71 ± 3 °C for a period of 42 h to simulate extreme dryness (drought) or high temperature.

After 42 h the dry samples were removed, and their masses recorded. Thereafter, one sample was set aside and tested for CBR (drying cycle No 1) and the CBR value was recorded. This procedure constitutes one cycle (48 h) of wetting and drying [33,34]. The process was repeated until all 10 cycles were achieved to see the effect of wetting–drying cycles on the durability of the treated subgrades. The purpose of conducting wetting–drying test in this study was to establish a limiting number of cycles at which the CBR of the subgrade materials fall below allowable limits and hence the subgrades become unsuitable for use in road construction. A high-quality subbase material will typically have a CBR value between 80% and 100% and a subgrade with a CBR value < 2% is unacceptable for use in road construction and would require modification or treatment [12].

3. Results and Discussion

3.1. Wetting and Drying Cycles

From the results of wetting–drying cycle on subgrade samples, deeper cracks were observed for subgrade 2 (extremely high plasticity index), which was composed of 75% of bentonite compared to subgrade 1 (25% bentonite). This observation was due to the high shrinkage potential because of the high amount of bentonite content in the mix as shown in Figure 1. Weight loss in samples was observed for all mix designs as the number of cycles increased [44]. A significant loss in mass of the samples was observed from cycles 4 to 10 for both wet and dry testing conditions. This weight loss was a result of the weakening of the interparticle bonds due to the constant swelling and shrinking caused by cyclic wetting and drying. Particles of treated soil were constantly disintegrating and falling-off as the wetting–drying process progressed. This behaviour is expected and has been observed in binders used in subgrade stabilisation when exposed to wetting and drying [45]. The study by [46], showed adverse effect of increasing temperatures on stabilised soil with increasing mass loss of soil. It was observed that increased heating conditions led to a decrease in sample sizes. A drastic drop in sample mass after oven drying was observed. However, samples regained some amount of mass during the wetting processes (soaking) due to water filling the pore spaces in the samples. The occurrence of mass loss was due to the evaporation of moisture out of the sample during oven drying at 71 ± 3 °C leading to the development of microcrack which expanded following water ingress during soaking with associated swelling. The loss in bond strength and expansion of the samples during the wetting phase further expanded the cracks and increased the void ratio of the samples thereby reducing the mass of the sample as shown in Figure 1.

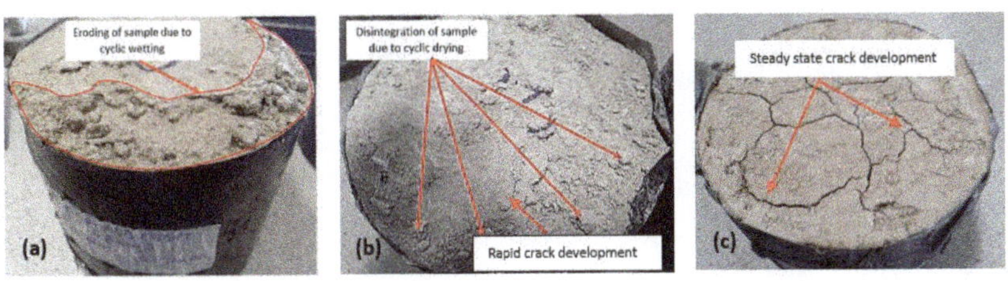

Figure 1. (**a**) Eroded subgrade sample due to cyclic wetting (**b**) Rapid crack development stage with loose particle following breakdown of interparticle bonds due to high temperature (**c**) Oven dry sample exhibiting steady-state crack development similar to a typical natural dry expansive subgrade.

Cracking after different numbers of wetting–drying cycles was obtained in the subgrade [47]. According to the British lime association [48], more moisture loss was observed through evaporation due to heat generated during the hydration and pozzolanic reactions synonymous with soil stabilisation using cement, lime and GGBS. Two stages of crack development at the end of the drying process of the high-plasticity-index subgrade were observed-the stage of rapid crack growth and the stage of steady crack development, these stages affect the strength of the soil while reducing the soil mass [47]. This behaviour is very common in naturally existing high-plasticity-index subgrade materials when subjected to high temperatures, especially in tropical areas. When oven-dry samples were submerged in water during the wetting cycle process, water, with a higher density, quickly occupied the pores, expelling the air from within the sample hence increasing the mass of the sample as shown in Figure 1a–c. Figures 2 and 3 show the gradual reduction in sample mass for subgrade materials as the number of cycles increases.

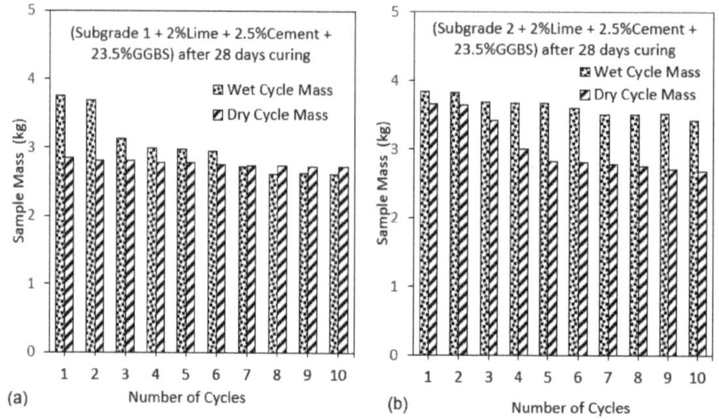

Figure 2. (**a**) Sample mass for subgrade 1 composed of GGBS after 28 days of curing (**b**) Sample mass for subgrade 2 composed of GGBS after 28 days of curing.

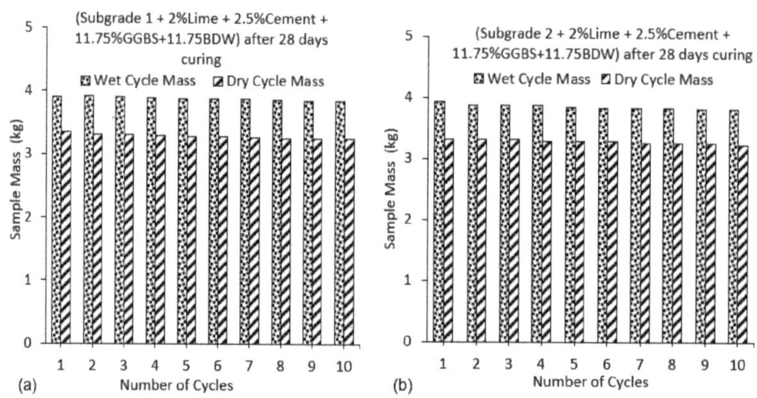

Figure 3. (**a**) Sample mass for subgrade 1 composed of GGBS and BDW after 28 days of curing (**b**) Sample mass for subgrade 2 composed of GGBS and BDW after 28 days of curing.

The effect of the addition of 11.75% of BDW to the binder mix can be seen from Figure 3a,b. The increase in mass loss from the use of 2% lime, 2.5%cement, and 23.5% GGBS with increasing cycles was significantly reduced from almost 10% to less than 2% for the wet cycles when 11.75% BDW was introduced into the binder. The gradual loss in mass of the samples resulted in lower moisture absorption. This was because a reduction in the mass of the stabilised soil from disintegration, led to an overall reduction in the void ratio as the wet–dry cycles progressed. The overall reduction in the void ratio also resulted in lower mass of the samples during the wetting cycle as the mass of the absorbed water was lowered in the process. The reduction in mass loss indicates further increase in the strength of the stabilised expansive soils from the addition of BDW. BDW is known for its high pozzolanic properties and have been reported to increase the strength of stabilised soils [49].

In addition, BDW known for its high specific area is very reactive as a highly pozzolanic material. The finer particle size of the BDW filled the interparticle pores, which resulted in a denser and more compact soil-binder matrix with enhanced strength and durability.

In the current study, the effect of 11.75% BDW is seen in enhancing the durability of the treated soils. From Figure 3a it seen that at the 10th cycle, the sample mass is relatively stable compared to the samples stabilised with 23.5% GGBS. This might indicate increased durability when BDW is utilised and suggests that BDW might be a useful additive in increasing the durability of stabilised samples.

3.2. *The California Bearing Ratio (CBR)*

The CBR values used in this study was the average CBR value of two samples to ensure validity and accuracy of the results. After conducting wetting–drying cycle on the treated subgrade samples, it was observed that the CBR of the subgrade samples was decreasing with an increase in the number of cycles as expected. This observation confirms the findings by [1], that subjecting soil to wetting–drying cycles caused the decrease in soil strength. The CBR value of subgrade 2 was less compared to subgrade 1. The reduction in CBR values was observed for all subgrade samples with an increase in wetting–drying cycles. The reduction in CBR value as wetting–drying increased could be attributed to the repeated swelling and shrinking due to wetting–drying resulting in loss of binder strength within the samples [1]. Very high CBR values more than twice that of wet samples were recorded for dry cycle samples. The extremely high CBR observed in dry cycles was due to the ability of high-plasticity-index clays to harden during drying under elevated temperatures. The hardening of the clay increased the CBR of the soil hence, the high CBR values were recorded for drying-cycle samples. The low CBR values recorded for wet-cycle samples compared with dry-cycle samples could be because high plasticity clay is weak in the presence of water [50]. Overall, the CBR recorded for subgrade 2 were lower compared with subgrade 1 as shown in Figures 4 and 5.

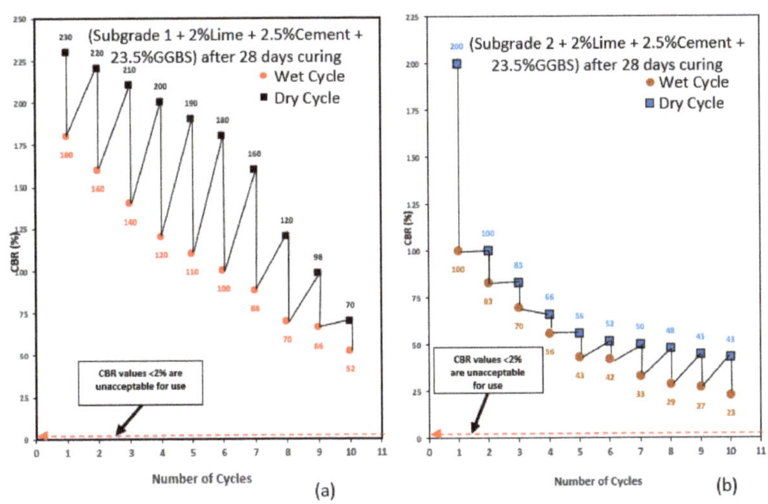

Figure 4. (**a**) Wetting–drying cycle results for subgrade 1 composed of GGBS after 28 days of curing. (**b**) Wetting–drying cycle results for subgrade 2 composed of GGBS after 28 days of curing.

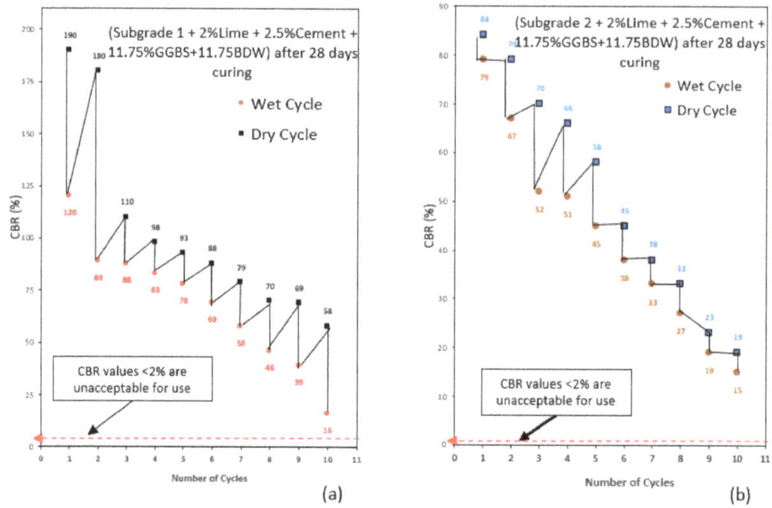

Figure 5. (**a**) Wetting–drying cycle results for subgrade 1 composed of GGBS and BDW after 28 days of curing. (**b**) Wetting–drying cycle results for subgrade 2 composed of GGBS and BDW after 28 days of curing.

This was due to the high amount of bentonite content in the mix, making it expansive and weak in compression. According to [12], subgrade materials with high bentonite content have low CBR values. Subgrade mix design composed of high GGBS content (23.5%) recorded the highest CBR values for wetting–drying cycles, making it the best performing mix in this study. The high strength values achieved with the addition of high amounts of GGBS in the mix supports some of the results presented by [12], which is because the relatively high calcium content in GGBS, the main reaction product of CSH gel is responsible for the increase in strength in a mix [51]. Studies have shown that the higher the amount of GGBS blends, the greater the strength and durability of the mix [52]. This shows that subgrade material treated with 23.5% GGBS can withstand many wetting–drying cycles and still maintain very high CBR values required for use in road construction. The addition of equal proportions of GGBS and BDW exhibited good CBR values usable in road construction at the end of the wetting–drying cycle process. BDW are pozzolanic materials with high alumina/silica content which in the presence of lime can form cementitious gel to increase strength in a mix [53]. Previous studies have also shown that the addition of brick dust increased soil strength by 1.7 to 2.3-fold [51,54]. The lowest CBR value recorded for subgrade 1 composed of 23.5% GGBS was 70% after 10 drying cycles and 52% after 10 wetting cycles. Subgrade 2 composed of 23.5% GGBS recorded a CBR value of 43% after 10 drying cycles and 23% after 10 wetting cycles. Subgrade 1 composed of equal amounts of 11.75% GGBS and 11.75% BDW recorded a CBR value of 58% after 10 drying cycles and 16% after 10 wetting cycles. Subgrade 2 composed of equal amounts of 11.75% GGBS and 11.75% BDW recorded a CBR value of 19% after 10 drying cycles and 15% after 10 wetting cycles. The result shows that CBR values achieved for all subgrades after 10 wetting–drying cycles are very good exceeding minimum 2%, making them suitable for use in road construction. Hence, expansive subgrade materials treated with GGBS and BDW are highly durable and can withstand harsh weather conduction without losing their strength. Figures 4 and 5 show CBR results for subgrade materials after wetting–drying cycles.

3.3. The Resilient Modulus

The resilient modulus is considered one of the critical parameters in the design of flexible road pavements as prescribed by some important design guidelines [55,56]. The resilient modulus is often used to estimate pavement layers' (subgrade, base and sub-base material) behaviour especially when subjected to conditions of cyclic loading. It is also a fundamental mechanical property of materials utilised to explain non-linear stress–strain characteristic of subgrades under the influence of repeated loading [57,58]. According to AASHTO, the resilient modulus is defined as the ratio of the applied cyclic axial stress to the resilient axial strain [59].

Several empirical and mathematical relationships have been proposed to estimate the resilient modulus of pavement materials based on the CBR of these materials as tested in the laboratory [55,59–64]. In this study, the resilient modulus of natural and stabilised subgrade materials subjected to cycles of wetting and drying derived from previous research as shown in Table 2 is compared as given in Figure 6.

Table 2. Empirical Relationship for Resilient Modulus Derivation.

SN	Expression	Reference
1	$10.33 \times \text{CBR}$	[62]
2	$38 \times (\text{CBR})^{0.711}$	[61]
3	$18 \times (\text{CBR})^{0.64}$	[63]
4	$21 \times (\text{CBR})^{0.65}$	[55]
5	$17.6 \times (\text{CBR})^{0.64}$	[59]
6	$49.37 \times (\text{CBR})^{0.59}$	[60]
7	$1.2 \times \text{CBR}$	[64]

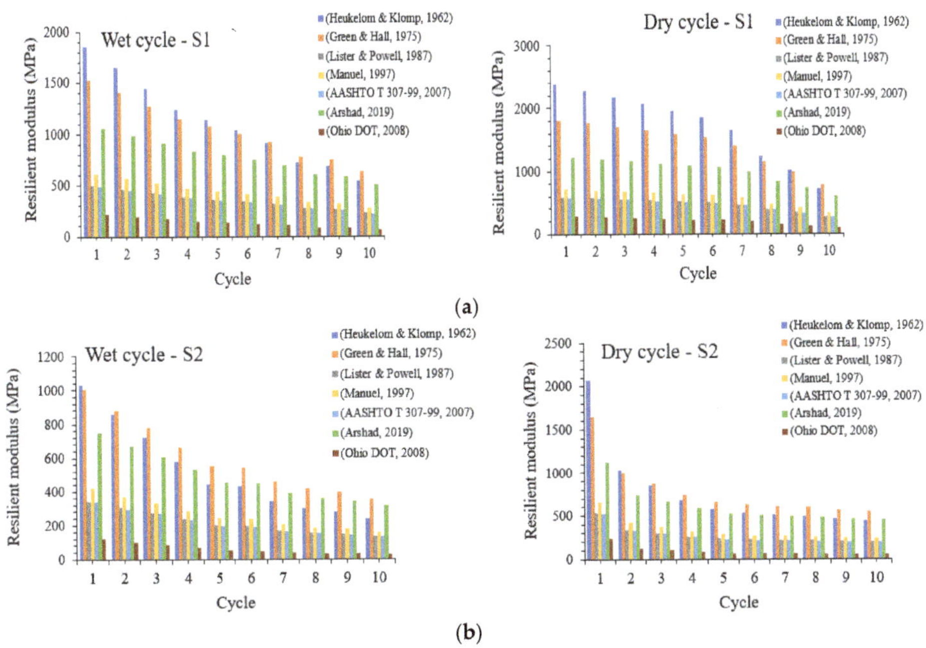

Figure 6. (**a**) Resilient modulus results for subgrade 1 composed of GGBS after 28 days of curing (**b**) Wetting–drying cycle results for subgrade 2 composed of GGBS after 28 days of curing [55,59–64].

As indicated in Figure 6, the resilient modulus seems to be generally decreasing with increasing cycles of wetting and drying for stabilised subgrades 1 and 2 without BDW. Across the estimated resilient modulus, values for subgrade 1 are all greater than those for subgrade 2. Moreover, for both wetting and drying cycles, the expression proposed by [62,64] seem to give upper and lower bound values of the resilient modulus, respectively. Meanwhile, the resilient modulus proposed by [61] seem almost indistinguishable. It could also be noticed that the resilient modulus derived from [55,59–64] all seem to remain unchanged with the increasing number of wetting and drying cycles especially for stabilised subgrade 1. A similar decreasing trend in the resilient modulus with increasing cycles of wetting and drying is observed for stabilised subgrades 1 and 2 with the inclusion of BDW as shown in Figure 7a,b.

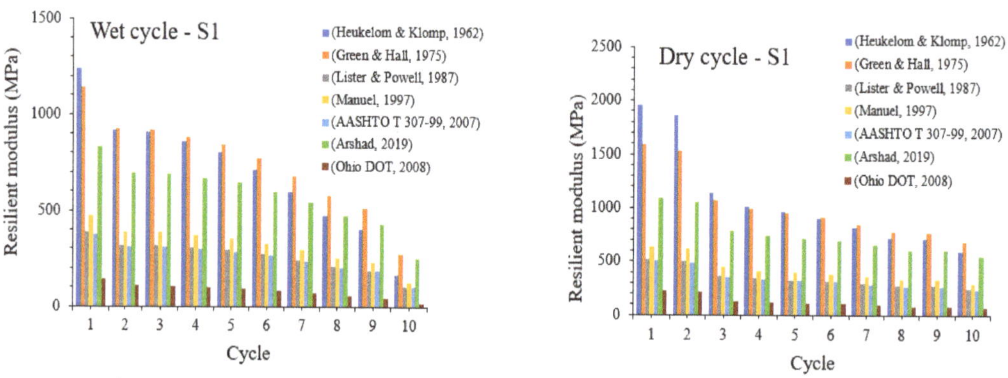

(**a**)

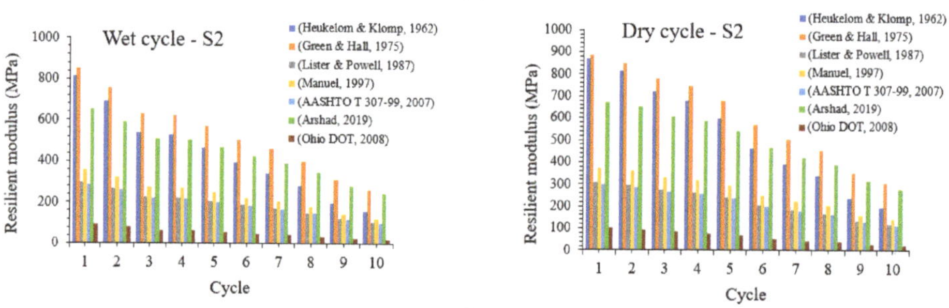

(**b**)

Figure 7. (**a**) Resilient modulus results for subgrade 1 composed of GGBS and BDW after 28 days of curing. (**b**) Wetting–drying cycle results for subgrade 2 composed of GGBS and BDW after 28 days of curing [55,59–64].

However, some differences are also noticed in the behaviour of the resilient modulus curves when comparing soils stabilised with and without the inclusion of BDW. One of the points of differences is generally noticed with the upper-bound resilient modulus, which is in accordance with the expressions proposed by [61,62] for soils stabilised with BDW included unlike those stabilised without the inclusion of BDW under both wetting and drying cycles. Additionally, for stabilised soil 1 under the drying cycles, the resilient modulus is slightly higher for the first two cycles but does remain unchanged until the 10th cycle. On the other hand, the lower-bound resilient modulus [55,59,64] for stabilised soil 2 with BDW included seems to demonstrate a clear decreasing trend with increasing wetting and drying cycles unlike for stabilised soil 2 without the inclusion of BDW. The results all show the impact of the fluctuating moisture levels from erratic environmental conditions on the performance of stabilised subgrade materials and highlight the improvements gained by blending pozzolanic materials such as BDW and GGBS in enhancing the strength and durability of the stabilised soils. A BDW and GGBS blend is encouraged to increase the durability of the stabilised subgrades.

3.4. Microstructural Properties

Microstructural analysis conducted showed a high formation of the calcium silicate hydrate (CSH) gel responsible for the increase in strength. Recent studies [53,65] have pointed out that the addition of GGBS and BDW increases the formation of CSH gel due to the rich calcium and pozzolanic characteristic of GGBS and BDW combination. From the result of energy dispersive x-ray analysis, as shown in Figure 8a–d, a calcium (Ca) content of 44.87% was recorded for subgrade 1 treated with the addition of 23.5% GGBS after 28 days of curing. Subgrade 1 composed of 11.75% GGBS and blended with 11.75% BDW recorded 32.73% calcium (Ca) after 28 days of curing. Subgrade 2 composed of 11.75% GGBS and blended with 11.75% BDW recorded 36.78% calcium (Ca) after 28 days of curing. Detailed results of SEM and EDX analysis conducted in this study are shown in Figure 8a–d. From the EDX results, the higher calcium content of subgrades 1 and 2 treated with 23.5% GGBS in combination with lime and cement is logical since GGBS has a higher calcium content, which was further increased with the addition of 2% lime and 2.5% cement. The increase in calcium, however, might not have sustained the increase in strength as calcium alone is insufficient for production of CHS obviously. Reasonable quantities of Si and Al are required to enhance the production of cementitious products. This is seen in the results of the samples with the addition of 11.75% BDW and is an indication of the growth of more cementitious products as shown in Figure 8a,b.

The surface morphology and reduced pores of GGBS- and BDW- treated samples, as shown in the micrographs in Figure 8a,b, align with the CBR results in Section 3.2 which also shows higher strengths for samples treated with GGBS blended with BDW at the end of the 10th cycle. Lower pores are synonymous with increased particle coating ability, which is a function of more CHS flakes as shown in the micrographs of Figure 8c,d. The increased interparticle bonds from a higher pozzolanic activity combined with the initial hydration products from the lime and cement enhanced the durability of the subgrades 1 and 2 when BDW was introduced into the mix. The mineralogical results from EDX also capture a spike in the Al and Si content from 2.37% Al and 4.54% Si to 8.60% Al and 19.01% Si for subgrade 1 when BDW was introduced as shown in Figure 8a,c. Additionally, for subgrade 2, this was from 6.74% Al and 19.70% Si to 10.27% Al and 20.59% Si when BDW was introduced into the binder mix as shown in Figure 8b,d. This strength enhancement is also seen in the lower mass loss recorded from the wetting–drying cycles discussed in Section 3.1. The SEM and EDX results corroborate the fact that the addition of GGBS blended with BDW might be a better combination in dealing with subgrade degradation from cyclic wetting and drying which has become a norm in the light of the rapidly changing environmental condition. The use of GGBS blended with BDW might be considered as an option in the development of resilient road pavement infrastructure to withstand the adverse effects of changing climatic conditions.

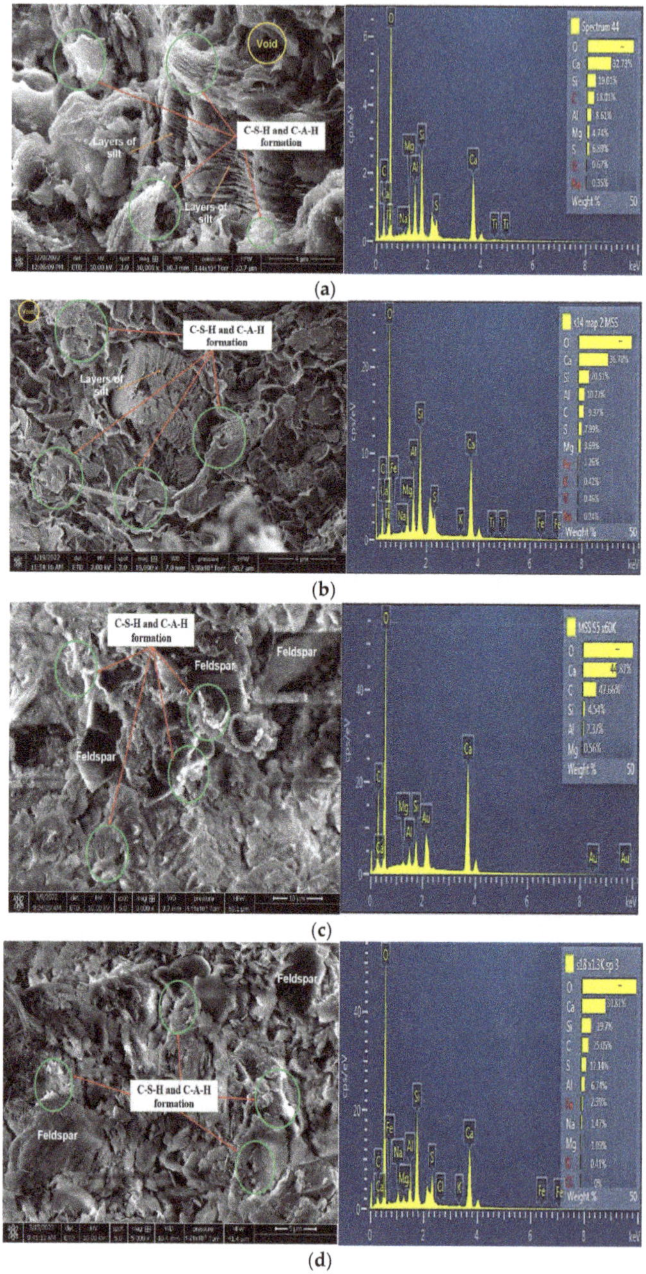

Figure 8. (**a**) SEM and EDX results for subgrade 1 composed of 11.75% GGBS and 11.75% BDW after 28 days curing. (**b**) SEM and EDX results for subgrade 2 composed of 11.75% GGBS and 11.75% BDW after 28 days curing. (**c**) SEM and EDX results for subgrade 1 composed of 23.5% GGBS after 28 days curing. (**d**) SEM and EDX results for subgrade 2 composed of 23.5% GGBS after 28 days curing.

4. Conclusions

This study focused on the effects of introducing sustainable binders (GGBS blended with BDW) in improving the strength and durability of lime- and cement-treated expansive subgrade for road pavement construction. The bearing capacity of the treated soils has been studied via CBR, under constant wetting and drying cycles. Additionally, microstructural studies have been undertaken to evaluate the changes in the fabric of the stabilised soil at the microlevel. The results of the treated samples containing BDW blends when compared with that of samples without BDW showed strength enhancement and increased durability. Based on the experimental results, the following conclusion can be drawn.

1. Expansive subgrade materials treated with a blend of GGBS and BDW showed higher resistance against degradation at the end of 10 cycles of wetting and drying. The higher strength at the end of the wetting–drying periods indicates that the addition of 11.75% BDW to the binder mix enhanced the resistance of the treated soils against expansion and shrinkage cracks.
2. The strength reduction over the wet–dry cycles was lower for samples treated with a blend of BDW. The addition of BDW increase the production of CHS gels due to the additional pozzolanic activity which increased interparticle bond under the wet–dry cycles.
3. The result of mass loss analysis of the treated soils aligns with the bearing capacity results and the microstructural characteristics indicates that the addition of 11.75% BDW blends into the binder mix is a useful means of lowering the mass loss which occurs through constant swelling and shrinking leading to the breakdown of the bonds. Lower mass loss indicates that subgrade treatment with a blend of BDW has higher resilience and could maintain strength in the event of rapidly fluctuating environmental conditions.

Author Contributions: Conceptualization, S.Y.O.A. and S.J.A.; methodology, S.Y.O.A. and S.J.A.; software, J.J.J., S.J.A., C.A.B. and E.U.E.; validation, S.Y.O.A., S.J.A. and C.A.B.; formal analysis, S.Y.O.A., S.J.A., E.U.E. and J.J.J.; investigation, S.Y.O.A. and S.J.A.; resources, S.J.A., C.A.B. and E.U.E.; data curation, S.J.A., S.Y.O.A., E.U.E. and J.J.J.; writing—original draft preparation, S.Y.O.A., E.U.E., J.J.J. and S.J.A.; writing—review and editing, S.Y.O.A., E.U.E., J.J.J. and S.J.A.; visualization, S.Y.O.A., S.J.A., C.A.B. and E.U.E.; supervision, S.JA. and C.A.B.; project administration, S.J.A. and C.A.B. All authors have read and agreed to the published version of the manuscript.

Funding: This research received no external funding.

Institutional Review Board Statement: Not applicable.

Informed Consent Statement: Not applicable.

Data Availability Statement: Data can be obtained from corresponding author upon reasonable request.

Conflicts of Interest: The authors declare no conflict of interest.

References

1. Hu, Z.; Peng, K.; Li, L.; Ma, Q.; Xiao, H.; Li, Z.; Ai, P. Effect of Wetting-Drying Cycles on Mechanical Behaviour and Electrical Resistivity of Unsaturated Subgrade Soil. *Adv. Civ. Eng.* **2019**, *2019*, 3465327. [CrossRef]
2. Chen, R.; Ng, C. Impact of wetting–drying cycles on hydro-mechanical behavior of an unsaturated compacted clay. *Appl. Clay Sci.* **2013**, *86*, 38–46. [CrossRef]
3. Zhang, J.; Jiang, Q.; Zhang, Y.; Dai, L.; Wu, H. Nondestructive Measurement of Water Content and Moisture Migration of Unsaturated Red Clays in South China. *Adv. Mater. Sci. Eng.* **2015**, *2015*, 542538. [CrossRef]
4. Farulla, C.A.; Ferrari, A.; Romero, E. Mechanical Behaviour of Compacted Scaly Clay During Cyclic Controlled-Suction Testing. In *Experimental Unsaturated Soil Mechanics*; Schanz, T., Ed.; Springer Proceedings in Physics; Springer: Berlin/Heidelberg, Germany, 2007; Volume 112. [CrossRef]
5. Holtz, W.G. Expansive Clays—Properties and Problems. *Q. Colo. Sch. Mines* **1973**, *54*, 89–125.
6. Soltani, A.; Deng, A.; Taheri, A.; O'Kelly, B. Intermittent swelling and shrinkage of a highly expansive soil treated with polyacrylamide. *J. Rock Mech. Geotech. Eng.* **2022**, *14*, 252–261. [CrossRef]

7. Zhang, M.; Tian, Z.; Zhu, Q.; Chen, J. In-situ assessment of soil shrinkage and swelling behavior and hydro-thermal regimes with a thermo-time domain reflectometry technique. *Soil Tillage Res.* **2023**, *227*, 105617. [CrossRef]
8. Gens, A.; Sánchez, M.; Sheng, D. On constitutive modelling of unsaturated soils. *Acta Geotech.* **2006**, *1*, 137–147. [CrossRef]
9. Kim, J.; Hwang, W.; Kim, Y. Effects of hysteresis on hydro-mechanical behavior of unsaturated soil. *Eng. Geol.* **2018**, *245*, 1–9. [CrossRef]
10. Chertkov, V. Physical modeling of the soil swelling curve vs. the shrinkage curve. *Adv. Water Resour.* **2012**, *44*, 66–84. [CrossRef]
11. Fernandes, M.; Denis, A.; Fabre, R.; Lataste, J.-F.; Chrétien, M. In situ study of the shrinkage-swelling of a clay soil over several cycles of drought-rewetting. *Eng. Geol.* **2015**, *192*, 63–75. [CrossRef]
12. Amakye, S.Y.O.; Abbey, S.J.; Booth, C.A.; Oti, J. Road Pavement Thickness and Construction Depth Optimization Using Treated and Untreated Artificially-Synthesized Expansive Road Subgrade Materials with Varying Plasticity Index. *Materials* **2022**, *15*, 2773. [CrossRef]
13. Jeremiah, J.J.; Abbey, S.J.; Booth, C.A.; Kashyap, A. Behaviour and Microstructural Characteristics of Lime-GGBS-Treated Kaolin Clay Contaminated with Gypsum. *Materials* **2023**, *16*, 874. [CrossRef]
14. Tripathy, S.; Rao, K.S.S. Cyclic Swell–Shrink Behaviour of a Compacted Expansive Soil. *Geotech. Geol. Eng.* **2009**, *27*, 89–103. [CrossRef]
15. Navagire, O.P.; Sharma, S.K.; Rambabu, D. Stabilization of black cotton soil with coal bottom ash. *Mater. Today Proc.* **2022**, *52*, 979–985. [CrossRef]
16. Amakye, S.Y.; Abbey, S.J.; Booth, C.A.; Mahamadu, A. Enhancing the engineering properties of subgrade materials using processed waste: A review. *Geotechnics* **2021**, *1*, 307–329. [CrossRef]
17. Tiwari, N.; Satyam, N.; Puppala, A.J. Strength and durability assessment of expansive soil stabilized with recycled ash and natural fibers. *Transp. Geotech.* **2021**, *29*, 100556. [CrossRef]
18. Abbey, S.J.; Eyo, E.U.; Oti, J.; Amakye, S.Y.; Ngambi, S. Mechanical Properties and Microstructure of Fibre-Reinforced Clay Blended with By-Product Cementitious Materials. *Geosciences* **2020**, *10*, 241. [CrossRef]
19. Khadka, S.D.; Jayawickrama, P.W.; Senadheera, S.; Segvic, B. Stabilization of highly expansive soils containing sulfate using metakaolin and fly ash based geopolymer modified with lime and gypsum. *Transp. Geotech.* **2020**, *23*, 100327. [CrossRef]
20. Pandey, A.; Schwab, P.; Little, D.N. Optimization of magnesium phosphate cement: Stabilization of a kaolinitic soil. *Transp. Geotech.* **2022**, *37*, 100854. [CrossRef]
21. Mohanty, S.; Roy, N.; Singh, S.P.; Sihag, P. Strength and durability of flyash, GGBS and cement clinker stabilized dispersive soil. *Cold Reg. Sci. Technol.* **2021**, *191*, 103358. [CrossRef]
22. Jiang, N.; Du, Y.; Liu, K. Durability of lightweight alkali-activated ground granulated blast furnace slag (GGBS) stabilized clayey soils subjected to sulfate attack. *Appl. Clay Sci.* **2018**, *161*, 70–75. [CrossRef]
23. Peethamparan, S.; Olek, J.; Lovell, J. Influence of chemical and physical characteristics of cement kiln dusts (CKDs) on their hydration behavior and potential suitability for soil stabilization. *Cem. Concr. Res.* **2008**, *38*, 803–815. [CrossRef]
24. Eyo, E.U.; Ng'ambi, S.; Abbey, S.J. Incorporation of a nanotechnology-based additive in cementitious products for clay stabilisation. *J. Rock Mech. Geotech. Eng.* **2020**, *12*, 1056–1069. [CrossRef]
25. Miller, G.A.; Azad, S. Influence of soil type on stabilization with cement kiln dust. *Constr. Build. Mater.* **2000**, *14*, 89–97. [CrossRef]
26. Abbey, S.J.; Ngambi, S.; Coakley, E. Effect of cement and by-product material inclusion on plasticity of deep mixing improved soils. *Int. J. Civ. Eng. Technol.* **2016**, *7*, 265–274.
27. Abbey, S.J.; Ngambi, S.; Olubanwo, A.O.; Tetteh, F.K. Strength and hydraulic conductivity of cement and by-product cementitious materials improved soil. *Int. J. Appl. Eng. Res.* **2018**, *13*, 8684–8694.
28. Eyo, E.U.; Ng'Ambi, S.; Abbey, S.J. Performance of clay stabilized by cementitious materials and inclusion of zeolite/alkaline metals-based additive. *Transp. Geotech.* **2020**, *23*, 100330. [CrossRef]
29. Rao, S.M.; Reddy, B.V.V.; Muttharam, M. The impact of cyclic wetting and drying on the swelling behaviour of stabilized expansive soils. *Eng. Geol.* **2001**, *60*, 223–233. [CrossRef]
30. Consoli, N.C.; Samaniego, Q.R.A.; Gonz´alez, L.E.; Bittar, E.J.; Cuisinier, O. Impact of Severe Climate Conditions on Loss of Mass, Strength, and Stiffness of Compacted Fine-Grained Soils-Portland Cement Blends. *J. Mater. Civ. Eng.* **2018**, *30*, 04018174. [CrossRef]
31. Wassermann, A.; Abdallah, A.; Cuisinier, O. Impact of wetting and drying cycles on the mechanical behaviour of a cement-treated soil. *Transp. Geotech.* **2022**, *36*, 100804. [CrossRef]
32. Ying, Z.; Benahmed, N.; Cui, Y.; Duc, M. Wetting-drying cycle effect on the compressibility of lime-treated soil accounting for wetting fluid nature and aggregate size. *Eng. Geol.* **2022**, *307*, 106778. [CrossRef]
33. *ASTM D559/D559M-15*; Standard Test Methods for Wetting and Drying Compacted Soil-Cement Mixture1. ASTM: West Conshohocken, PA, USA, 2016.
34. *BS EN 13286-47*; Unbound and Hydraulically Bound Mixtures—Test Methods for the Determination of California Bearing Ratio, Immediate Bearing Index and Linear Swelling. British Standards Institution: London, UK, 2021.
35. *BS EN 197-1*; Cement Composition, Specifications and Conformity Criteria for Common Cements. British Standards Institution: London, UK, 2011.
36. *BS EN 459-1*; Building Lime Definitions, Specifications and Conformity Criteria. British Standards Institution: London, UK, 2015.
37. *BS EN 771-1:2011+A1:2015*; Specification for Masonry Units. Clay Masonry Units. British Standards Institution: London, UK, 2015.

38. *BS EN 15167-1:2006*; Ground Granulated Blast Furnace Slag for Use in Concrete, Mortar and Grout Definitions, Specifications and Conformity Criteria. British Standards Institution: London, UK, 2006.
39. *BS EN 13286-2-2012*; Unbound and Hydraulically Bound Mixtures—Test Methods for Laboratory Reference Density and Water Content. Proctor Compaction. British Standards Institution: London, UK, 2012.
40. *BS EN ISO 17892-1-2021*; Geotechnical Investigation and Testing. Laboratory Testing of Soil—Determination of Water Content. British Standards Institution: London, UK, 2021.
41. *BS 1377-4:1990*; Methods of Test for Soils for Civil Engineering Purposes—Compaction—Related Tests. British Standards Institution: London, UK, 1990.
42. Amakye, S.Y.O.; Abbey, S.J.; Booth, C.A. DMRB Flexible Road Pavement Design Using Re-Engineered Expansive Road Subgrade Materials with Varying Plasticity Index. *Geotechnics* **2022**, *2*, 395–411. [CrossRef]
43. Amakye, S.Y.O.; Abbey, S.J.; Booth, C.A.; Oti, J. Performance of Sustainable Road Pavements Founded on Clay Subgrades Treated with Eco-Friendly Cementitious Materials. *Sustainability* **2022**, *14*, 12588. [CrossRef]
44. Harichane, K.; Ghrici, M.; Khebizi, W.; Missoum, W. Effect of the Combination of Lime and Natural Pozzolana on the Durability of Clayey Soils. *Electron. J. Geotech. Eng.* **2010**, *15*, 1194–1210. Available online: https://www.researchgate.net/publication/228501953_Effect_of_the_Combination_of_Lime_and_Natural_Pozzolana_on_the_Durability_of_Clayey_Soils (accessed on 8 July 2022).
45. Ahmadi, S.; Ghasemzadeh, H.; Changizi, F. Effects of thermal cycles on microstructural and functional properties of nano treated clayey soil. *Eng. Geol.* **2021**, *280*, 105929. [CrossRef]
46. Zihms, S.G.; Switzer, C.; Karstunen, M.; Tarantino, A. Understanding the Effect of High Temperature Processes on the Engineering Properties of Soils. In Proceedings of the 18th International Conference on Soil Mechanics and Geotechnical Engineering, Paris, France, 2–6 September 2013. Available online: https://www.cfms-sols.org/sites/default/files/Actes/3427-3430.pdf (accessed on 8 July 2022).
47. Tu, Y.; Zhang, R.; Zhong, Z.; Chai, H. The strength behaviour and desiccation crack development of silty clay subjected to wetting-drying cycles. *Front. Earth Sci. Sec. Geohazard Georisks* **2022**, *10*, 852820. [CrossRef]
48. British Lime Association. Soil Stabilisation with Lime, Cement and Other Binders. 2015. Available online: https://britishlime.org/documents/Plantworx_Presentation_Britpave_Plantworx_Stabilisation_Chemistry_Presentation_2015_2.pdf (accessed on 8 July 2022).
49. Kinuthia, J.M.; Nidzam, R.M. Towards zero industrial waste: Utilisation of brick dust waste in sustainable construction. *Waste Manag.* **2011**, *31*, 1867–1878. [CrossRef]
50. Amakye, S.Y.O.; Abbey, S.J.; Booth, C.A. Road pavement defect investigation using treated and untreated expansive road subgrade materials with varying plasticity index. *Transp. Eng.* **2022**, *9*, 100123. [CrossRef]
51. Rogers, S.B. Evaluation and Testing of Brick Dust as a Pozzolanic Additive LIME Mortars for Architectural Conservation. Master's Thesis, University of Pennsylvania, Philadelphia, PA, USA, 2011. Available online: https://repository.upenn.edu/cgi/viewcontent.cgi?article=1178&context=hp_theses (accessed on 16 July 2022).
52. Hewlett, P. *Lea's Chemistry of Cement and Concrete*, 4th ed.; Butterworth-Heinemann: Oxford, UK, 2003; pp. 637–678, ISBN 978-0-7506-6256-7.
53. Amakye, S.Y.; Abbey, S.J.; Olubanwo, A.O. Consistency and mechanical properties of sustainable concrete blended with brick dust waste cementitious materials. *SN Appl. Sci.* **2021**, *3*, 420. [CrossRef]
54. Hidalgo, C.; Carvajal, G.; Muñoz, F. Laboratory Evaluation of Finely Milled Brick Debris as a Soil Stabilizer. *Sustainability* **2019**, *11*, 967. [CrossRef]
55. Manuel, A. *Development of a Rational Probabilistic Approach for Flexible Pavement Analysis*; University of Maryland: College Park, MD, USA, 1997; Volume 2.
56. van Til, C.J.; Mccullough, B.F.; Vallerga, B.A.; Hick, R.G. *Evaluation of AASHTO Interim Guide for the Design of Pavement Structures*; National Cooperative Highway Research Program: Washington, DC, USA, 1972.
57. Shaqlaih, A.; White, L.; Zaman, M. Resilient Modulus Modeling with Information Theory Approach. *Int. J. Géoméch.* **2013**, *13*, 384–389. [CrossRef]
58. Zaman, M.; Solanki, P.; Ebrahimi, A.; White, L. Neural Network Modeling of Resilient Modulus Using Routine Subgrade Soil Properties. *Int. J. Géoméch.* **2010**, *10*, 1–12. [CrossRef]
59. *AASHTO T 307-99*; Standard Method of Test for Determining the Resilient Modulus of Soils and Aggregate Materials. American Association of State and Highway Transportation Officials: Washington, DC, USA, 2007.
60. Arshad, M. Development of a Correlation between the Resilient Modulus and CBR Value for Granular Blends Containing Natural Aggregates and RAP/RCA Materials. *Adv. Mater. Sci. Eng.* **2019**, *2019*, 1–16. [CrossRef]
61. Green, J.L.; Hall, J.W. *Nondestructive Vibratory Testing of Airport Pavements. Volume 1. Experimental Test Results and Development of Evaluation Methodology and Procedure*; U.S. Department of Transportation: Washington, DC, USA, 1975.
62. Heukelom, W.; Klomp, A. Dynamic Testing as a Means of Controlling Pavements during and after Construction. In Proceedings of the International Conference on the Structural Design of Asphalt Pavements, University of Michigan, Ann Arbor, MI, USA, 20–24 August 1962.
63. Lister, N.W.; Powell, D. Design practices for bituminous pavements in the United Kingdom. In Proceedings of the 6th International Conference on the Structural Design of Asphalt Pavements, Ann Arbor, MI, USA, 13–17 July 1987.

64. Ohio Department of Transportation. *Pavement Design Manual*; Ohio Department of Transportation: Columbus, OH, USA, 2008.
65. Abbey, S.J.; Eyo, E.U.; Jeremiah, J.J. Experimental Study on Early Age Characteristics of Lime-GGBS-Treated Gypseous Clays under Wet–Dry Cycles. *Geotechnics* **2021**, *1*, 402–415. [CrossRef]

materials

MDPI

Review

Recent Developments on the Effects of Micro- and Nano-Limestone on the Hydration Process, Products, and Kinetics of Cement

Xin Li and Mingli Cao *

School of Civil Engineering, Dalian University of Technology, Dalian 116024, China; lixin_@mail.dlut.edu.cn
* Correspondence: minglic@dlut.edu.cn

Abstract: Limestone is commonly used in cement concrete due to its unique nature and type. It has physical effects (nucleation effect and dilution effect) and chemical effects on the hydration process of cement. This paper reviews the effects of three representative limestone materials on the hydration process, hydration products, and hydration kinetics. In the hydration process, the reaction was delayed by limestone powder with a particle size larger than 20 μm and calcium carbonate whiskers due to their dilutive effect. On the other hand, limestone powder with a particle size smaller than 20 m and calcium carbonate nanoparticles facilitated the reaction through nucleation and chemical effects. Limestone has a similar effect on hydration products, promoting the production of C-S-H through nucleation. The mechanism of action for this nucleation effect depends on the differences in crystalline form and particle size of the three types of micro- and nano-calcium. Chemical effects impact the amount of AFt produced, with the generation of new products being the main reaction influenced by the limestone admixture.

Keywords: cement hydration; calcium whisker; limestone powder; nano-calcium carbonate

1. Introduction

Concrete is the most widely used building material globally due to the easy availability of raw materials and low cost. However, the production of cement emits significant amounts of CO_2 and consumes large amounts of energy, which is harmful to the environment [1,2]. Therefore, researchers aim to decrease carbon emissions and energy consumption in cement production by reducing the amount of cement used in concrete [3,4]. To reduce the amount of cement used, supplementary cementitious materials (SCMs) are commonly employed as an alternative to cement, with fly ash being the most common admixture [5–7], silica fume [8], slag [9–11], rock dust [12,13], limestone [14–17], and et al. These SCMs can also enhance various properties of concrete [18]. Limestone is a calcium carbonate sedimentary rock with calcite as its principal component. Calcium carbonate is a compound that includes chalcopyrite, aragonite, and calcite [19]. Among the various forms of calcium carbonate ($CaCO_3$), calcite is the most stable at room temperature and pressure. Its crystal system is tripartite, its space group is $R\bar{3}c$, and its most stable cleavage surface is $(10\bar{1}4)$ [20,21]. This cleavage surface is particularly prevalent in geological environments and can be exploited in the production of cement clinkers [22]. Aragonite is a high-pressure phase of calcium carbonate crystals that exhibits an orthorhombic crystal system with a space group of *Pmcn* and a most stable cleavage surface of (001) [21]. It is usually widely distributed as a component of the shells of marine organisms. Depending on the differences in their size [23,24], crystalline shape [25,26], and morphology [27,28], they can play different roles in concrete, such as aggregates [29], fillers [30], and microfibers [31]. Numerous studies have shown that the addition of limestone improves various properties of concrete [14,32]. When used as an aggregate, it can effectively improve the compressive and splitting tensile strength of concrete [33,34], and when used as a filler, it can effectively improve the compactness of the matrix and enhance the performance of concrete [35], including compressive strength and durability [10,36]. Calcium carbonate whiskers (CWs), like other microfibers, can prevent the expansion of microscopic cracks [37,38]. Limestone particles smaller than 1 mm can form a ternary system with other materials that have volcanic ash activity (such as fly ash), which contributes to the hydration reaction of the cement [39]. This, in turn, improves the mechanical properties and microstructure of the concrete [40].

Numerous studies have been conducted to explain the mechanisms by which limestone affects cement hydration. The possible mechanisms that are widely accepted include dilution, nucleation, and chemical action [41]. Dilutive effects are caused by the addition of admixtures, which increase the interparticle distance and dilute the concentration of cement in a certain area. This results in a slower rate of hydration in that area [42]. The nucleation effect refers to the fact that the limestone surface is more likely to provide nucleation sites, attract free Ca^{2+} for nucleation and growth, and facilitate the production of hydration products [20,43,44]. Chemical effects mean that calcium carbonate can react with C_3A to form monocarbonaluminate and monocarbonaluminate (Hc/Mc) and participate in the hydration reaction [17]. The three types of action described above work together in the limestone–cement system to determine the role of calcium carbonate in the hydration process.

Numerous studies have investigated the impact of calcium carbonate on the hydration process and products. This paper reviews the role and factors influencing the hydration reaction and products of cement by limestone with different morphologies, particle sizes, and crystal types. Additionally, the literature summary presents some expectations for future research. There are still many deficiencies in the studies on the role of micro- and nano-calcium in the hydration of cement. Few studies have investigated the influence of CWs on the phase composition and morphology of the products during the hydration process. There are no systematic studies on the quantitative characterization of the influencing factors of micro- and nano-limestone in the hydration process.

As micro- and nano-calcium are micro-active dopants, the additional nucleation area provided by them is not quantitatively characterized in relation to the enhancement of nucleation rate, which also leads to the lack of accurate quantitative representation of the role of micro- and nano-limestone in the hydration kinetics. The role of calcium carbonate whiskers in hydration kinetics and product morphology has not been specifically analyzed and concluded, and can be studied in more depth in future research.

2. Influence Mechanism

Limestone affects the hydration process of cement through dilution, nucleation, and chemical effects. These effects are often interdependent in hydration reactions, and the factors influencing them are complex. This section discusses the modes of action and influencing factors of these effects.

2.1. Dilution Effect

The dilutive effect is widespread in mineral admixtures [42,45]. The addition of mineral admixtures typically results in a reduction in cement content, which dilutes the silicate cement content in certain areas [30,42]. This increases the space available for the formation of hydration products, leading to a decrease in the supersaturation of the pore solution required for C-S-H generation [46]. As a result, the cement is more readily hydrated under the influence of the dilutive effect [30]. However, the dilutive effect results in a less compact cement matrix due to the need for higher pore solution supersaturation during hydration in smaller-diameter pores [30,47]. The dilutive effect is solely physical and is influenced primarily by particle size and dosage [32]. Increasing the dosage and particle size enhances the dilutive effect [32,48], as shown in Figure 1. When the average particle size of limestone is larger than that of the cement clinker, the hydration rate is drastically reduced, in which case the dilutive effect does not promote the early hydration reaction [32].

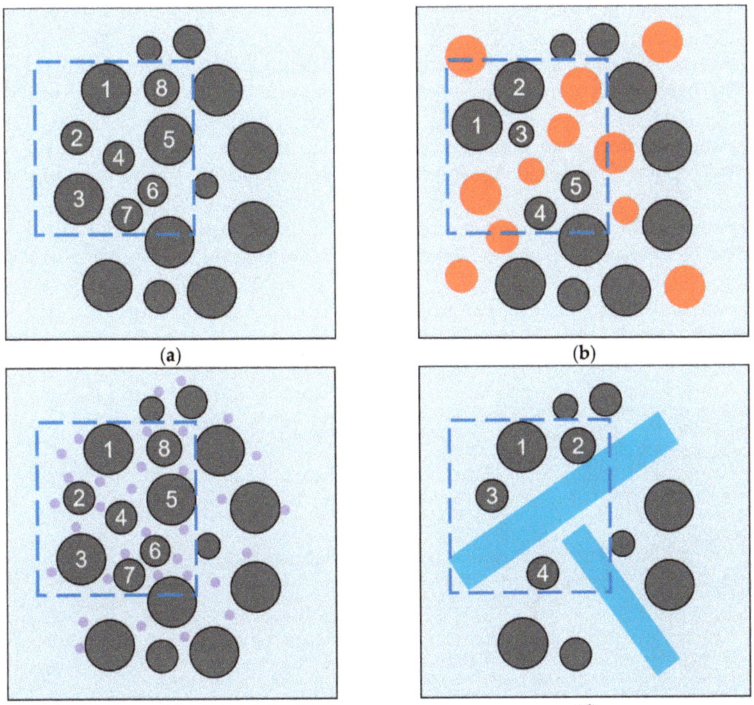

Figure 1. Schematic representation of the dilutive effects for the different blended systems; numbers 1–8 represent the number of cement clinker particles in a fixed area. Orange represents limestone, purple represents calcium carbonate nanoparticles, and blue represents calcium carbonate whiskers. (**a**) Cement; (**b**) Cement–limestone; (**c**) Cement–nano-limestone; (**d**) Cement–carbonate calcium whisker.

2.2. Nucleation Effect

The impact of nucleation on the hydration reaction is typically facilitated in two ways. Firstly, calcite (10$\bar{1}$4) has a surface atomic arrangement similar to that of C-S-H, which enables it to adsorb Ca^{2+} and allow C-S-H to nucleate and grow on its surface [49,50]. Secondly, limestone promotes nucleation by dissolving Ca^{2+} so that C-S-H reaches the required degree of supersaturation for nucleation [32,51]. The addition of calcite enhances the dissolution of alite, which, in turn, promotes the nucleation of C-S-H [32,41]. The two approaches described above work together in the hydration reaction. Calcite promotes C-S-H nucleation, as demonstrated by the calculation of its saturation index. The saturation index of C_3S is lower than 0.04, while the saturation index of calcite is $\geq$0.25. This suggests that C-S-H tends to nucleate on the surface of C_3S only when the SI is lower than 0.04 [50,52]. When the water–cement ratio is between 0.35 and 0.5, the pH range of the pore solution makes C-S-H more likely to undergo heterogeneous nucleation [50,52,53] rather than secondary nucleation on C-S-H or C_3S surfaces [50]. Therefore, during the experimental process, C-S-H shows a preference for heterogeneous nucleation on the calcite surface. When C-S-H nucleates heterogeneously on the calcite surface, the required degree of supersaturation for C-S-H nucleation is reduced, shortening the induction period [51,54]. Additionally, the presence of limestone provides a larger heterogeneous nucleation surface for the hydrated products, facilitating their nucleation and growth [51].

The nucleation effect mechanism suggests that the crystalline form of limestone is the primary factor affecting the nucleation effect. The calcium ions of calcite are in the same surface area as the oxygen ions atoms, so that the arrangement of both types of atoms can be observed simultaneously in the (10$\bar{1}$4) solvation surface, as shown in Figure 2a,b. Calcite has a similar atomic arrangement to that of C-S-H and, therefore, has a more pronounced nucleation [55]. On the other hand, the calcium ions of aragonite are not in the same plane as the oxygen ions, so only the calcium ion arrangement can be observed on the (001) solvation surface and the oxygen ion arrangement cannot be observed, and, therefore, aragonite does not have a specific facilitating effect on C-S-H nucleation [21]. The pseudo-hexagonal pattern in Figure 2d is inconsistent with all the patterns in calcite, which also leads to the very different crystal properties of aragonite and calcite. However, mineral admixtures, such as aragonite, can still facilitate C-S-H nucleation by providing a larger area for hydration product nucleation [56]. In addition to the differences in crystal structure, the amount of limestone doping also affects the nucleation effect. As the amount of limestone doping increases, the nucleation density of C-S-H increases while the size of C-S-H decreases, as shown in Figure 3. Additionally, the amount of C-S-H generated also increases with the increase in doping [49,54].

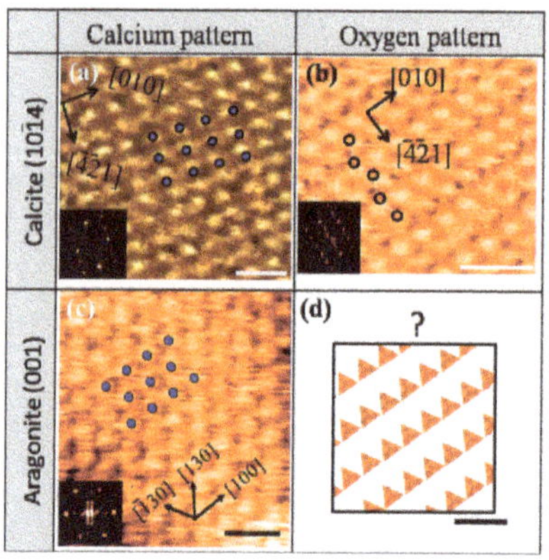

Figure 2. Comparison of atomically resolved and Fourier-transformed images of calcite cleavage (10$\bar{1}$4) and aragonite (001) faces. On the calcite cleavage plane, (**a**) there is a latticed calcium layer and (**b**) a zigzag oxygen pattern, and (**c**) the calcium layer was only observed on the (001) face of aragonite [21], (**d**) oxygen pattern iobserved on the (10$\bar{1}$4) face of aragonite.

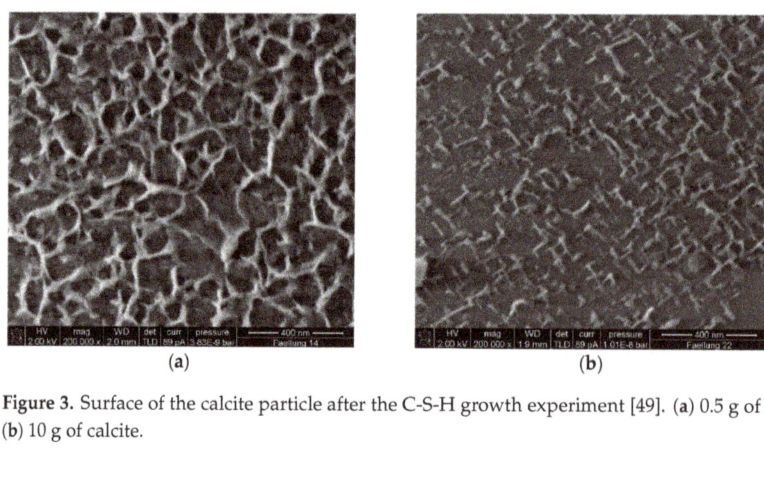

(**a**) (**b**)

Figure 3. Surface of the calcite particle after the C-S-H growth experiment [49]. (**a**) 0.5 g of calcite; (**b**) 10 g of calcite.

2.3. Chemical Effect

In systems containing limestone, which is slightly soluble in water [57], excessive tricalcium aluminate (C_3A) can react with it to form hemicarbonaluminate (Hc) and monocarbonaluminate (Mc) [17]. The specific reactions are illustrated in Equations (1)–(3) [58].

$$C_3A + CC + 11H \rightarrow C_4ACH_{11} \quad (1)$$

$$C_3A + 0.5CC + 0.5CH + 11.5H \rightarrow C_4ACH_{0.512} \quad (2)$$

$$C_3A + CaCO_3 \rightarrow Mc/Hc \text{ (mono/hemicarboaluminate)} \quad (3)$$

Equation (1) shows that the generation of Hc can consume calcium hydroxide and promote the hydration reaction by reducing the amount of hydration products. Additionally, the presence of limestone allows excessive C_3A to participate in the reaction to produce Hc and Mc, which prevents the conversion of ettringite (Aft) to AFm and stabilizes Aft [32]. The limited improvement of limestone's reaction to C_3A during the later stages of the hydration process [32] is primarily due to the fact that the majority of the C_3A in pure cement undergoes hydration. The chemical reaction is affected by two main factors: the amount of limestone and C_3A in the system and the reaction kinetics [39]. The chemical reactions are minimally affected by the properties of the limestone itself, and differences in crystal type and particle size do not have a major effect on the chemical effects [26,32,55]. The aluminum phase content in the system is typically increased using mineral admixtures that are rich in alumina and blended with limestone. This promotes chemical effects [58,59]. The inclusion of the aluminum phase from the mineral admixtures resulted in a significant increase in the production of AFt. Additionally, the production of Mc also increased [30], as illustrated in Figure 4.

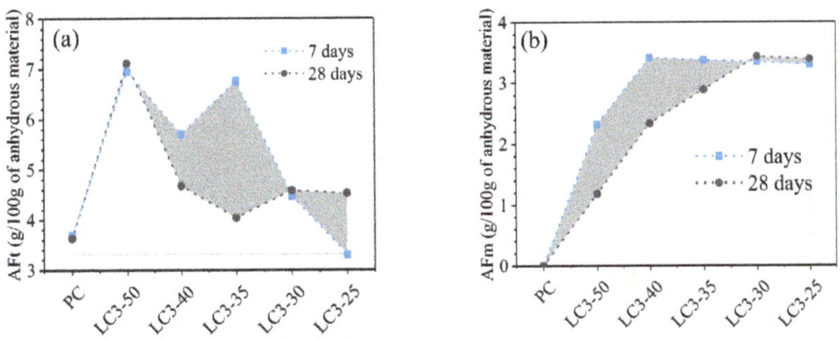

Figure 4. Quantified AFt (**a**) and AFm (**b**) phases of the investigated mixtures based on XRD Rietveld refinement, the gray area represents the amount of change in AFt and AFm from 7 to 28 days [30].

2.4. Discussion

The impact of the three types of limestone on hydration products was comparable. Nucleation effects promoted the production of C-S-H, while chemical effects influenced the amount of AFt and the production of new products. Although limestone powder and nano-calcium carbonate have similar effects on hydration products, the difference in particle size between the two leads to a difference in the mechanism of influence. The particle size of limestone powder is similar to that of C_3S. As a result, C-S-H attaches to its surface and grows vertically. This promotes the dissolution of C_3S and the nucleation of C-S-H. In contrast, nano-calcium carbonates are much smaller than C_3S particles. Therefore, they can contribute to the secondary nucleation of C-S-H by adhering to the surface of C_3S and C-S-H, which also results in C-S-H usually having a higher density in the systems containing nano-calcium carbonate.

3. Influence of Limestone on Hydration Reactions

Limestone with a particle size below 1 mm generally plays a diluting, nucleating, and chemical role in the cement hydration process [59,60]. Dilution usually slows down the hydration rate and delays the exothermic hydration starting [26,32]. The nucleation effect usually results in an increase in the hydration rate with an increase in the peak exothermic value of hydration [16,61,62].

3.1. Limestone Powder (LP)

Limestone powder (LP) typically has a particle size between 10 and 100 μm, which is smaller than the average particle size of a cement clinker [16]. LP can, therefore, affect the hydration reaction through physical and chemical effects. Table 1 presents a review of the impact of varying sizes of limestone powder (LP) on the cumulative heat release, peak hydration heat release, and peak occurrence time.

Table 1. Exothermic changes in hydration.

Reference	W/C	Particle Size	Dosage	Cumulative Heat Release	Peak Heat of Hydration	Appearance Time
[16]	0.5	10–100 μm	5 wt%	Reduced by 6%	-	About 0.6 h ahead of schedule
			15 wt%	Reduced by 18%	-	About 0.8 h ahead of schedule
			25 wt%	Reduced by 17%	-	About 1.1 h ahead of schedule
[45]	0.4	6.5 μm	10, 20, 30, 40, 50 wt%	85% increase at 30 wt%	About 28% improvement at 30 wt%	About 2.5 h ahead of time
		3.5 μm		71% increase at 30 wt%	About 30% improvement at 30 wt%	About 3.5 h ahead of schedule
		9 μm		Approx. 63% improvement at 30 wt%		About 2.5 h ahead of time
		3 μm		-	About 5% improvement at 10% dosing	
		15 μm		-	Improvement of about 1% at 10% dosing	Approximately 1% ahead of schedule
[32]	0.4	2 μm	20 wt%	-	Improvement of about 11%	Approximately 13% ahead of schedule
		130 μm		-	Reduction of about 19%	Delayed by about 16%
[63]	0.4	1–10 μm	25 wt%	Improvement of about 28%	Improvement of about 32%	Approximately 19% ahead of schedule
			50 wt%	Improvement of about 19%	Improvement of about 42%	Approximately 44% ahead of schedule

Table 1 shows that dosage and particle size are the main factors influencing the exothermic hydration in LP. For LP with a particle size below 20 μm, appropriate doping can increase the total amount of exothermic hydration, increase the peak rate of exothermic hydration, and advance the time of peak exothermic hydration [7,45,64]. Aqel et al. [64] demonstrated that the heat of hydration increased with the addition of LP, and this increase was inversely proportional to the particle size. The hydration reaction was promoted by LP, resulting in a shorter induction time, mainly due to the nucleating effect, leading to the second exothermic peak advancing [65,66]. However, for LP with a particle size larger than 20 μm, the addition will significantly reduce the hydration exotherm [16,26,32], and variations in the above results are mainly due to LP particle size selection. Briki et al. [32] conducted a study on the effect of LPs with different finenesses on early hydration. Results showed that the packing effect of LPs with a particle size of 2 μm compensates for the dilution effect, which is attributed to the fact that finer LPs increase the undersaturation of alite, which results in a faster dissolution rate of alite and promotes the nucleation effect of C-S-H on the surface of LPs [41,52], leading to higher exothermic hydration of alite.

After the dosage of LP exceeded 20 wt%, the peak rate of the exothermic rate of cement hydration decreased, and the time of peak appearance was delayed [16,25,66], as shown in Figure 5. This reduction in the exothermic rate of hydration was mainly due to the diluting effects of LP [25,39]. Although LP provides more nucleation sites for C-S-H with increasing doping, the addition of LP also increases the effective water–cement ratio [25]. This dilutive effect outweighs the nucleation effect [32]. However, increasing the dosage also promotes chemical effects and the production of Hc and Mc, which favors the formation of hydration products [16].

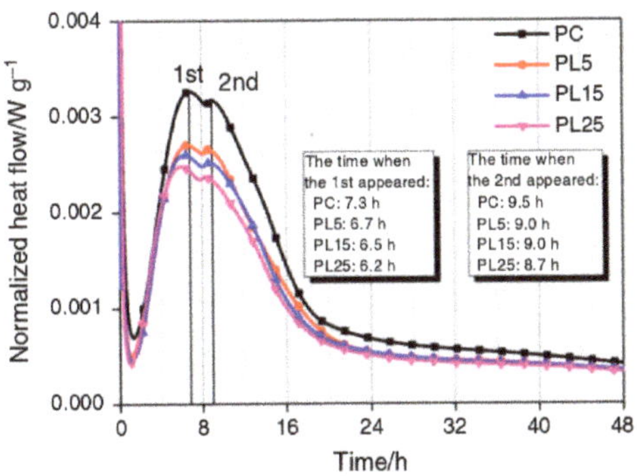

Figure 5. Hydration heat flow curve of the LP Portland cement (PC) binary system [16].

3.2. *Calcium Carbonate Whiskers (CWs)*

Calcium carbonate whiskers are typically fibrous, with diameters ranging from 0.5 to 2 μm and lengths from 20 to 30 μm [67]. The crystalline form of these whiskers is usually aragonite, which can undergo a transformation to calcite at 450 °C [68,69]. Calcite whiskers were obtained by Li et al. through high-temperature treatment of the CW. The temperature was raised to 450 °C, kept for 2 h, and then cooled naturally [26]. For the reaction peak of alite, the calcite CW accelerated the appearance time of its peak and increased the intensity of the exothermic peak of the C_3S reaction compared to the aragonite CW. However, for the second exothermic peak, the peak intensities and appearance times of the two were similar.

It has been demonstrated that, similar to larger LPs, the addition of CWs mainly has a dilutive effect and reduces the exothermic heat of hydration. This suggests that the length dimension of the CWs is the main factor controlling the variation of the heat of hydration. Li et al. [26] used XRD to determine the chemical products, Hc and Mc, as shown in Figure 6. The peak intensities of both were found to be essentially the same, which is consistent with the results of the hydration exotherm. Therefore, it can be concluded that crystallographic differences primarily affect the hydration reaction through physical effects, which influence the nucleation of C-S-H on the CW surface. This suggests that differences in crystallographic shape have minimal impact on the chemical effects [26,30].

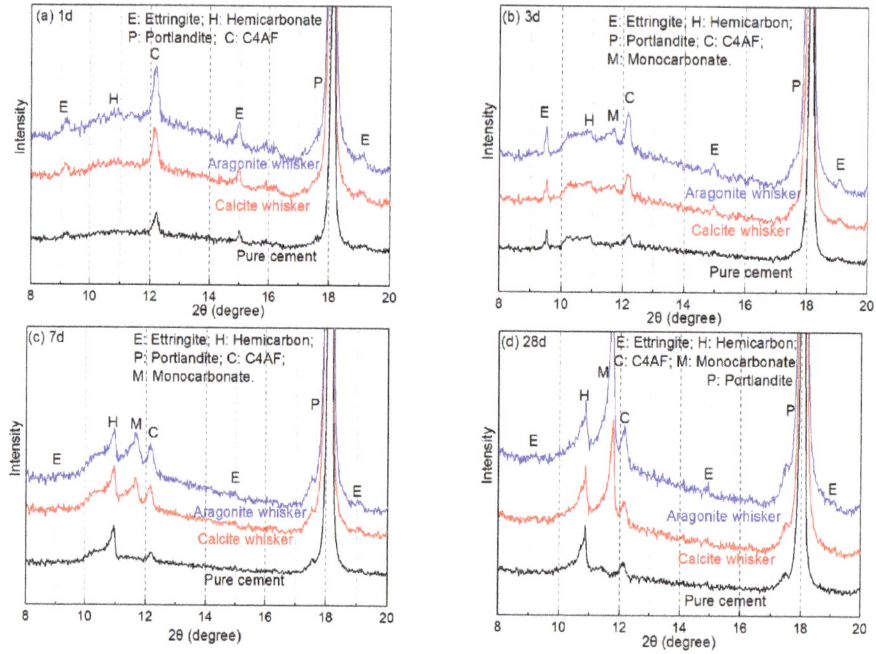

Figure 6. XRD patterns of aragonite and calcite whisker-reinforced cement paste [26]. (**a**) 1 d of hydration; (**b**) 3 d of hydration; (**c**) 7 d of hydration; (**d**) 28 d of hydration.

3.3. Nano-Calcium Carbonate (NC)

The size of calcium carbonate nanoparticles typically ranges from 10 to 105 nm, with irregular shapes and an average size of approximately 50 nm. The particles are typically cubic or connected in agglomerated chains [24,70], and XRD test results indicate that the primary constituent is calcite [24]. Currently, some researchers have chosen to pass gaseous CO_2 during the mixing process to produce calcium carbonate nanoparticles. These are known as in situ grown calcium carbonate nanoparticles (in situ NC), and in addition to solid-state calcium carbonate nanoparticles [71], they increase the density of hydration products [72]. However, the addition of CO_2 only slightly enhances cement hydration exotherm, which suggests that in situ NC does not significantly affect hydration exotherm [73]. This may be due to the fact that in situ NC generation occurs simultaneously with the hydration reaction.

The addition of NC can dramatically increase the exothermic rate of hydration while advancing the appearance of the alite reaction peak and shortening the induction period [74]. NC has a high surface energy, which can promote ion migration by adsorbing Ca^{2+} released from C_3S dissolution and thus shorten the induction period [24,74,75]. The incorporation of NC results in a reduction of approximately four hours in the dormant period [75]. Additionally, NC can act as a nucleation site for C-S-H, promoting its growth and playing a nucleation effect [74]. However, the high surface energy of NC makes it prone to agglomeration in the cement matrix, which negatively affects the properties of cementitious materials [9,76]. Differences in the mode of dispersion also led to changes in the early hydration kinetics, with ultrasonically dispersed NC facilitating the hydration reaction [77]. When NC is more uniformly dispersed, it can promote early hydration, increase the hydration rate, and increase the degree of hydration (Doh) [78]. Currently, the recommended dosage of NC in cementitious materials ranges from 1 wt% to 4 wt% [9,79]. Excessive NC dosage can result in agglomeration, preventing proper dispersion and leading to lower exothermic values [61,75]. It has been demonstrated that ultrasonication-dispersed NC can enhance hydration by providing additional nucleation sites [80]. Furthermore, NC can also have a chemical effect by dissolving CO_3^{2-} to react with C_3A [24]. Wu et al. [81] also found that chemical effects can advance the second exothermic peak and increase the exothermic values of hydration.

3.4. Comparison of the Effect of Different Calcium Carbonates on Hydration Reactions

Table 2 presents a detailed comparison of the primary functions of various forms and grain sizes of limestone in the process of hydration. The presence of the nucleation effect in limestone is widespread when the particle size is below 100 μm. The surface composition of limestone is more similar to that of C-S-H gels [20,62] compared to quartz, making it more susceptible to the adsorption of hydration products for nucleation [32]. Additionally, the inclusion of limestone promotes the easier dissolution of C_3S, which, in turn, enhances hydration reactions [32]. The dilution effect typically occurs in LP and CWs when using high dosages and large particle sizes. According to Ahmed et al. [30], the promotion of the hydration reaction through filling and nucleation effects is offset by the dilution effect when replacing more than 30wt% of the cement proportion. When the average particle size of LP is larger than that of the cement clinker, its dilutive effect becomes more apparent in the hydration reaction, resulting in a decrease in the rate of the reaction [32]. The atomic arrangement of aragonite and calcite has a significant impact on the hydration reaction, with the crystalline form of limestone also contributing to the hydration reaction, which is mainly due to differences in the atomic arrangement of the two decisions. Calcite has a Ca and O atom arrangement more similar to that of the C-S-H gel surface, which can attract free Ca^{2+} to nucleate and grow and promote the generation of C-S-H [20,43,44]. The effect of aragonite on the hydration reaction proceeds mainly through the dissolution of CO_3^{2-} for chemical reactions [26,55].

Table 2. Effect of micro- and nano-calcium carbonate on cement hydration.

Reference	Type	Particle Size	Dosage (wt%)	Mechanism of Action
Cao et al. [37]	CW	length 20–30 μm, diameter 0.5–2 μm	5%, 10%, 15%, 20%	nucleation effect
Ming et al. [82]	CW	length 20–30 μm	10%	nucleation effect, chemical effect
Sato et al. [61]	NC	50–120 nm	10%, 20%	nucleation effect, chemical effect
Ouyang et al. [28]	LP	9 μm	30%	nucleation effect, dilutive effects
Aqel et al. [69]	LP	17 μm,12 μm, 3 μm	5%, 10%, 15%	dilutive effects
Berodier [83]	LP	2 μm, 15 μm,	40%	nucleation effect, dilutive effects
Zemei Wu et al. [81]	NC	15–105 nm	1.6%, 3.2%, 4.8%, 6.4%	nucleation effect
Li et al. [26]	aragonite CW calcite CW	length 20–30 μm, diameter 0.5–2 μm	5%, 10%	nucleation effect, chemical effect, dilutive effects

4. Effect of Calcium Carbonate on Hydration Products

The effects of calcium carbonate on the hydration reaction alter the type and morphology of the hydration products. The morphology of the hydration products has been influenced by physical effects. Nucleation effects have promoted the generation of hydration products [32], while dilutive effects have negatively impacted the compactness of the cement matrix [30]. Chemical effects generate Hc and Mc, which also grow in the cement matrix, altering the microscopic morphology of the hydration products [32,84].

4.1. Limestone Powder (LP)

The addition of LP may impact the hydration products in terms of physical and chemical effects. When the nucleation effect dominates, the production of C-S-H increases. The addition of LP refines the hydration products [84,85], resulting in shorter and coarser needles and rods of C-S-H [15]. In contrast, if the dilution effect dominates, the compactness of the cement matrix decreases, leading to a significant reduction in C-S-H production [30]. In contrast to the C_3S surface, C-S-H on the calcite surface exhibits oriented growth, as depicted in Figure 7. This growth is primarily attributed to the nucleation effect of calcite [51]. Furthermore, the addition of LP reduces the production of CH, which gradually decreases with increasing LP doping [35,86].

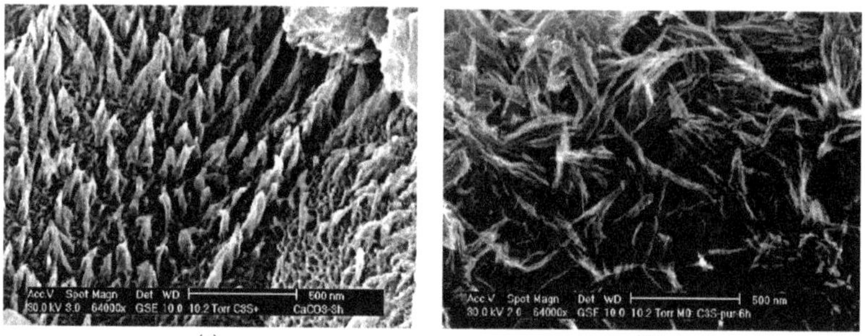

(**a**) (**b**)

Figure 7. The microstructure of C-S-H during the hydration of C_3S and calcite powder [51]. (**a**) Growth of C-S-H on calcite; (**b**) Growth of C-S-H on C_3S.

The chemical effects of LP result in the production of Hc and Mc, which consume excess C_3A [32]. However, it should be noted that Hc is not always stable and can decrease as the reaction age increases [16]. Additionally, the content of Hc can be stable when the content of blended calcite-type LP is lower than 2–3 wt% [84]. Mc and Hc are generated in the interfacial zone between the LP and the cement matrix, forming firmly consecutive crystalline aggregates [84]. Furthermore, the chemical effects stabilize the presence of AFt and prevent its conversion to AFm [30,87,88]. This is advantageous for the properties of cementitious materials, as AFt has a higher strength and larger solid phase volume. To stabilize AFt and promote chemical effects, researchers often elect to introduce alumina-rich admixtures, such as metakaolin, fly ash, and calcined clay, to partially replace cement by blending with LP [63,65,88]. In such ternary systems, the dilution effect is not obvious and, in addition, since the generation of Hc and Mc requires the consumption of CH, it will promote hydration reactions, thus increasing the hydration of the cement clinker [30] and making the cement matrix denser.

The incorporation of LP into hydration products resulted in a decline in compressive strength, which became more pronounced with an increase in LP dosage. The 28d compressive strength of the PC-LP system was observed to be inferior to that of the control when the LP dosage exceeded 20 wt% [14]. This phenomenon can be attributed to the dilution effect of a high dosage of LP, which plays a dominant role in the dilution process [30,64]. When SCMs with volcanic ash activity are co-mingled with LP to participate in the hydration reaction, a small increase in the early compressive strength of this ternary system is observed due to the synergistic effect of the nucleation surface provided by LP [7,11]. This happens while the SCMs increase the activity of LP and enhance the hydration process [10].

4.2. *Calcium Carbonate Whiskers (CWs)*

CW bridges and deflects microcracks in the cement matrix, resulting in a denser matrix [27,89]. Furthermore, the inclusion of the CW decreases the orientation index of CH, indicating that it restricts the area available for CH crystal growth and improves the density of the cement paste. When comparing the calcite CW with the aragonite CW, it can be observed that the calcite CW has a rougher surface with more hydration products than the surface of the aragonite CW [26], as shown in Figure 8. The CW also has chemical effects and can react with C_3A to produce Hc and Mc during hydration [26,55]. Additionally, aragonite, which is more soluble than calcite [69], can dissolve more CO_3^{2-} and participate in the chemical reactions. The CW also has synergistic effects with alumina-rich materials. It provides a calcium source for the volcanic ash reaction of fly ash, while fly ash also provides aluminates for the chemical effects of the CW [82].

(a) (b)

Figure 8. Microstructure of CW-reinforced hardened cement paste [26]. (**a**) 10% aragonite type CW; (**b**) 10% calcite type CW.

The addition of the CW has a small effect on C-S-H generation, and due to its larger volume, it mainly acts as a diluent and reduces the concentration of the cement clinker, as the compressive strength decreases with increasing CW dosage [26,37,82]. The calcite whiskers obtained after the treatment had the most obvious effect on compressive strength improvement [26]. When the CW addition was 40 vol%, the strength was reduced by about 18.65% [31]. A total of 10% of the CW can increase the 28d compressive strength of mortar by 13%, effectively improving the physical and mechanical properties of mortar [90]. The CW is more often used with the rest of the fibers to prepare multi-scale hybrid fiber-reinforced concrete (MHFRCC) [91]. The CW is more commonly used with other fibers in the production of multi-fiber hybrid-reinforced concrete (MHFRCC) [38]. The commonly used fibers include polyvinyl alcohol (PVA) fibers [91,92], steel fibers (SFs), and hybrid fibers (HFs) [93], and the combination of the CW and PVA fibers with steel fibers can effectively improve the mechanical properties of MHFRCC under static loading and also increase the compressive strength of the mortar [91,94]. For hybrid fibers, the CW has good synergy with hybrid fibers of different lengths, and 10% of the CW can increase the compressive strength by about 7% [93]. However, by adding 1% of the CW, 0.45% HF, and 0.36% SF, the compressive strength can be increased by 43% compared to PC [95].

4.3. *Nano-Calcium Carbonate (NC)*

The effect of NC on hydration products is similar to that of LP, which can stabilize the presence of AFt by chemically generating Hc and Mc and consuming CH to promote the hydration reaction to proceed [39]. At lower levels of NC agglomeration, NC promotes early hydration reactions and the generation of C-S-H [78]. The addition of NC also affects the growth of CH by decreasing the growth of CH{001} facets and accelerating the growth of CH{101} facets, leading to a decrease in hexagonal plates and an increase in the amount of prismatic CH [24]. On the other hand, the addition of NC increased the CO_3^{2-} content in the cement matrix, and the dissolution of CO_3^{2-} could replace SO_4^{2-} in AFt, forming a crystal structure similar to that of AFt [24].

NC is connected to the hardened cement paste with a smaller interfacial transition zone [9,96]. The nucleation of NC on the surface of C-S-H gels promotes the secondary generation of C-S-H gels [97]. This also allows the twice-generated C-S-H gels to be cross-coated with the hydration products of NC, resulting in the generation of ultra-high-density C-S-H gels [24]. This reduces the percentage of low-density C-S-H gels and improves the bulk density of the gels [98]. The effect is more pronounced in the early stages of cement hydration [24,99].

The addition of NC improved the compressive strength as it promoted the formation of ultra-high-density C-S-H gels [24]. With increasing dosage, the compressive strength showed a tendency to increase and then decrease, and when the dosage was greater than 4 wt%, the compressive strength of the NC cement system was lower than that of

PC [76,80,100]. When the dosage of NC was 3.2 wt%, its effect on compressive strength was most obvious [96,101], and it improved the strength growth rate by 1-3d [81,100]. Since the effect of NC on strength was mainly expressed by the increase in C-S-H density, it mainly changed the compressive strength in the pre-hydration period, and the compressive strength was basically unchanged after 28d [81].

By comparing NC and LP, it is possible to observe the difference in the effect of micron-sized limestone and nanosized limestone on hydration products, as illustrated in Figure 9. For LP, the particle size is comparable to that of C_3S. Therefore, C-S-H will adhere to its surface and grow vertically until the C-S-H size reaches a critical length of about 400 nm [51]. Because the surface of calcite is more prone to act as a nucleation site, the density of C-S-H growth on the surface of C_3S is lower, and the surface of C_3S is not fully covered by hydration products [49,51]. This makes it easier to dissolve. Compared to systems without limestone, the rate of hydration reaction is faster. The effect of NC on nucleation is mainly reflected in its ability to destroy the silicon-rich layer on the surface of C_3S while reducing the ionic concentration around the silicon-rich layer, thus shortening the induction period and promoting the hydration reaction, and at the same time, it can also adhere to the surface of C-S-H and promote the nucleation of C-S-H [24]. The addition of NC typically results in the growth of more C-S-H on its surface, with nucleation occurring on the C_3S surface. This leads to the formation of denser C-S-H gels.

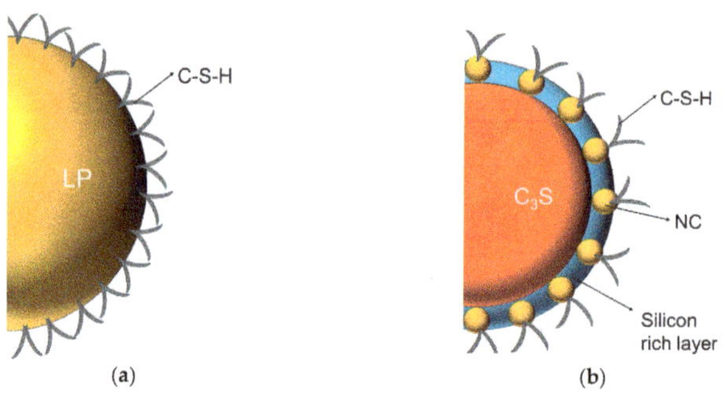

Figure 9. Schematic representation of the nucleation of C-S-H in limestone powder and nano-calcium carbonate. (**a**) Limestone powder (LP); (**b**) Nano carbonate calcium (NC).

4.4. *Effect of Limestone on Durability*

The primary component of limestone is calcium carbonate ($CaCO_3$), which results in limestone primarily influencing the sulfate resistance of cementitious materials [102]. Sulfate reacts with the C-S-H gel produced by cement hydration to form products, such as calomel and gypsum, in a humid environment in the presence of CO_3^{2-}, leading to expansion and cracking of cementitious materials [102,103]. Limestone provides carbonate ions for this process, and thus the addition of LP generally has a detrimental effect on the resistance of cementitious materials to sulfate attack [104]. When the LP dosage is increased to 15 wt%, it has a noticeable effect on the durability of the cementitious materials [103]. The expansion rate was faster, and the damage occurred earlier than in PC [103,104]. Low temperatures also result in a reduction in the sulfate attack resistance, which is facilitated by a slight increase in CO_2 at 0–5 °C, leading to an increase in the CO_3^{2-} content [36,103]. The sulfate attack resistance of the LP-PC system can be effectively enhanced by the incorporation of the remaining SCMs [105]. Following the addition of metakaolin, the concrete samples subjected to sulfate attack, and the overall properties and apparent morphology exhibited significant improvements in comparison to those of PC and LP-PC [106].

In addition, chloride ion permeability is an important index for evaluating the durability of cementitious materials. The increase in LP doping leads to a deterioration of the chloride ion permeability of the cementitious materials [106]. Nevertheless, when chloride ions and sulfate ions are present simultaneously, the presence of chloride ions reduces the extent of sulfate attack. The complication of the co-existence of Cl^- and SO_4^{2-} is mainly found in seawater erosion. Furthermore, chloride ions penetrate deeper into the matrix than sulfate ions [107]. In addition to chloride erosion, Nadelman et al. [108] investigated the impact of limestone on the physical salt erosion induced by the addition of LP at higher water–cement ratios (0.6). This resulted in the refinement of the pore structure and the observation of more expansion cracking due to salt crystallization. In addition to salt attack, the performance of cementitious materials under high-temperature conditions is also an important index for evaluating their suitability for such conditions [39]. The incorporation of NC and CWs can effectively improve the performance of cementitious materials after high temperatures [109,110].

5. Numerical Modeling of Hydration

Numerical simulation can be used to explore the hydration process by calculating the hydration reactions, determining the roles of each reactant in the process [111–114], identifying the factors that control the transformation of each hydration process [115], and understanding the mechanism by which external factors or admixtures influence the hydration process [40,48]. Thermodynamic calculations are the main method used today for the numerical modeling of limestone hydration.

5.1. The Thermodynamics of Hydration

Thermodynamic calculations can provide a reliable representation of the phase composition and the chemical composition of a system at a given temperature and pressure [116], characterize the effect of the external environment on the phase composition of the cement and hydration products, and determine the degree of cement hydration at the time of interest [115]. Currently, the main software used includes ① GEMs, and the corresponding databases include Cemdata07 [113,117], Cemdata14, and Cemdata18 [116]. ② PHREEQC version 3 software, as a commonly used geochemical calculation software, can also be used to calculate the thermodynamics of cement hydration [118,119], which mainly uses the PHREEQC database with the HATCHES database [120].

GEM computational simulations support the computational simulation of many hydration processes, including geopolymers, ordinary Portland cement [115], and Portland cement [121]. Matschei [17] carried out thermodynamic calculations on the hydration reaction system involving calcite and determined that calcite is involved in the system, as shown in Figure 10. Initially, the formation conditions of Hc and Mc were derived. Based on this study, some researchers have further investigated the specific generation conditions and existence states of Hc and Mc using GEMs [116,119], which showed that the chemical compositions of Hc and Mc are independent of the $CaCO_3$ content [116] and that Hc and Mc can be further converted to hydrogarnets [113].

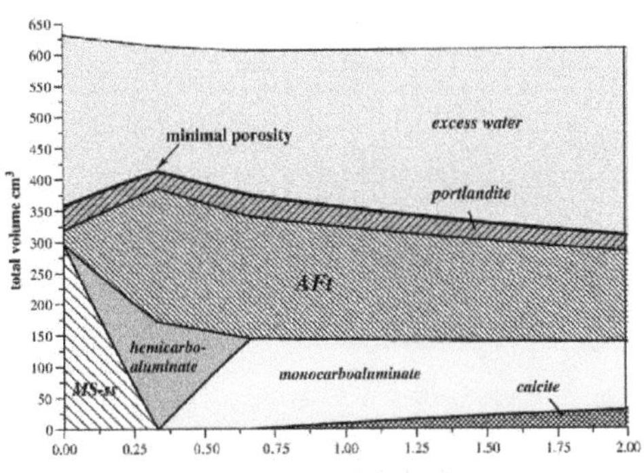

Figure 10. Variation of hydration products with the CO_2/Al_2O_3 ratio under constant total solid conditions [17].

5.2. Hydration Kinetics

Blending limestone with other SCMs that are rich in aluminum and partially replacing cement can result in significant performance gains due to the chemical effects of limestone [122]. Further research was conducted to investigate its effect on hydration kinetics and its role in ternary systems [30,40]. Kunther et al. [111,123] used ^{27}Al MAS NMR and ^{29}Si MAS NMR to determine the products at different reaction ages based on GEMS. They optimized the results of the thermodynamic tests and proposed to carry out the reaction kinetics calculations of the components in the ternary system. These calculations can be used to estimate the extent of the reaction, with the equations shown in Equations (4)–(6) [40,111]. Combining kinetic calculations with GEMs effectively improves the accuracy of thermodynamic calculations, and experimental results fit well with simulation results [123]. Figure 11 shows the kinetic calculations for ternary systems, which typically focus on the generation of hydration products at different reaction ages. Alite has the highest degree of hydration and the fastest rate of increase in hydration for the same hydration age when metakaolin (MK) and limestone (LS) masses are constant. Blite has a slower hydration reaction and a low degree of hydration [111]. The degree of hydration and the rate of reaction of MK, on the other hand, increased as the mass of LS increased, mainly because MK would react with LS to promote hydration [88,111]. The results indicate that when there is a sufficient alumina phase in the system, the hydration products are dominated by C-S-H gels. This is because the later reaction between the biotite and limestone consumes part of CH [86].

$$Q_i(t) = Q_{i0} + \bar{k}_i exp \cdot (-n_i/t) \tag{4}$$

$$Q_{i0} = q_i \cdot \alpha(t_1) \tag{5}$$

$$\bar{k}_i = q_i \cdot [\alpha(t_x) - \alpha(t_1)] \tag{6}$$

where i is the reactant, $Q_i(t)$ and Q_{i0} are the reactants i at time t and at the start of the reaction, respectively, k_i is used to limit the possible dissolved mass during the reaction, n_i is the hydration rate parameter, q_i is the initial dissolved amount, and $\alpha(t_x)$ is the degree of hydration at the end of the reaction.

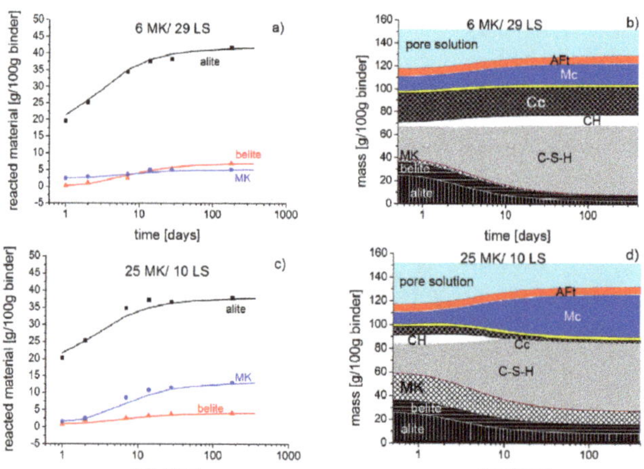

Figure 11. Comparison of the experimental (symbols) and modeled (lines) quantities of reacted alite, belite, and MK for 1 day to 182 days of hydration with (**a**) 6 wt% metakaolin and 29 wt% limestone powder, and (**c**) 25 wt% metakaolin and 10 wt% limestone powder. The composition of hydration products for 1 day to 182 days is shown in (**b**,**d**) [111].

6. Application Trend of Limestone

Currently, limestone is more commonly blended with clay and cement and used as LC3 instead of cement [124]. The preparation process of LC3 has lower carbon emissions [125], while the mechanical properties of LC3 are more excellent [126,127]. Furthermore, due to the lower content of cement in LC3, less CH is produced by hydration, which makes the durability performance of LC3 concrete better than that of cement concrete [126].

Theodore and colleagues demonstrated that the production cost of LC3 is approximately 10–20% lower than that of silicate cement, with the potential for further reduction if industrial wastes are employed in lieu of clay [128]. Additionally, the exceptional durability of LC3 will result in a reduction in the frequency of building repairs and lower maintenance costs [90]. In addition to LP, the application of the CW as a microfiber can also reduce the cost of multi-scale hybrid fiber concrete [92]. The CW is manufactured at a cost of approximately USD 200 per ton [129], so the CW can partially replace the more expensive PVA fiber with steel fiber, reducing costs [91]. Meanwhile, the inclusion of microfibers can stop the growth of microcracks in concrete in a limited way and improve the flexural properties of concrete [130]. In addition, the production cost of NC is lower compared to other nanomaterials, while NC is compatible with cementitious materials [131]. Therefore, NC is also a common choice when it comes to nano-enhancement [132]. However, nanomaterials need to be dispersed before they can exert a better reinforcing effect [79,97], and the engineering practical application is less at present.

The current application of limestone in engineering practice is primarily for use as an admixture to partially replace cement or as a component of LC3 [133,134]. Furthermore, LC3 is now also used in the construction field in the initial application. The model house in Jhansi, India, employs LC3 as a component of the cementitious material, utilizing 26.6 tons of industrial waste, a process that reduces CO_2 emissions by 15.5 tons. Additionally, in India, LC3 is utilized in the construction of road pavements [135]. In Latin America, LC3 is employed in a variety of settings, including buildings, offshore test sites, artistic sculptures, and pavements. The LC3 House in Santa Clara, Cuba, is an example of a structure that emits 30% less CO_2 during production compared to conventional concrete [135].

7. Conclusions and Outlook

7.1. Conclusions

This review examines the impact of limestone with varying morphologies, crystal types, and grain sizes on hydration reaction, products, and kinetics, as well as the factors that influence them. From this analysis, the following conclusions can be drawn:

(1) The effects of limestone on the hydration reaction can be divided into two categories: promotion through nucleation and chemical effects and delay through dilution. As the dosage and particle size of limestone powder increase, the dilutive effect becomes more pronounced. Calcium carbonate whiskers, due to their large size, primarily have a dilutive effect on the hydration reaction. Nano-calcium carbonate promotes the hydration reaction through nucleation and chemical effects due to its small particle size. The differences between the three types of limestone mentioned are attributed to variations in particle size and crystal type. Particle size affects the dilutive effect, while crystal type has a greater impact on the nucleation effect.

(2) With regard to LP, its addition primarily served to enhance the filling and nucleation effects, thereby facilitating the generation of C-S-H and improving the compressive strength. In the case of the CW, its principal role was that of a microfiber, which reduced the development of microcracks and improved the mechanical properties of cementitious materials. As NC can play a more significant filling role in smaller pores, it promotes the generation of C-S-H with CH and improves mechanical properties.

(3) Numerical simulations of hydration in multifaceted systems containing limestone typically employ thermodynamics, such as GEMs, to predict and simulate the composition of hydration products at different ages. Empirical formulas are still mostly

used for the kinetic calculation of the multivariate system, and the thermodynamic fitting results are more accurate after incorporating the kinetic calculation.

(4) There are three primary mechanisms by which limestone affects cement hydration. The first is the dilutive effect, which occurs when limestone is added to the mixture, reducing the cement content in a particular region, increasing the space available for the growth of hydration products, and promoting the hydration reaction and product formation. (2) The nucleation effect has two main aspects: ① the attraction of calcite to Ca^{2+} and ② the reduction in the supersaturation degree of the solution required for C-S-H nucleation. Therefore, the promotional effect of aragonite on C-S-H nucleation is much weaker than that of calcite. (3) The chemical effects of the reaction between limestone and excessive C_3A to produce Hc and Mc and to stabilize the AFt produced by hydration, the main factor influencing the chemical effects was the addition of limestone.

7.2. Outlook

The advent of contemporary technological innovations has led to a surge in the availability of micro- and nano-limestone. In situ NC is effectively dispersed, while the process utilizes CO_2 emitted from cement production, thereby enhancing the sustainability of the concrete preparation process. However, this approach is currently constrained to the production of calcite, and the process itself limits further investigation into the impact of nano-aragonite on the hydration reaction.

The main national standards and regulations for limestone cement set limits on the amount of limestone that can be used. However, these standards do not address the differences between different types of limestone in the application process, and future recommendations should be tailored to the different types of limestone.

Author Contributions: Writing—original draft preparation, X.L.; writing—review and editing, M.C. All authors have read and agreed to the published version of the manuscript.

Funding: This research received no external funding.

Data Availability Statement: The data in this article cannot be shared directly for article type reasons.

Conflicts of Interest: The authors declare no conflict of interest.

References

1. Ashraf, W.; Olek, J.; Jain, J. Microscopic features of non-hydraulic calcium silicate cement paste and mortar. *Cem. Concr. Res.* **2017**, *100*, 361–372. [CrossRef]
2. Yang, K.-H.; Jung, Y.-B.; Cho, M.-S.; Tae, S.-H. Effect of supplementary cementitious materials on reduction of CO_2 emissions from concrete. *J. Clean. Prod.* **2015**, *103*, 774–783. [CrossRef]
3. Selim, M.; Khalifa, R.; Elshamy, E.; Zaghlal, M. Structural Efficiency of Fly-Ash Based Concrete Beam-Column Joint Reinforced by Hybrid GFRP and Steel Bars. *Case Stud. Constr. Mater.* **2024**, *20*, e02927. [CrossRef]
4. Kanagaraj, B.; Kiran, T.; Gunasekaran, J.; Nammalvar, A.; Arulraj, P.; Gurupatham, B.G.A.; Roy, K. Performance of Sustainable Insulated Wall Panels with Geopolymer Concrete. *Materials* **2022**, *15*, 8801. [CrossRef] [PubMed]
5. Moudio, A.M.N.; Tchakouté, H.K.; Ngnintedem, D.; Andreola, F.; Kamseu, E.; Nanseu-Njiki, C.; Leonelli, C.; Rüscher, C. Influence of the synthetic calcium aluminate hydrate and the mixture of calcium aluminate and silicate hydrates on the compressive strengths and the microstructure of metakaolin-based geopolymer cements. *Mater. Chem. Phys.* **2021**, *264*, 124459. [CrossRef]
6. Miao, X.; Pang, X.; Li, S.; Wei, H.; Yin, J.; Kong, X. Mechanical strength and the degradation mechanism of metakaolin based geopolymer mixed with ordinary Portland cement and cured at high temperature and high relative humidity. *Chin. J. Chem. Eng.* **2023**, *60*, 118–130. [CrossRef]
7. Vance, K.; Aguayo, M.; Oey, T.; Sant, G.; Neithalath, N. Hydration and strength development in ternary portland cement blends containing limestone and fly ash or metakaolin. *Cem. Concr. Compos.* **2013**, *39*, 93–103. [CrossRef]
8. Muller, A.C.A.; Scrivener, K.L.; Skibsted, J.; Gajewicz, A.M.; McDonald, P.J. Influence of silica fume on the microstructure of cement pastes: New insights from 1H NMR relaxometry. *Cem. Concr. Res.* **2015**, *74*, 116–125. [CrossRef]
9. Hosan, A.; Shaikh, F.U.A.; Sarker, P.; Aslani, F. Nano- and micro-scale characterisation of interfacial transition zone (ITZ) of high volume slag and slag-fly ash blended concretes containing nano SiO_2 and nano $CaCO_3$. *Constr. Build. Mater.* **2020**, *269*, 121311. [CrossRef]
10. Arora, A.; Sant, G.; Neithalath, N. Ternary blends containing slag and interground/blended limestone: Hydration, strength, and pore structure. *Constr. Build. Mater.* **2016**, *102*, 113–124. [CrossRef]

11. Menéndez, G.; Bonavetti, V.; Irassar, E.F. Strength development of ternary blended cement with limestone filler and blast-furnace slag. *Cem. Concr. Compos.* **2003**, *25*, 61–67. [CrossRef]
12. Stempkowska, A.; Gawenda, T.; Chajec, A.; Sadowski, Ł. Effect of Granite Powder Grain Size and Grinding Time of the Properties of Cementitious Composites. *Materials* **2022**, *15*, 8837. [CrossRef] [PubMed]
13. Shamsabadi, E.A.; Ghalehnovi, M.; De Brito, J.; Khodabakhshian, A. Performance of Concrete with Waste Granite Powder: The Effect of Superplasticizers. *Appl. Sci.* **2018**, *8*, 1808. [CrossRef]
14. Ramezanianpour, A.M.; Hooton, R.D. A study on hydration, compressive strength, and porosity of Portland-limestone cement mixes containing SCMs. *Cem. Concr. Compos.* **2014**, *51*, 1–13. [CrossRef]
15. Nehdi, M.; Mindess, S. Opitimization of high strength limestonefiller cement mortars. *Cem. Concr. Res.* **1996**, *26*, 883–893. [CrossRef]
16. Ma, J.; Yu, Z.; Shi, H.; Zhang, Y.; Shen, X. Long-term hydration behavior and pore structure development of cement–limestone binary system. *J. Therm. Anal. Calorim.* **2021**, *143*, 843–852. [CrossRef]
17. Matschei, T.; Lothenbach, B.; Glasser, F. The role of calcium carbonate in cement hydration. *Cem. Concr. Res.* **2007**, *37*, 551–558. [CrossRef]
18. Madan, C.S.; Munuswamy, S.; Joanna, P.S.; Gurupatham, B.G.A.; Roy, K. Comparison of the Flexural Behavior of High-Volume Fly AshBased Concrete Slab Reinforced with GFRP Bars and Steel Bars. *J. Compos. Sci.* **2022**, *6*, 157. [CrossRef]
19. Hay, R.; Peng, B.; Celik, K. Filler effects of $CaCO_3$ polymorphs derived from limestone and seashell on hydration and carbonation of reactive magnesium oxide (MgO) cement (RMC). *Cem. Concr. Res.* **2023**, *164*, 107040. [CrossRef]
20. Rode, S.; Oyabu, N.; Kobayashi, K.; Yamada, H.; Kühnle, A. True Atomic-Resolution Imaging of (10$\bar{1}$4) Calcite in Aqueous Solution by Frequency Modulation Atomic Force Microscopy. *Langmuir* **2009**, *25*, 2850–2853. [CrossRef]
21. Araki, Y.; Tsukamoto, K.; Oyabu, N.; Kobayashi, K.; Yamada, H. Atomic-Resolution Imaging of Aragonite(001)Surface in Waterby Frequency Modulation Atomic Force Microscopy. *Jpn. J. Appl. Phys.* **2012**, *51*, 08KB09. [CrossRef]
22. Addadi, L.; Weiner, S. Control and Design Principles in Biological Mineralization. *Angew. Chem. Int. Ed.* **1992**, *31*, 153–169. [CrossRef]
23. Camiletti, J.; Soliman, A.M.; Nehdi, M.L. Effects of nano- and micro-limestone addition on early-age properties of ultra-high-performance concrete. *Mater. Struct.* **2012**, *46*, 881–898. [CrossRef]
24. Fu, Q.; Zhang, Z.; Zhao, X.; Xu, W.; Niu, D. Effect of nano calcium carbonate on hydration characteristics and microstructure of cement-based materials: A review. *J. Build. Eng.* **2022**, *50*, 104220. [CrossRef]
25. Wang, D.; Shi, C.; Farzadnia, N.; Shi, Z.; Jia, H.; Ou, Z. A review on use of limestone powder in cement-based materials: Mechanism, hydration and microstructures. *Constr. Build. Mater.* **2018**, *181*, 659–672. [CrossRef]
26. Li, L.; Mingli, C.; Hong, Y. Comparative roles between aragonite and calcite calcium carbonate whiskers in the hydration and strength of cement paste. *Cem. Concr. Compos.* **2019**, *104*, 1003350. [CrossRef]
27. Gong, P.; Zhang, C.; Wu, Z.; Zhang, G.; Mei, K.; Gao, Q.; Cheng, X. Study on the effect of $CaCO_3$ whiskers on carbonized self-healing cracks of cement paste: Application in CCUS cementing. *Constr. Build. Mater.* **2022**, *321*, 126368. [CrossRef]
28. Ouyang, X.; Koleva, D.A.; Ye, G.; Van Breugel, K. Understanding the adhesion mechanisms between C S H and fillers. *Cem. Concr. Res.* **2017**, *100*, 275–283. [CrossRef]
29. Menadi, B.; Kenai, S.; Khatib, J.; Aït-Mokhtar, A. Strength and durability of concrete incorporating crushed limestone sand. *Constr. Build. Mater.* **2009**, *23*, 625–633. [CrossRef]
30. Ahmed, A.H.; Nune, S.; Liebscher, M.; Köberle, T.; Willomitzer, A.; Noack, I.; Butler, M.; Mechtcherine, V. Exploring the role of dilutive effects on microstructural development and hydration kinetics of limestone calcined clay cement (LC3) made of low-grade raw materials. *J. Clean. Prod.* **2023**, *428*, 139438. [CrossRef]
31. Cao, M.; Wei, J. Microstructure and mechanical properties of $CaCO_3$ whisker-reinforced cement. *J. Wuhan Univ. Technol. Sci. Ed.* **2011**, *26*, 1004–1009. [CrossRef]
32. Briki, Y.; Zajac, M.; Haha, M.B.; Scrivener, K. Impact of limestone fineness on cement hydration at early age. *Cem. Concr. Res.* **2021**, *147*, 106515. [CrossRef]
33. Juenger, M.C.G.; Siddique, R. Recent advances in understanding the role of supplementary cementitious materials in concrete. *Cem. Concr. Res.* **2015**, *78*, 71–80. [CrossRef]
34. Alhozaimy, A.M. Effect of absorption of limestone aggregates on strength and slump loss of concrete. *Cem. Concr. Compos.* **2009**, *31*, 470–473. [CrossRef]
35. He, Z.; Cai, R.; Chen, E.; Tang, S. The investigation of early hydration and pore structure for limestone powder wastes blended cement pastes. *Constr. Build. Mater.* **2019**, *229*, 116923. [CrossRef]
36. Sotiriadis, K.; Mazur, A.; Tolstoy, P.; Frankeová, D. Chloride effect on sulfate attack in hydrated Portland-limestone cement assessed by 29Si NMR spectroscopy and thermal analysis. *Mater. Today Proc.* 2023; *in press*. [CrossRef]
37. Cao, M.; Zhang, C.; Lv, H.; Xu, L. Characterization of mechanical behavior and mechanism of calcium carbonate whisker-reinforced cement mortar. *Constr. Build. Mater.* **2014**, *66*, 89–97. [CrossRef]
38. Khan, M.; Cao, M.; Ali, M. Effect of basalt fibers on mechanical properties of calcium carbonate whisker-steel fiber reinforced concrete. *Constr. Build. Mater.* **2018**, *192*, 742–753. [CrossRef]
39. Cao, M.; Ming, X.; He, K.; Li, L.; Shen, S. Effect of Macro-, Micro- and Nano-Calcium Carbonate on Properties of Cementitious Composites—A Review. *Materials* **2019**, *12*, 781. [CrossRef] [PubMed]
40. Karkhaneh, S.; Tarighat, A.; Jahromi, S.G. Kinetics behavior of delayed ettringite in limestone calcined clay cement (LC3) by thermodynamic approach and consideration of the time factor. *Constr. Build. Mater.* **2023**, *367*, 129143. [CrossRef]

41. Berodier, E.; Scrivener, K. Understanding the Filler Effect on the Nucleation and Growth of C-S-H. *J. Am. Ceram. Soc.* **2014**, *97*, 3764–3773. [CrossRef]
42. Zhang, Z.; Chen, W.; Han, F.; Yan, P. A new hydration kinetics model of composite cementitious materials, Part 2: Physical effect of SCMs. *J. Am. Ceram. Soc.* **2020**, *103*, 3880–3895. [CrossRef]
43. Pourchet, S.; Pochard, I.; Brunel, F.; Perrey, D. Chemistry of the calcite/water interface: Influence of sulfate ions and consequences in terms of cohesion forces. *Cem. Concr. Res.* **2013**, *52*, 22–30. [CrossRef]
44. Scrivener, K.L.; Juilland, P.; Monteiro, P.J. Advances in understanding hydration of Portland cement. In Proceedings of the 14th International Congress on the Chemistry of Cement (ICCC 2015), Beijing, China, 13–16 October 2015.
45. Khan, R.I.; Ashraf, W. Effects of ground wollastonite on cement hydration kinetics and strength development. *Constr. Build. Mater.* **2019**, *218*, 150–161. [CrossRef]
46. Myers, R.J.; Geng, G.; Li, J.; Rodríguez, E.D.; Ha, J.; Kidkhunthod, P.; Sposito, G.; Lammers, L.N.; Kirchheim, A.P.; Monteiro, P.J.M. Role of Adsorption Phenomena in Cubic Tricalcium Aluminate Dissolution. *Langmuir* **2016**, *33*, 45–55. [CrossRef]
47. Briki, Y.; Avet, F.; Zajac, M.; Bowen, P.; Ben Haha, M.; Scrivener, K. Understanding of the factors slowing down metakaolin reaction in limestone calcined clay cement (LC3) at late ages. *Cem. Concr. Res.* **2021**, *146*, 106477. [CrossRef]
48. Termkhajornkit, P.; Barbarulo, R. Modeling the coupled effects of temperature and fineness of Portland cement on the hydration kinetics in cement paste. *Cem. Concr. Res.* **2012**, *42*, 526–538. [CrossRef]
49. Bellmann, F.; Scherer, G.W. Analysis of C-S-H growth rates in supersaturated conditions. *Cem. Concr. Res.* **2018**, *103*, 236–244. [CrossRef]
50. Jönsson, B.; Nonat, A.; Labbez, C.; Cabane, B.; Wennerström, H. Controlling the Cohesion of Cement Paste. *Langmuir* **2005**, *21*, 9211–9221. [CrossRef]
51. Stark, J.; Möser, B.; Bellmann, F. Nucleation and growth of C-S-H phases on mineral admixtures. *Adv. Constr. Mater.* **2007**, *2007*, 531–538.
52. Zajac, M.; Skocek, J.; Lothenbach, B.; Mohsen, B.H. Late hydration kinetics: Indications from thermodynamic analysis of pore solution data. *Cem. Concr. Res.* **2020**, *129*, 105975. [CrossRef]
53. Andalibi, M.R.; Kumar, A.; Srinivasan, B.; Bowen, P.; Scrivener, K.; Ludwig, C.; Testino, A. On the mesoscale mechanism of synthetic calcium–silicate–hydrate precipitation: A population balance modeling approach. *J. Mater. Chem. A* **2018**, *6*, 363–373. [CrossRef]
54. Garrault-Gauffinet, S.; Nonat, A. Experimental investigation of calcium silicate hydrate (C-S-H) nucleation. *J. Cryst. Growth* **1999**, *200*, 565–574. [CrossRef]
55. Yeşilmen, S.; Al-Najjar, Y.; Balav, M.H.; Şahmaran, M.; Yıldırım, G.; Lachemi, M. Nano-modification to improve the ductility of cementitious composites. *Cem. Concr. Res.* **2015**, *76*, 170–179. [CrossRef]
56. Friebert, M. Der Einfluss von Betonzusatzstoffen auf die Hydratation und Dauer-haftigkeit selbstverdichtender Betone. Ph.D. Thesis, Bauhaus-Universität Weimar, Weimar, Germany, 2005.
57. Bentz, D.P.; Sato, T.; de la Varga, I.; Weiss, W.J. Fine limestone additions to regulate setting in high volume fly ash mixtures. *Cem. Concr. Compos.* **2012**, *34*, 11–17. [CrossRef]
58. De Weerdt, K.; Haha, M.B.; Le Saout, G.; Kjellsen, K.O.; Justnes, H.; Lothenbach, B. Hydration mechanisms of ternary Portland cements containing limestone powder and fly ash. *Cem. Concr. Res.* **2011**, *41*, 279–291. [CrossRef]
59. Kakali, G.; Tsivilis, S.; Aggeli, E.; Bati, M. Hydration products of C3A, C3S and Portland cement in the presence of $CaCO_3$. *Cem. Concr. Res.* **2000**, *30*, 1073–1077. [CrossRef]
60. Darweesh, H.H.M. Limestone as an accelerator and filler in limestone-substituted alumina cement. *Ceram. Int.* **2004**, *30*, 145–150. [CrossRef]
61. Sato, T.; Beaudoin, J.J. Effect of nano-$CaCO_3$ on hydration of cement containing supplementary cementitious materials. *Adv. Cem. Res.* **2011**, *23*, 33–43. [CrossRef]
62. Mikanovic, N.; Khayat, K.; Pagé, M.; Jolicoeur, C. Aqueous $CaCO_3$ dispersions as reference systems for early-age cementitious materials. *Colloids Surf. A Physicochem. Eng. Asp.* **2006**, *291*, 202–211. [CrossRef]
63. Drissi, S.; Shi, C.; Li, N.; Liu, Y.; Liu, J.; He, P. Relationship between the composition and hydration-microstructure-mechanical properties of cement-metakaolin-limestone ternary system. *Constr. Build. Mater.* **2021**, *302*, 124175. [CrossRef]
64. Aqel, M.; Panesar, D.K. Hydration kinetics and compressive strength of steam-cured cement pastes and mortars containing limestone filler. *Constr. Build. Mater.* **2016**, *113*, 359–368. [CrossRef]
65. Hu, J. The Hydration Mechanism and the Improvement of Early Properties of Ternary Cement Blends with Limestone Powder and Metakaolin. Master's Thesis, Southeast University, Nanjing, China, 2021.
66. Zhou, S.-C. Study on Evolutionary Regularity and Micro Mechanisms of Macro Properties of Composite Limestone Powder-Fly Ash-Slag Concrete. Ph.D. Thesis, China University of Mining and Technology, Xuzhou, China, 2019.
67. Saulat, H.; Cao, M.; Khan, M.; Khan, M.M.; Rehman, A. Preparation and applications of calcium carbonate whisker with a special focus on construction materials. *Constr. Build. Mater.* **2020**, *236*, 117613. [CrossRef]
68. Yoshioka, S.; Kitano, Y. Transformation of aragonite to calcite through heating. *Geochem. J.* **1985**, *19*, 245–249. [CrossRef]
69. Bentz, D.P.; Ardani, A.; Barrett, T.; Jones, S.Z.; Lootens, D.; Peltz, M.A.; Sato, T.; Stutzman, P.E.; Tanesi, J.; Weiss, W.J. Multi-scale investigation of the performance of limestone in concrete. *Constr. Build. Mater.* **2015**, *75*, 1–10. [CrossRef]
70. Luan, C.; Zhou, Y.; Liu, Y.; Ren, Z.; Wang, J.; Yuan, L.; Du, S.; Zhou, Z.; Huang, Y. Effects of nano-SiO_2, nano-$CaCO_3$ and nano-TiO_2 on properties and microstructure of the high content calcium silicate phase cement (HCSC). *Constr. Build. Mater.* **2022**, *314*, 125377. [CrossRef]

71. Goodbrake, C.J.; Young, J.F.; Berger, R.L. Reaction of Beta-Dicalcium Silicate and Tricalcium Silicate with Carbon Dioxide and Water Vapor. *J. Am. Ceram. Soc.* **1979**, *62*, 168–171. [CrossRef]
72. Monkman, S.; Lee, B.E.J.; Grandfield, K.; MacDonald, M.; Raki, L. The impacts of in-situ carbonate seeding on the early hydration of tricalcium silicate. *Cem. Concr. Res.* **2020**, *136*, 106179. [CrossRef]
73. Monkman, S.; Sargam, Y.; Raki, L. Comparing the effects of in-situ nano-calcite development and ex-situ nano-calcite addition on cement hydration. *Constr. Build. Mater.* **2022**, *321*, 126369. [CrossRef]
74. Sato, T.; Diallo, F. Seeding Effect of Nano-$CaCO_3$ on the Hydration of Tricalcium Silicate. *Transp. Res. Rec. J. Transp. Res. Board* **2010**, *2141*, 61–67. [CrossRef]
75. Ding, Y.; Liu, J.-P.; Bai, Y.-L. Linkage of multi-scale performances of nano-$CaCO_3$ modified ultra-high performance engineered cementitious composites (UHP-ECC). *Constr. Build. Mater.* **2020**, *234*, 117418. [CrossRef]
76. Xu, Z.; Zhou, Z.; Du, P.; Cheng, X. Effects of nano-limestone on hydration properties of tricalcium silicate. *J. Therm. Anal. Calorim.* **2017**, *129*, 75–83. [CrossRef]
77. Douba, A.; Hou, P.; Kawashima, S. Hydration and mechanical properties of high content nano-coated cements with nano-silica, clay and calcium carbonate. *Cem. Concr. Res.* **2023**, *168*, 107132. [CrossRef]
78. Nguyen, V.T.; Lee, S.Y.; Kim, D.J. Simulation of the effect of nano-$CaCO_3$ agglomeration on the hydration process and microstructural evolution of cement paste. *Case Stud. Constr. Mater.* **2023**, *19*, e02612. [CrossRef]
79. Shaikh, F.U.; Supit, S.W. Mechanical and durability properties of high volume fly ash (HVFA) concrete containing calcium carbonate ($CaCO_3$) nanoparticles. *Constr. Build. Mater.* **2014**, *70*, 309–321. [CrossRef]
80. Kawashima, S.; Seo, J.W.T.; Corr, D.; Hersam, M.C.; Shah, S.P. Dispersion of $CaCO_3$ nanoparticles by sonication and surfactant treatment for application in fly ash–cement systems. *Mater. Struct.* **2013**, *47*, 1011–1023. [CrossRef]
81. Wu, Z.; Shi, C.; Khayat, K.; Wan, S. Effects of different nanomaterials on hardening and performance of ultra-high strength concrete (UHSC). *Cem. Concr. Compos.* **2016**, *70*, 24–34. [CrossRef]
82. Ming, X.; Cao, M.; Lv, X.; Yin, H.; Li, L.; Liu, Z. Effects of high temperature and post-fire-curing on compressive strength and microstructure of calcium carbonate whisker-fly ash-cement system. *Constr. Build. Mater.* **2020**, *244*, 118333. [CrossRef]
83. Berodier, E.M.J. Impact of the Supplementary Cementitious Materials on the Kinetics and Microstructural Development of Cement Hydration. Ph.D. Thesis, Swiss Federal Institute of Technology Lausanne, Lausanne, Switzerland, 2019.
84. Zajac, M.; Rossberg, A.; Le Saout, G.; Lothenbach, B. Influence of limestone and anhydrite on the hydration of Portland cements. *Cem. Concr. Compos.* **2014**, *46*, 99–108. [CrossRef]
85. Zajac, M.; Durdzinski, P.; Stabler, C.; Skocek, J.; Nied, D.; Haha, M.B. Influence of calcium and magnesium carbonates on hydration kinetics, hydrate assemblage and microstructural development of metakaolin containing composite cements. *Cem. Concr. Res.* **2018**, *106*, 91–102. [CrossRef]
86. Antoni, M.; Rossen, J.; Martirena, F.; Scrivener, K. Cement substitution by a combination of metakaolin and limestone. *Cem. Concr. Res.* **2012**, *42*, 1579–1589. [CrossRef]
87. Dhandapani, Y.; Santhanam, M.; Kaladharan, G.; Ramanathan, S. Towards ternary binders involving limestone additions—A review. *Cem. Concr. Res.* **2021**, *143*, 106396. [CrossRef]
88. Tang, J.; Wei, S.; Li, W.; Ma, S.; Ji, P.; Shen, X. Synergistic effect of metakaolin and limestone on the hydration properties of Portland cement. *Constr. Build. Mater.* **2019**, *223*, 177–184. [CrossRef]
89. Li, L.; Cao, M.; Xie, C.; Yin, H. Effects of $CaCO_3$ whisker, hybrid fiber content and size on uniaxial compressive behavior of cementitious composites. *Struct. Concr.* **2018**, *20*, 506–518. [CrossRef]
90. Cao, M.; Khan, M.; Ahmed, S. Effectiveness of Calcium Carbonate Whisker in Cementitious Composites. *Period. Polytech. Civ. Eng.* **2020**, *64*, 265–275. [CrossRef]
91. Cao, M.; Xie, C.; Li, L.; Khan, M. Effect of different PVA and steel fiber length and content on mechanical properties of $CaCO_3$ whisker reinforced cementitious composites. *Mater. Constr.* **2019**, *69*, 200. [CrossRef]
92. Cao, M.; Li, L.; Khan, M. Effect of hybrid fibers, calcium carbonate whisker and coarse sand on mechanical properties of cement-based composites. *Mater. Constr.* **2018**, *68*, 156. [CrossRef]
93. Khan, M.; Cao, M.; Ai, H.; Hussain, A. Basalt Fibers in Modified Whisker Reinforced Cementitious Composites. *Period. Polytech. Civ. Eng.* **2022**, *66*, 344–354. [CrossRef]
94. Cao, M.; Khan, M. Effectiveness of multiscale hybrid fiber reinforced cementitious composites under single degree of freedom hydraulic shaking table. *Struct. Concr.* **2020**, *22*, 535–549. [CrossRef]
95. Khan, M.; Cao, M.; Xie, C.; Ali, M. Effectiveness of hybrid steel-basalt fiber reinforced concrete under compression. *Case Stud. Constr. Mater.* **2022**, *16*, e00941. [CrossRef]
96. Li, W.; Huang, Z.; Cao, F.; Sun, Z.; Shah, S.P. Effects of nano-silica and nano-limestone on flowability and mechanical properties of ultra-high-performance concrete matrix. *Constr. Build. Mater.* **2015**, *95*, 366–374. [CrossRef]
97. Papatzani, S.; Paine, K.; Calabria-Holley, J. A comprehensive review of the models on the nanostructure of calcium silicate hydrates. *Constr. Build. Mater.* **2015**, *74*, 219–234. [CrossRef]
98. Jinchang, P.; Ronggui, L. Improvement of performance of ultra-high performance concrete based composite material added with nano materials. *Frat. Ed Integrità Strutt.* **2016**, *10*, 130–138. [CrossRef]
99. Svenum, I.-H.; Ringdalen, I.G.; Bleken, F.L.; Friis, J.; Höche, D.; Swang, O. Structure, hydration, and chloride ingress in C-S-H: Insight from DFT calculations. *Cem. Concr. Res.* **2020**, *129*, 105965. [CrossRef]

100. Wu, Z.; Khayat, K.H.; Shi, C.; Tutikian, B.F.; Chen, Q. Mechanisms underlying the strength enhancement of UHPC modified with nano-SiO_2 and nano-$CaCO_3$. *Cem. Concr. Compos.* **2021**, *119*, 103992. [CrossRef]
101. Assaedi, H.; Alomayri, T.; Kaze, C.R.; Jindal, B.B.; Subaer, S.; Shaikh, F.; Alraddadi, S. Characterization and properties of geopolymer nanocomposites with different contents of nano-$CaCO_3$. *Constr. Build. Mater.* **2020**, *252*, 119137. [CrossRef]
102. Irassar, E.F. Sulfate attack on cementitious materials containing limestone filler—A review. *Cem. Concr. Res.* **2009**, *39*, 241–254. [CrossRef]
103. Ramezanianpour, A.M.; Hooton, R.D. Thaumasite sulfate attack in Portland and Portland-limestone cement mortars exposed to sulfate solution. *Constr. Build. Mater.* **2013**, *40*, 162–173. [CrossRef]
104. Yang, Z.; Zhang, W.; Zhu, H.; Chen, Y.; Xu, L.; Wang, P.; Lai, Y. Thaumasite form of sulfate attack in ettringite rich-ternary systems: Effects of limestone filler, etching solutions and exposure temperature. *Dev. Built Environ.* **2023**, *15*, 100208. [CrossRef]
105. Gunjal, S.; Turkane, S.D.; Patankar, S.; Kondraivendhan, B. Effect of magnesium sulphate and sulphuric acid attack on limestone calcined clay cement concrete. *Mater. Today Proc.* 2023; *in press*. [CrossRef]
106. Sun, J.; Chen, Z. Influences of limestone powder on the resistance of concretes to the chloride ion penetration and sulfate attack. *Powder Technol.* **2018**, *338*, 725–733. [CrossRef]
107. Rashad, A.M.; Ezzat, M.; ElNagar, A.M.; El-Nashar, M. Valorization of limestone powder as an additive for fly ash geopolymer cement under the effect of the simulated tidal zone and seawater attack. *Constr. Build. Mater.* **2023**, *369*, 130616. [CrossRef]
108. Nadelman, E.; Kurtis, K. Durability of Portland-limestone cement-based materials to physical salt attack. *Cem. Concr. Res.* **2019**, *125*, 105859. [CrossRef]
109. Kanagaraj, B.; Nammalvar, A.; Andrushia, A.D.; Gurupatham, B.G.A.; Roy, K. Infuence of Nano Composites on the Impact Resistance of Concrete at Elevated Temperatures. *Fire* **2023**, *6*, 135. [CrossRef]
110. Cao, M.; Yuan, X.; Ming, X.; Xie, C. Effect of High Temperature on Compressive Strength and Microstructure of Cement Paste Modified by Micro- and Nano-calcium Carbonate Particles. *Fire Technol.* **2022**, *58*, 1469–1491. [CrossRef]
111. Kunther, W.; Dai, Z.; Skibsted, J. Thermodynamic modeling of hydrated white Portland cement–metakaolin–limestone blends utilizing hydration kinetics from 29Si MAS NMR spectroscopy. *Cem. Concr. Res.* **2016**, *86*, 29–41. [CrossRef]
112. Liu, X.; Luo, Q.; Xie, H.; Li, S.; Zhang, J.; Xia, C.; Ding, Y.; Chen, Y.; Gao, R.; Wei, Z.; et al. Effect of calcium alumina silicate hydrate nano-seeds on the hydration of low clinker cement. *J. Build. Eng.* **2023**, *66*, 105844. [CrossRef]
113. Matschei, T.; Lothenbach, B.; Glasser, F.P. Thermodynamic properties of Portland cement hydrates in the system $CaO–Al_2O_3–SiO_2–CaSO_4–CaCO_3–H_2O$. *Cem. Concr. Res.* **2007**, *37*, 1379–1410. [CrossRef]
114. Naber, C.; Bellmann, F.; Sowoidnich, T.; Goetz-Neunhoeffer, F.; Neubauer, J. Alite dissolution and C-S-H precipitation rates during hydration. *Cem. Concr. Res.* **2019**, *115*, 283–293. [CrossRef]
115. Schöler, A.; Lothenbach, B.; Winnefeld, F.; Ben Haha, M.; Zajac, M.; Ludwig, H.-M. Early hydration of SCM-blended Portland cements: A pore solution and isothermal calorimetry study. *Cem. Concr. Res.* **2017**, *93*, 71–82. [CrossRef]
116. Lothenbach, B.; Kulik, D.A.; Matschei, T.; Balonis, M.; Baquerizo, L.; Dilnesa, B.; Miron, G.D.; Myers, R.J. Cemdata18: A chemical thermodynamic database for hydrated Portland cements and alkali-activated materials. *Cem. Concr. Res.* **2019**, *115*, 472–506. [CrossRef]
117. Lothenbach, B.; Winnefeld, F. Thermodynamic modelling of the hydration of Portland cement. *Cem. Concr. Res.* **2006**, *36*, 209–226. [CrossRef]
118. Parkhurst, D.L.; Appelo, C.A.J. Description of Input and Examples for PHREEQC Version 3—A Computer Program for Speciation, Batch-Reaction, One-Dimensional Transport, and Inverse Geochemical Calculations. *US Geol. Surv. Tech. Methods* **2013**, *6*, 497.
119. Damidot, D.; Lothenbach, B.; Herfort, D.; Glasser, F. Thermodynamics and cement science. *Cem. Concr. Res.* **2011**, *41*, 679–695. [CrossRef]
120. Elizalde, M.; Aparicio, J. Current theories in the calculation of activity coefficients—II. Specific interaction theories applied to some equilibria studies in solution chemistry. *Talanta* **1995**, *42*, 395–400. [CrossRef] [PubMed]
121. Lothenbach, B.; Le Saout, G.; Gallucci, E.; Scrivener, K. Influence of limestone on the hydration of Portland cements. *Cem. Concr. Res.* **2008**, *38*, 848–860. [CrossRef]
122. Wang, Y.; Shui, Z.; Wang, L.; Gao, X.; Huang, Y.; Song, Q.; Liu, K. Alumina-rich pozzolan modification on Portland-limestone cement concrete: Hydration kinetics, formation of hydrates and long-term performance evolution. *Constr. Build. Mater.* **2020**, *258*, 119712. [CrossRef]
123. Kunther, W.; Dai, Z.; Skibsted, J. Thermodynamic modeling of Portland cement-metakaolin-limestone blends. In Proceedings of the 14th International Congress on the Chemistry of Cement (ICCC 2015), Beijing, China, 13–16 October 2015.
124. Her, S.; Im, S.; Liu, J.; Suh, H.; Kim, G.; Sim, S.; Wi, K.; Park, D.; Bae, S. Exploring the potential of pulverized oyster shell as a limestone substitute in limestone calcined clay cement (LC3) and its implications for performance. *Constr. Build. Mater.* **2024**, *425*, 135918. [CrossRef]
125. Fakhri, R.S.; Dawood, E.T. Limestone powder, calcined clay and slag as quaternary blended cement used for green concrete production. *J. Build. Eng.* **2023**, *79*, 107644. [CrossRef]
126. Mañosa, J.; Calderón, A.; Salgado-Pizarro, R.; Maldonado-Alameda, A.; Chimenos, J.M. Research evolution of limestone calcined clay cement (LC3), a promising low-carbon binder—A comprehensive overview. *Heliyon* **2024**, *10*, e25117. [CrossRef] [PubMed]
127. Frías, M.; Guerrero, A.; Monasterio, M.; Insignares, Á.; de Rojas, M.I.S. Viability of using limestone concrete waste from CDW to produce ternary cements type LC3. *Constr. Build. Mater.* **2024**, *411*, 134362. [CrossRef]

128. Hanein, T.; Thienel, K.-C.; Zunino, F.; Marsh, A.T.M.; Maier, M.; Wang, B.; Canut, M.; Juenger, M.C.G.; Ben Haha, M.; Avet, F.; et al. Clay calcination technology: State-of-the-art review by the RILEM TC 282-CCL. *Mater. Struct.* **2021**, *55*, 3. [CrossRef]
129. Cai, J.M.; Pan, J.L. Using Calcium Carbonate Whisker in Engineered Cementitious Composites. In Proceedings of the 9th International Conference on Fracture Mechanics of Concrete and Concrete Structures, Berkeley, CA, USA, 29 May–1 June 2016.
130. Xie, C.; Cao, M.; Si, W.; Khan, M. Experimental evaluation on fiber distribution characteristics and mechanical properties of calcium carbonate whisker modified hybrid fibers reinforced cementitious composites. *Constr. Build. Mater.* **2020**, *265*, 120292. [CrossRef]
131. Kim, G.; Cho, S.; Moon, J.; Suh, H.; Her, S.; Sim, S.; Bae, S. Investigation of the hydrate formation and mechanical performance of limestone calcined clay cement paste incorporating nano-$CaCO_3$ and nano-SiO_2 as partial limestone substitutes. *Constr. Build. Mater.* **2024**, *418*, 135335. [CrossRef]
132. Meng, S.; Ouyang, X.; Fu, J.; Ma, Y.; Ye, G. New insights into the role of MWCNT in cement hydration. *Mater. Struct.* **2021**, *54*, 238. [CrossRef]
133. Zhang, R.; Scott, A.N.; Panesar, D.K. Carbonation and CO_2 reabsorption of cement-based materials: Influence of limestone filler and ground-granulated blast-furnace slag. *Constr. Build. Mater.* **2024**, *416*, 135166. [CrossRef]
134. Kim, G.; Kurtis, K.E. Early-stage assessment of drying shrinkage in Portland limestone cement concrete using nonlinear ultrasound. *Constr. Build. Mater.* **2022**, *342*, 128099. [CrossRef]
135. Barbhuiya, S.; Nepal, J.; Das, B.B. Properties, compatibility, environmental benefits and future directions of limestone calcined clay cement (LC3) concrete: A review. *J. Build. Eng.* **2023**, *79*, 107794. [CrossRef]

MDPI AG
Grosspeteranlage 5
4052 Basel
Switzerland
Tel.: +41 61 683 77 34

Materials Editorial Office
E-mail: materials@mdpi.com
www.mdpi.com/journal/materials

www.ingramcontent.com/pod-product-compliance
Lightning Source LLC
Chambersburg PA
CBHW040803250926
51850CB00007BA/1069
* 9 7 8 3 7 2 5 8 6 4 3 6 2 *